ESSENTIALS OF ENGINEERING DESIGN

ESSENTIALS OF ENGINEERING DESIGN

Joseph W. Walton

WEST PUBLISHING COMPANY
ST. PAUL • NEW YORK • SAN FRANCISCO • LOS ANGELES

COPYRIGHT © 1991 By WEST PUBLISHING COMPANY
50 W. Kellogg Boulevard
P.O. Box 64526
St. Paul, MN 55164-1003

All rights reserved

Printed in the United States of America

98 97 96 95 94 93 92 91   8 7 6 5 4 3 2 1 0

Library of Congress Cataloging-in-Publication Data

Walton, Joseph W.
   Essentials of engineering design/Joseph W. Walton.
     p.   cm.
   Includes index.
   ISBN 0-314-76550-6 (soft)
   1. Engineering design.   I. Title.
TA174W.W266   1991
620′.0042—dc20

90-40253
CIP

Production Credits
Copyediting: Michael Michoud
Illustrations: Scientific Illustrators
Text Design: Geri Davis, Quadrata, Inc.
Cover Design: Pollock Design Group
Cover Image: FPG International
Composition: Trigraph Incorporated

Photo Credits

14, H. Armstrong Roberts; **15**, H. Armstrong Roberts; **16**, H. Armstrong Roberts; **17**, H. Armstrong Roberts; **18**, Stock-Boston; **19**, H. Armstrong Roberts; **28**, H. Armstrong Roberts; **28**, H. Armstrong Roberts; **29**, Camerique; **30**, Camerique; **31**, H. Armstrong Roberts; **32**, Camerique; **33**, H. Armstrong Roberts; **34**, H. Armstrong Roberts; **35**, H. Armstrong Roberts; **36**, Camerique; **37**, H. Armstrong Roberts; **38**, Paul Fortin, Stock-Boston; **40**, Courtesy of Numonics Corporation; **42**, Courtesy of Deere & Co.; **51**, H. Armstrong Roberts; **52**, H. Armstrong Roberts; **53**, H. Armstrong Roberts; **200**, Courtesy of Xavier Park Unit of Mercy Health Center, Dubuque, Iowa; **208**, H. Armstrong Roberts

Contents

Preface xiv
Introduction for Students xvii
List of Symbols xix

SECTION ONE ENGINEERING PROBLEM SOLVING PROCEDURES 2

1 Introduction To Engineering 4
1.1 Engineering Experience 5
1.2 Sources of Design Failure 10
1.3 Unsolved Problems 13
1.4 References 19
1.5 Exercises 21

2 Engineering Activities 24
2.1 Primary Job 24
2.2 Necessary Skills 25
2.3 Engineering Developments and Advancements 26
2.4 Responsibilities 38
2.5 Day-to-day Concerns 44
2.6 Constraints and Limited Resources 46
2.7 Project Planning 52
2.8 Communication 57
2.9 Personal Characteristics and Abilities 64
2.10 References 65
2.11 Exercises 67

3 Problem Solving 74

- 3.1 Sources of Problems 74
- 3.2 Creativity 76
- 3.3 Roadblocks To Creative Problem Solving 76
  CASE STUDY #1: Door Holder-opener 79
- 3.4 Problem Solving Activities 80
- 3.5 References 83
- 3.6 Exercises 84

4 Problem Definition 88

- 4.1 Methods of Problem Definition 88
  CASE STUDY #2: Parts Moving Problem 89
- 4.2 Related Activities 96
- 4.3 Cautions 102
- 4.4 System Problems 103
- 4.5 References 104
- 4.6 Exercises 105

5 Problem Solution Idea Generation 108

- 5.1 Specific Considerations 109
- 5.2 Physical Laws 117
- 5.3 Brainstorming 119
  CASE STUDY #3: Idea generation 121
- 5.4 Work Simplification 124
- 5.5 Feasibility Study 125
- 5.6 Summary 127
- 5.7 References 128
- 5.8 Exercises 128

6 Refinement and Analysis 134

- 6.1 Levels of Analysis 135
- 6.2 Related Activities 138
  CASE STUDY #4: Drawing Board Table Adjustment 141
  CASE STUDY #5: Room Cooling 147
- 6.3 Summary 154
- 6.4 References 154
- 6.5 Exercises 155

7 Decision and Implementation 166

7.1 Idea Selection 168
  CASE STUDY #6: Baseball Field Revision 171
7.2 Implementation 174
7.3 References 176
7.4 Exercises 177

Design Problems 182

Group I—Incomplete Problem Identification 182
Group II—Preliminary Problem Solution Given 186

SECTION TWO DETAILS OF REFINEMENT AND ANALYSIS 196

8 Engineering Models 198

8.1 Types of Models 198
  CASE STUDY #7: Tree Age Analysis 202
8.2 Boundary Conditions 205
8.3 Full-Scale Models 206
8.4 Scale Models 207
8.5 Dimensional Analysis 209
  CASE STUDY #8: Wind Force Modeling 212
8.6 Experimental Data Analysis 216
  CASE STUDY #9: Empirical Formula Refinement 222
8.7 Experimentation 228
8.8 Additional Study Topics 232
8.9 References 232
8.10 Exercises 232

9 Optimization 240

9.1 Calculus Techniques 241
  CASE STUDY #10: Tool Life Optimization 246
  CASE STUDY #11: Pipe Packing Optimization 251
  CASE STUDY #12: Cost Ratio Analysis 253
9.2 Non-calculus Techniques 256
9.3 Linear Programming 266
9.4 Additional Concerns 272
  CASE STUDY #13: Balcony Seating Optimization 272
  CASE STUDY #14: Spherical Storage Tank Optimization 276

9.5   Experimental Optimization 279
CASE STUDY #15: Experimental Optimization 281
9.6   Additional Study Topics 287
9.7   References 287
9.8   Exercises 288

10 Ethics 298

10.1   Definition 298
10.2   Sources of Problems 299
10.3   Codes 300
10.4   Specific Situations 305
CASE STUDY #16: Personal Conflict 306
CASE STUDY #17: Corporate Conflict 307
CASE STUDY #18: Government Ethics 308
10.5   Study Cases 309
10.6   References 309
10.7   Exercises 310

Appendices 313

A-1   Basic quantities and units. 314
A-2   Basic quantities and units. 315
B   SI Prefixes and multiplying factors. 316
C   Selected conversion factors. 316
D   Energy values of common fuels. 317
E   Element data. 318
F-1   Properties of selected materials. 319
F-2   Properties of selected materials. 321
F-3   Properties of cast iron. 323
F-4   Properties of selected aluminum alloys. 325
F-5   Properties of selected magnesium alloys. 326
F-6   Properties of selected brasses. 327
F-7   Properties of selected bronzes. 328
F-8   Properties of selected nickel alloys. 329
F-9   Properties of selected titanium alloys. 330
F-10   Properties of selected ceramics. 331
F-11   Properties of concrete. 332
F-12   Selected values of the modulus of elasticity. 332
F-13   Steel alloy numbering system. 333
F-14   Magnetic alloys. 334
F-15   Aluminum identification coding system. 335

Contents

F-16	Selected commercial glasses.	336
F-17	Variation in properties of wood.	336
F-18	Ultimate tensile and compressive strength of selected brittle materials.	337
F-19	Selected values of Poisson's ratio.	337
F-20	Mechanical properties of selected engineering materials.	338
F-21	Selected values of quasi-static toughness and yield strength.	339
F-22	Static and dynamic properties of selected materials.	340
F-23	Density and mechanical properties of bulk materials and filaments.	341
F-24	Selected electrical resistivities.	342
F-25	Density and mechanial properties of selected composite materials.	342
G	Normal distribution.	343
H	t distribution.	345
I	F distribution.	347
J	Chi-squared distribution.	350
K	Effect of surface conditions on the fatigue strength of steel.	351
L	Stress concentration factors.	352
M	Stress intensity factors.	356
N	Copper wire data.	361
O	Beam loading formulas.	362
P	Properties of selected geometric shapes.	369
Q-1	Wide flange structural section properties.	370
Q-2	Standard flange structural section properties.	371
Q-3	Standard channel structural section properties.	372
Q-4	Standard angle structural section properties.	373
R	R-values of insulating materials.	374

Answers To Selected Exercises 376

Index 394

Preface

Essentials of Engineering Design contains 10 of the 15 chapters of *Engineering Design: From Art to Practice*. *Essentials* is an overview of fundamental methods and procedures for solving engineering design problems. It includes discussion of and emphasis on performing the essential, though sometimes routine, details of analyzing and implementing design solutions, while stressing flexibility and adaptability. It emphasizes analysis and synthesis techniques, for these allow an engineer to tackle new unsolved problems. Examples that illustrate various concepts are drawn primarily from my personal experiences in engineering. Engineers don't spend most of their time solving earth-shaking revolutionary design problems; day-to-day mundane-seeming problems occupy most of their time. It is not usual for an engineer to solve calculus integrals after graduation, the more important problem being to determine how the best ideas can be manufactured economically into reliable products.

This book is my solution to a problem: to have a text for an engineering design course that contains a variety and quantity of problems without suggesting that there is a final word on any problem. Technology and state of the art changes require that engineers be prepared to keep up with the pace of change and be aware of the various interrelationships that exist between the different areas of engineering.

During my 18 years of teaching, one of the most common student comments has been that textbooks are hard to read. I have tried to write in a style that reads easily while keeping the readers aware of details. Many authors write for the experts, not for the students. The result is that students often do not read for learning or understanding. I also think that students learn as much from example as from theory, so I include examples where others might include theory.

The amount of time spent on extended design problems will influence how much time is spent on the exercises in each chapter. Multiple, shorter projects can serve the same purpose as a semester long project, depending on personal objectives. Projects should be undertaken to broaden the experience of the student and the teacher, and to facilitate accurate and detailed analysis on the part of the student. Detailed formal reports, both written and oral, are part of the project work. Suggested technical prerequisite courses for this text are Graphics, Calculus, Chemistry, Physics, Statics, Strength of Materials, Material Science, and Computer Programming.

Chapters 1–7 survey the engineering problem solving procedure. Chapters 1 and 2 illustrate the range of problems and activities in which engineers are involved. Chapter 3 presents an overview of problem-solving procedures, and Chapters 4 through 7 present some of the details of tasks engineers perform while solving design problems. It is assumed that student projects will be assigned and developed as the topics of the design procedure are studied. Some engineering students have never been exposed to the range of activities that complement the problem-solving task. Chapters 4 through 7 cover these activities and offer examples for illustration. Chapter 5 lists questions an engineer should ask to insure an optimum final design solution. The first seven chapters contain over 200 exercises, ranging from those appropriate for group discussion to those a student could spend a week or more investigating. The exercises stress the common, but essential, considerations necessary for good design. Many exercises have no single answer, which is a true engineering predicament. Answers the students provide will vary depending on their background, personal experience, and available references. Class discussion of the exercises is important so that ideas and experiences can be shared and the assumptions and compromises needed for solutions can be agreed on. Cooperation, understanding, and communication between specialty areas of engineering is a real engineering requirement.

A variety of design problems are proposed at the end of Chapter 7 for student work. The length of time spent on a particular problem depends on the depth of analysis desired, how many students work together on a project, and whether working models are built. Students can be assigned a design project from Chapter 7 or other sources at the beginning of the semester, and solutions can be developed as Chapters 4 through 7 are studied. To make the exercises on experimentation meaningful, solutions must be proposed, experiments run, and data collected and analyzed.

Chapters 8 and 9 cover modeling and optimization procedures to assist the engineer in design analysis. Exercises in Chapters 8 and 9 are more apt to have one answer because they are intended for single concept discussion and practice. They are smaller, but important, details of the overall design process. Chapter 10 raises ethical questions that an engineer may face from time to time, the non-mathematical problems that need more than a calculator to answer.

As a guide for the support for chapter discussions, references are listed at the end of each chapter and in Section 4.2, neither of which is meant to be exhaustive. Parenthetical notation, (2), used in the text refers to the numbered references at the end of the chapter. English to metric conversions were done using Theodore Wildi's *Units*, 2nd ed., published by Volta Inc., Canada, 1972.

Acknowledgements

I want to thank the following reviewers who have offered constructive criticism during the development of this text.

Joseph R. Baumgarten, Iowa State University

George Bibel, University of Akron

James L. Blechschmidt, Arizona State University

Larry Clemen, Iowa State University

Annette Fetter, Howard R. Green Consulting Engineers

Gary A. Gabriele, Rensselaer Polytechnic Institute

R. A. Gardner, Washington University in Saint Lewis

Timothy K. Hight, Santa Clara University

Paul P. Lin, Cleveland State University

K. Ross Johnson, Michigan Technological University

Charles Reif Hammond, Dover, NH

Barry W. McNeil, Arizona State University

Steven P. Nichols, The University of Texas at Austin

John V. Perry, Jr. Texas A&M University

George Saliba, University of Pittsburgh

A. Samson, University of California at Los Angeles

Robert F. Steidel, Jr., University of California at Berkeley

Donald H. Thomas, Drexel University

David G. Ullman, Oregon State University

John R. Zimmerman, University of Delaware

Introduction For Students

For students of engineering, or any discipline, understanding the levels of intellectual activity needed to enhance learning will assist with developing proficiency in using the problem solving procedures. A summary and brief discussion of these main intellectual activities will aid in your understanding of the role they play in your development as an engineer. The key words for the activities that affect the intellectual thought process are:

1. Know
2. Comprehend
3. Apply
4. Analyze
5. Organize
6. Synthesize
7. Evaluate
8. Create

Basic knowledge is important. Facts and figures must be *known*, but knowing is not enough. Knowing that the yield strength of a material is 30,000 psi is not of much value unless one *comprehends* what yield strength is. Once a concept is known and comprehended, it can be *applied* to a specific case. Example: A physical member made from a material with an allowable yield strength of 30,000 psi will withstand a load of 15,000 lbf before deforming permanently if it has a cross section of 0.50 in^2.

A problem can be *analyzed* by breaking it into component parts, each of which can be studied separately, as well as in the context of assembly. Fracture and other failure modes for each part are studied and the components designed accordingly. If the analysis results are *organized*, then the problem solver can *synthesize*. Synthesis includes the prediction of what will happen to the overall design solution when a change is made in a component. If the location of a component is changed, or its method of attachment to the next member is altered, or its material is changed, the size and shape of other components and the life of the product may also be affected. Synthesis is not complete until the prediction is *evaluated*, perhaps with a computer model, a scale model, or a full size model. If the evaluation results do not meet expectations, then alternative solutions must be *created* and the test process repeated. Iteration of ideas, analysis, and test is continued until the results satisfactorily solve the design problem.

The purpose of this book is to provide you with ideas, suggestions, and exercises to help develop your skills so you can perform design problem-solving activities in an efficient way. Reading about how to solve problems is only the beginning; practice is essential. Exercises are provided so you can practice the concepts and improve your problem-solving skills. Some of the exercises will be so easy you may wonder what the catch is. Others will be so complex you may wonder if a solution is possible. Open-ended design exercises at the end of Chapter 7 give you the opportunity to practice the steps required to solve a problem, including how to go about obtaining necessary support information. There is no single best answer to a design problem, only a compromise after you consider relevant and restricting factors. Discussion with your instructor and your fellow students is good learning; reading theory and examples is not enough. Example and practice are as valuable as theory.

Make good use of what you have learned in previous courses: calculus, differential equations, chemistry, material science, statics, dynamics, strength of materials, or computer science. Engineering problems don't appear with a footnote telling you what solution technique to use. Be creative and develop new solution techniques along the way.

The real problems of day-to-day engineering cannot be itemized in a checklist. Your ability to work and communicate with people, combined with your resourcefulness and creativity, will determine the success or failure of your projects. Do not be discouraged with failures. Learn from them and make the next solution better.

List of Symbols

A	=	Area, generally perpendicular to applied force.
B	=	Intercept on semilog and log-log graph paper.
b	=	Intercept on linear coordinate graph paper.
D	=	Circle diameter.
δ	=	Beam deflection.
E	=	Modulus of elasticity.
f_n	=	Natural frequency.
I	=	Area moment of inertia.
K_c	=	Fracture toughness.
m	=	Slope of a line on rectangular graph paper.
μ	=	Absolute viscosity of a fluid.
P_c	=	Critical load for column buckling.
ρ	=	Material density.
R	=	Thermal resistance coefficient, modulus of resilience.
r	=	Radius of a circle; Coefficient of correlation.
S_{ut}, S_u	=	Ultimate tensile strength.
S_{yt}	=	Yield tensile strength.
S_{yc}, σ_y	=	Yield compressive strength.
S_{ys}	=	Yield shear strength.
x or **y**	=	Mean of a set of data.

ESSENTIALS OF ENGINEERING DESIGN

SECTION ONE

ENGINEERING PROBLEM SOLVING PROCEDURES

Engineering design and problem solving procedures are the lifeblood of practicing engineers. Problems are encountered at every stage of an engineering project and engineers must know how to deal with them. Some problems repeat themselves, resulting in specialized fields of engineering. Engineering has become so specialized that different courses of study and disciplines are available. Engineers become trained to solve a variety of problems and needs of society: nuclear, computer, chemical, mechanical, electrical, agricultural, and oceanographic, to name just a few. Some problems, because of constant changes in the state of the art, do not repeat themselves, but still must be solved. Finding a way to repair or replace a distorted space telescope mirror, and finding a reliable, economical source of non-polluting energy are examples of problems yet to be solved as this book is written.

Regardless of which speciality a student engineer wishes to master, or which uncharted problem area is going to be searched for answers to the world's problems, there is common ground to cover, common information to learn, and common procedures to guide in the quest for acceptable solutions to problems. Section I of this text provides procedures, study material, and learning exercises for problem solving. These procedures can then be applied to any engineering (and also non-engineering) problem. The guidelines for problem solving presented and discussed do not guarantee that a utopian solution will be found, only the "best" within your limits. Specific data on material and material failure theories, as well as solutions to problems that have already been solved, are of little value unless an engineer uses the information in a logical and systematic way, hence the need for guidelines to the problem-solving procedure. Section I includes hints, examples, and illustrations on how to sort and sift through information, whether it comes from handbooks or from your own creativity.

All the guidelines and data in the world, however, will not be of value to you unless they are tempered with experience, unbiased analysis, and good judgment. Engineering provides many stories of peculiar, surprising, and sometimes unfortunate events. These events teach us to be careful because they are unforgiving if we make a mistake. The bridge fails, the ship sinks, or the building collapses in the storm. Even as engineers work within their own area of specialty they must be constantly alert for the unusual and the unexpected. Following the guidelines for problem solving will only help reach an optimum solution, they will not provide a guarantee of success.

Examples and exercises in Section I are specific, even though the procedures which are presented are universally applicable and generally can be used on any problem. Section II also contains specific examples and exercises which use specific procedures. However, these procedures are not universally applicable.

1 Introduction To Engineering

"The trouble with the future is that it usually arrives before we're ready for it."
✦ ARNOLD H. GLASOW

Engineers are professional people who work to solve the problems of today's society and also plan ahead to prevent problems in the future. The purpose of this book is to help develop the skills necessary to perform problem solving activities effectively. This is done by using theory, examples, and exercises to illustrate that engineering problems in general have alternate acceptable solutions. Solutions are a compromise, a balance between conflicting restrictions that are overcome using personal judgment based on experience and theory. It is unfortunate that so much of engineering training is connected with "getting the right answer," because the conclusion many students have is that a problem has only one correct answer. For some exercise problems this may be so. Single answer problems are required so that students can study new concepts and become familiar with elementary procedures and methods. Engineers polish their skills by practicing simple techniques and then applying them to more complex issues. Few, if any, real problems have only one answer. What, you ask, can be required of the problem $2x = 4$ other than solving for the answer $x = 2$? The difference is that $2x = 4$ is not an engineering problem. It is a mathematical model that might be used to help solve an engineering problem. The model $2x = 4$ may be but one solution to dividing up a space of 4 feet, namely into 2 equal pieces. Perhaps a better model to help solve the real problem is $3x = 4$, dividing the space into 3 equal pieces. More generally the model $aX = 4$ is applicable, using the parameter a and investigating many solutions.

1.1 Engineering Experience

Engineering decisions are based on theories studied in classes, tempered with the experience of making theories work, and adjusted for the actual restrictive conditions encountered. This is one reason why state of the art and related procedures are built up from past successes and failures. Some of the best learning an engineer will ever get occurs when a proposed solution is tried and it does not work as planned. It is hoped that the failure will not cost lives or too much money, and only personal pride will suffer. On the other hand a design may be successful for a reason other than the one assumed critical by the engineer. History has taught the engineering profession many things. Some of these are obvious in today's technical world; others have been learned more recently and were not obvious at all. Following are summaries of cases that illustrate some of the lessons that have been learned the hard way.

Case #1: In the 1850s, when cast iron railroad bridges in Pennsylvania were failing, the order went out to replace them with wooden bridges. Wood held up under the same conditions, but no one knew for sure why the cast iron did not. Today we know why cast iron was not a good choice for the bridges; the combination of tension loads at stress concentrations is more than brittle cast iron can withstand. When the proposed change in the state of the art from wood to cast iron was attempted and failed, the engineering decision was to revert to a tried and proved method, wood bridges.

Case #2: In December, 1928, one leaf of the bridge over the Hackensack River in New Jersey failed after less than two years of service. The bridge was a two-leaf drawbridge that used a 770-ton counterweight on each leaf. The moving counterweight, accelerating and decelerating, coupled with component friction, caused forces and vibrations that the supporting structure could not withstand. The dynamics of the bridge had not been properly understood by the designers. Strain gage tests and transit measuring verification of the problem were made on the leaf that had not collapsed. The bridge was strengthened with additional tower supports and was returned to service. The theoretical model analysis had not been substantiated with real time testing.

Case #3: In March, 1937, a high school in Texas, just four years old, converted from burning methane city gas to natural gas obtained less expensively from a nearby oil field. Shortly after conversion, the school exploded, killing 455 people. Leaks in the system, which apparently were there prior to the fuel switch, caused the natural gas, denser than the methane, to accumulate in the basement and explode when an electrical switch was used. Investigation concluded that the area through which the pipes ran was inadequately ventilated. The investigation report also recommended that malodorants be added to all natural gas so leaks can be more easily detected. Lessons are often learned the hard way, but learning and reacting are essential for technological advancements.

Case #4: On November 7, 1940, the Tacoma Narrows Bridge failed during a moderate 40 mph cross wind. Other bridges, notably the Golden Gate Bridge, had been designed with the same basic state of the art, but had not failed. Analysis done after the bridge collapsed showed that the bridge was not deep enough for its length. It was aerodynamically unstable and simply "flew" apart. No model wind testing for this phenomenon had been done prior to construction. Aerodynamic instability had never before been recognized as a cause of bridge failure.

Case #5: In 1943, Liberty ships were failing when used in the cold environment of the North Sea. Liberty ships were transport vessels, fabricated with welded rather than riveted seams, and used during World War II to ship soldiers and supplies. Many of the ships were cracking, and some sank because of the cracks. Engineers realized that welded construction did not allow for the suppression of crack growth as did riveted construction. The failures confirmed stress concentration theory because many of the cracks started at the square corners of hatchways. The failures also demonstrated that stress concentrations that are not critical when metal is above the transition temperature, become critical below the transition temperature. The materials react brittlely, and cracks grow rapidly and unexpectedly. Fracture mechanics analysis is now used regularly by design engineers.

Case #6: In the early 1950s, a de Havilland Comet airplane failed by exploding in flight. The prototype had received a test load at twice the expected load pressure, and all seemed safe; the source of the failure was a puzzle. It is now known that the overload testing resulted in a residual compressive stress at stress concentration points. A component containing residual compressive stresses will withstand more cycles of tensile loads than will a part that does not contain residual compressive stresses. Comet airplanes which had not been tested exploded in flight when a crack formed at the corner of a window. The internal cabin pressure caused the crack to grow at a high rate which in turn caused the fuselage to explode.

Case #7: In August, 1953, a fire in the General Motors transmission plant in Lavonia, Michigan made clear the need to design large buildings with fire walls, protection for steel roof framing, and sprinkler systems, even though the contents are not considered a fire hazard.

Case #8: In 1963, the Vaiont Dam in Italy withstood unusual forces in an impressive, yet disastrous way. A landslide caused a wave of water to pass *over* the dam and kill more than 2000 people in the valley below. The dam was strong enough to withstand the high forces, but the project was a disaster. The dam remains filled with 300 million cubic yards of earth and is useless for storing water or generating electricity. A design engineer must always consider the effect and consequences of designs on the surrounding environment; analyze the overall system.

Case #9: In 1966, a 14-foot long, earth-filled pocket was discovered in an eight-foot diameter concrete shaft during the construction of the John Hancock Center in Chicago. The shaft's top had shifted during construction, and an inspection uncovered the problem. Worker failure? Sabotage? Ignorance? Good design won't succeed with poor construction techniques.

Case #10: A rule of thumb occasionally heard is "If the building holds together while being built, it will hold together when it's complete." One reason behind this statement is that structures often have stresses placed on them during construction that exceed the stress placed on them after construction is complete. One such concern is the load caused by wet concrete and the low strength of uncured concrete. This was made clear in a dramatic fashion in March, 1973 when a 26-story building under construction in Washington, D.C. collapsed killing 14 workers. Support shores had been removed before poured concrete had become strong enough to hold up the ensuing construction. The collapse started on the 22nd floor, and all 22 floors collapsed to the ground.

Bridge construction is another area where construction loads and stresses may govern the sizes of many components. Many members hang in space as cantilever beams until other members are built up to them and properly attached.

Case #11: In May, 1976, the South Carolina Canadian Bridge fell, and two semitrailer truck drivers were killed. An expansion joint had locked, and all the bridge contraction due to temperature change occurred at the one joint that would move. One joint did not have enough safe travel to accommodate the contraction, and it pulled itself off the supporting structure. Poor maintenance on the joints? Poor design on the joint surfaces? Poor design for not considering the contraction accumulation at one joint? All lessons learned the hard way.

Case #12: In January, 1978, the roof of the Hartford Civic Center collapsed under the weight of snow. The roof structure had been designed with the help of computer analysis. The program, however, had been written with assumptions that did not "hold up." Even during construction it was noted by contractors that the roof was sagging. No concern was given to this observation for the designers felt confident the design was correct because a computer was used for the analysis. Computers can't make right what engineers don't start right.

Case #13: In 1979, a DC-10 airplane lost an engine in flight and crashed. Design and test data were reviewed, and no design problems were uncovered. After more than one accident occurred, further checking discovered that a normal engine service operation was causing the problem. The engines were periodically dismounted, serviced, and remounted. Mounting holes were becoming elongated during the service procedures and were the source of failure.

Case #14: In 1980, the Alexander L. Kielland, an offshore oil rig, broke up in normal North Sea weather. When components of the failed structure were analyzed, a three-inch crack was found in a part near a weld joint. Paint in the crack revealed that the crack had been there before the unit went into service. Poor material? Poor inspection? Tragic results.

Case #15: In 1981, the Kansas City Hyatt Regency Hotel had a skywalk collapse during use just after opening. The skywalk had been designed to hang from the overhead structure so that the open area below was not cluttered with support columns. After the accident, analysis revealed that a change had been made during construction. The change had not been checked out with proper calculations and had placed more load on a hanging joint than it had originally been designed to carry. When the live load of use was added, the total was more than the structure could carry and it collapsed. The Hyatt lobby was rebuilt using a proven design, that of placing columns under the structure.

Case #16: Also in 1981, Grumman Flexible buses used in New York started developing fatigue cracks in their frames. The design had used a low safety factor because of design restrictions. The buses had to be fuel efficient, smooth riding, maneuverable, lightweight, accessible for the handicapped, air conditioned, and low in cost. The number of conflicting requirements resulted in a compromised design that was too fragile for the use environment.

Case #17: In 1983, the Mianus River Bridge in Connecticut collapsed after 38 years of service when a section hanger cracked and failed. How many cycles had the hanger absorbed in 38 years? Post-failure analysis indicated that a combination of questionable design and poor maintenance caused the problem. Periodic bridge inspections had been made. Was the design done in 1945 consistent with the state of the art in that year? Don't judge an old design by today's standards, judge them by standards of the time they were built. If state of the art changes are significant, a structure may require strengthening or replacement.

Case #18: In 1986, Joe Walton built a log rack in Iowa to store firewood off the ground. He built the rack from standard two-inch water pipe and water-resistant treated wood. The pipes were driven into the ground for legs. The rack fell over during the first rainstorm because the low compressive strength of wet soil had not been accounted for.

Case #19: In 1987, a private Iowa college allocated money to update and rebuild its outdoor track facilities. Track widening and resurfacing were the primary objectives. As work on the project progressed, it was discovered that the original track surface was not level. In fact the enclosed football field and track area sloped nearly three feet from one end to the other. The project reportedly ran 33 percent over its budget of $150,000 just to fill in and make the track level. Why was the critical project step of defining the problem and collecting adequate data for planning and scheduling omitted?

Case #20: In April, 1988, the lesson of metal fatigue was again brought to attention when a Boeing 737-200 lost its cabin roof during flight. A preliminary report indicated that six metal fatigue cracks were found emanating from rivet holes in the aluminum skin. Inadequate test? Inadequate inspection? Improper use? More problems for the engineer.

Case #21: In the summer of 1988, beaches in New York and New Jersey were closed to the public when hospital waste washed ashore. Tests on the waste showed evidence of the AIDS virus. Accidental? Illegal dumping? Is the world population to succumb to its own waste? Pollution prevention and control is a full time job for an environmental engineer, and a concern of every engineer.

Case #22: In November, 1988 an arm on a carnival ride in Florida broke off and one person died and seven others were injured. The ride had been recently inspected and no problems were found. The arm failed at a crack that had been painted over and was not visible. The crack had been there for some time as demonstrated by the rust found under the paint. Poor inspection procedures? Painted to hide flaws? Fatigue continues to control the life of physical structures.

Case #23: Lest the impression is made that all engineering problems are structural and material oriented, consider the collision of two trains in August, 1988. The collision caused the evacuation of 1500 area residents because one of the trains had two derailed cars of denatured alcohol that caught fire. Three human errors contributed to the collision and the dangerous conditions.

1. The collision occurred 16 minutes before one of the trains was scheduled to leave the train yard. Someone made a bad decision on starting time.

2. The two cars containing alcohol were the second and third cars of the train, though company rules required them to be at the end of the train. Someone violated company rules.

3. The tracks where the collision occurred had no wayside signals because the tracks were regional tracks with limited traffic. Someone had decided the cost of the signals was not warranted.

It is unfortunate that some things have to be learned from mistakes. Mistakes and ignorance cost money, time, and often lives. That is why efforts between engineers with different specialties are commonplace today. There are so many aspects to large commercial projects that it would be impossible for one person to be able to be responsible for all of them. Even seemingly small projects such as creating a strain gage pressure-sensing system for roller bearing races requires the joint efforts of mechanical, electrical, metallurgical, and manufacturing engineers. A project such as a large building certainly needs the cooperation of many specialists such as structural, electrical, safety, civil, and mechanical engineers.

Similarly, the design of nuclear power plants requires the joint efforts of nuclear engineers and material engineers. Nuclear power plants use pipe made from stainless steel because of its resistance to corrosion. However, salt from any source in the presence of moisture will corrode stainless steel. Salt can come from the air or from the sweat of the workers who weld the pipe. Pipes are routinely inspected for cracks so they can be repaired or replaced before serious leakage occurs. Materials used in nuclear power plants are also exposed to neutron bombardment. When the plants were first designed, it was not known what effect this would have on the structural materials. It is now known that continuous exposure to neutrons will raise the transition temperature of the metals. The result is that sudden cooling of the material can cause it to fail in a brittle manner.

The more complicated a problem, the less likely it is that direct application of any one theory will solve it. The restrictions and assumptions made during theory development are one source of possible conflict. The problem at hand may have variations and options present that were not known or considered during theory development. The following examples illustrate how poor understanding of a new situation and improper application of a theory may also contribute to inappropriate analysis.

EXAMPLE 1.1:
Theory development assumptions.

During the development of many material deformation theories, a mathematical simplification about angles is usually made. If the angle of deformation is small enough, the angle in radians is equal to the tangent of the angle. This simplification is fine for "small" angles, but is inadequate as the size of the deformation increases. If you forget the theory development assumption you may incorrectly apply the theory.

EXAMPLE 1.2:
Temperature variation affect.

Recall that the strength of materials decreases as the temperature increases. If you forget this, trouble can occur. An unplanned temperature increase can occur if an object is placed in a confined space under the sun's rays. Temperature problems can also occur to a part that is ordinarily convection cooled if the cooling fan stops or the air flow passages become blocked. Unplanned events may happen if an object made from a thermoplastic material is left too close to an incandescent light bulb.

Your engineering training has placed emphasis on theories and known scientific principles. It is up to you to see that they are applied correctly. Many problems often arise because of decisions that "seemed like a good idea at the time." Always consider the total effect of what you do, not just present conditions.

Engineers have learned many lessons the hard way, but they have also achieved great successes. Following are some of the notable engineering achievements that are often taken for granted by many people:

1. The Brooklyn Bridge, which opened for service in 1883, was the longest suspension bridge in the world until 1931. The bridge is still in service, safely carrying traffic from Brooklyn to Manhattan.
2. The Panama Canal, a 51-mile navigable link between the Pacific and Atlantic Oceans, was built in the period 1905-1914, and required the movement of 268 million cubic yards of dirt.
3. The Mackinac suspension bridge, a four-lane, 26,444 foot connection between the upper and lower peninsulas of Michigan, replaced car ferry service in November, 1957.
4. The series of 29 locks and dams built on the upper Mississippi River controls water flow to maintain a nine foot channel depth necessary for river barge traffic from the Ohio River to Minneapolis-St. Paul, Minnesota.
5. San Francisco buildings built after 1906 to be earthquake resistant survived the October, 1989 earthquake.
6. The spaceship Voyager 2, launched in August, 1977, is powered by the heat from the radioactive decay of plutonium oxide. The craft sent images and data on the makeup of planets back to earth during its 12-year trip, culminating with a close up look at Neptune in August, 1989.

1.2 Sources Of Design Failure

There are many reasons why products fail to live up to expectations. Failures can range from a catastrophe like a bridge collapsing, to an inconvenience such as a door opener that reacts too slowly. Engineers must be aware of the reasons why these things happen and try to avoid the problems. Following is a summary of 14 of the more common sources of engineering errors, omissions, and failures.

The *problem to be solved may not be understood* because the originator of the problem did not convey information correctly. Associated with a possible poor problem description may be a lack of questioning. Even if you think you know what the problem is, ask questions. Review your impression of the problem with the originator to be sure you understand it. Added discussion can only help clarify the situation. Don't be in too big a hurry to get started on the solution. A bad start to a design problem almost always assures a bad ending.

The design may be based on *erroneous data*. If the design of a dam is based on a high water level of 15 feet, but the measurement was taken from an undefined reference point, the final design probably won't do the intended job. For example, a problem with a surveying job was solved by clarifying a reference point. Two different surveys located the boundary of a lot in different places. The difference was 40 feet, half a lot width. Additional checking showed that the two surveys were started at opposite ends of the section of land, and that the section was too big. An original survey had made the section half a lot too big.

Design errors can be made because of *invalid or overextended assumptions*. Errors of invalid assumptions are very difficult to trace because often a designer doesn't record why a particular design decision was made. Get in the habit of recording all your work, not just the conclusions. If a storm sewer design was based on average rainfall, but this was not documented, it would be hard to verify why the system flooded from time to time.

Faulty reasoning from good assumptions can also cause problems. Bad reasoning can occur when the premise is wrong, or if the premise is applied to the wrong condition. Assuming that 1010 steel is always ductile is a bad assumption. If heated and quenched, the 1010 steel might be harder than desired; if an impact load is applied, the material might react in a brittle fashion. Concluding that the sine $0.25_R = 0.25$ could be a bad conclusion depending on the accuracy required. See Exercises 1.7 and 1.8.

An *incorrectly stated principle* can lead to many design problems. The statement "A feather and a ball bearing will fall to the ground from equal heights in equal time" is true if the surrounding medium is a vacuum, but it is not true in normal atmosphere.

Improper experiments are often a source of problems. An experiment can be set up so that preconceived results will almost be guaranteed. If no control sample is run at the same time, the results may be attributable to the wrong variable. It might be possible that a variation was present that was not recorded, like line voltage variation in an electrical experiment, or temperature variation in a chemical experiment.

Poor data collection from a valid experiment can also result in invalid conclusions. Data recorded in experiments are subject to errors. Errors include the incorrect reading of instruments or incorrect recording of correct readings. Included in this can be the transposition of figures or illegible handwriting. Errors in copying can also occur when data are collected from handbooks or charts. The wrong row or column can be entered, or a critical footnote might be overlooked. In very critical conditions reading and recording by more than one person is appropriate. Often errors can go unnoticed for a long time as in this example.

EXAMPLE 1.3:
Data error.

Isaac Newton recorded and used an incorrect value for the distance from the earth to the sun and calculated a mass for the earth inconsistent with the value used in other of his calculations. The error was not discovered until 1987, 300 years later.

Good data can be made bad with *errors in calculation*. Incorrect use of a calculator or computer program will give errors, and the possibility is strong that an error will not be obvious. Practice with orders of magnitude review will help. Repeating calculations or verifying them graphically is suggested for critical designs.

Once a design has been determined, drawings of the parts to be manufactured or system flow charts must be created because drawings are the communication link between the design engineer and manufacturing. A *drawing error* that causes a breakdown in communication can create many problems, as the following example illustrates.

> **EXAMPLE 1.4:**
> **Drawing error.**
>
> A layout drawing of a tank to hold liquid detergent and parts to be washed was made at half size by engineer A. Engineer A went on vacation and engineer B was called in to complete the drawings. Engineer B was not familiar with the project and dimensioned the drawing as if it had been drawn full size. The result was a nice half-size model of the wash tank, but the tank was of no value to the production shop.

Many other drawing errors can occur: incorrect threads to fit a purchased part, wrong sizes because of a layout measuring error, incorrect tolerances for function, and errors in basic view representation.

Parts drawn correctly must be manufactured, and *incorrect manufacture* can cause a good design to turn bad. A part that was machined to size for test purposes, but flame cut to size for production, failed because of surface irregularity stress concentrations. Manufacturing control is just as important as using the correct design failure criteria during product refinement and analysis. *Incorrect assembly* can also cause properly machined parts to fail. A part that was spread and forced into assembly over two dowel pins failed because of the pre-use tensile load placed on a critical part of a casting.

A well made part must be shipped to the customer and protected from damage. *Errors in packaging and shipping* can cause even the best made products to fail. Parts being shipped are often subjected to loads and vibrations that are not present when the product is in real use. Springs and other structural members can accumulate many cycles of stress before the product ever reaches the customer.

> **EXAMPLE 1.5:**
> **Shipping problem.**
>
> One problem related to shipping was an axle bearing that became out of round due to transportation-induced vibration. The vibration peened a flat spot on the bearing. The load was being applied to the same place on the bearing with each cycle of vibration because the vehicle was tied down. The bearing joint design was changed, and modifications to the shipping procedure were made. Air pressure in the rubber tires was increased after tie down to reduce the spring effect of the tires.

Products that do not go directly to the customer must be stored safely so they do not deteriorate. *Incorrect storage* can cause products to deteriorate and fail when placed into service. Unprotected steel parts must be stored in low humidity areas, or coated with rust inhibitor to avoid rust. High temperature can also cause materials to deteriorate. Parts must be stored in such a way that creep distortion does not occur because of inadequate support.

A successful product must be used properly by the consumer, but *improper use by the consumer* is very difficult to prevent. Good owner manuals and safety labels are minimal protection against improper use. Over time, many products have become laden with safety features because of improper owner use. Examples include lawn mowers with blade-stopping interlock mechanisms, and snow blowers with auger-stopping interlock mechanisms. Improper use by consumers includes failure to perform regular maintenance. Engineers try to design around this problem by creating "maintenance-free" products such as cars that do not need chassis lubrication, motors with self-lubricating sintered bronze bearings, and house siding that never needs painting.

Designing and building products that meet all these potential problems and still allow for a profit is indeed a challenge. It points out the large responsibility that engineers have; engineering is not a profession to enter into lightly.

Engineering design principles do not change when problems are solved in areas other than your specialty. Principles don't change; only the problems and the applications change. Every job and profession has its own set of problems. Although there are some very specific considerations made in this book, such as the fatigue life of rotating shafts, a new problem may require that a known procedure be used in a new way. Whatever your job, and whatever type of business you find yourself in, the problem-solving approach can be used. You should have the potential to solve any problem that comes along. Even if you don't know anything about underwater agriculture or making parts on the moon, if the problem-solving procedure is followed and analysis is done properly, problems in these areas are not insurmountable. This is not to say that someone more knowledgeable than you won't solve the problems faster and arrive at more efficient solutions, but problem solving and engineering design activities will provide an acceptable answer if you perform them with purpose and desire.

1.3 Unsolved Problems

Many unsolved problems require creativity and inventiveness to solve. As technology advances, new levels of performance are required. People and products are extended. Rotating parts need to spin faster, resistance to higher operating temperatures is required, and equipment needs to operate in the weightlessness of outer space. Supersonic flight has its own set of problems, as does nuclear power. Larger and stronger components are needed for taller buildings, longer bridges, and higher-payload airplanes. Other products need to be smaller, such as electronic components and artificial human implants. Systems to control the latest innovations have to be more responsive to people and the variations that can occur.

The problems you work on as an engineer may be less conspicuous than those mentioned. You may work to make a failing part function as it was intended, or simply make it more reliable. Perhaps the problem is that one part of a system is interfering with another part of the system. Recall that some of the first computer systems interfered with radio and TV signals when units were too close together. You might have to redesign parts to make them lighter or smaller, or so they can be assembled more easily.

There are many areas of science and industry that have problems that have not been solved satisfactorily, or solved at all. If you are searching for problem areas in which to try out your problem-solving skills, you will have many opportunities.

Environmental pollution is one of society's largest unsolved problems. As the world population grows, it places demands on the supplies of food, clothing, and shelter,

FIGURE 1.1

Worn-out and discarded product disposal pollution.

and it also creates waste disposal problems as illustrated in Figures 1.1, 1.2 and 2.17. Some of the by-products of our society that must be handled include human body wastes (sewage systems), product use waste (garbage, both biodegradable and non-biodegradable), manufacturing wastes (coal ashes, spent nuclear fuel rods, and waste chemicals), and old buildings. Environmental problems must concern engineers as well as the general public, and require combined efforts to solve. These problems include human wastes, mining wastes, worn out motor vehicles, industrial heating and power needs, water supply, radioactivity (natural and human-made), power plant by-products (heat, air pollutants, and fuel waste), animal waste, foundry waste, home heating and ventilation, refuse disposal, use of pesticides and weed killers, fertilizer residue, detergent residue, carbon monoxide, sulfur oxides, hydrocarbons from combustion processes, and

FIGURE 1.2

Garbage disposal pollution.

nitrous oxides. Some of these problems affect our water supply, as shown in Figure 1.3, while others affect the atmosphere. Breathable air can be in short supply in areas where industry and people crowd together and cause air toxicity. An average person needs 1000 gallons of breathable air per day. Where does it come from? Figures 1.4 and 1.5 are examples of environmental air pollution.

Other real and immediate problems exist in the urban areas of the world. Food, water, clothing, and shelter are not always available to the people who need them. Engineers are involved in the systems and technology required to meet the demands.

Problems concerning **energy** need attention. Fossil fuels such as coal, oil, and natural gas are being used up and sources depleted. Coal, which is the most abundant fossil fuel, is higher in by-product contaminants than the less abundant gas and oil.

FIGURE 1.3

Industrial water pollution.

Special equipment is required to burn coal and control air pollution. Special safety precautions are required for transporting and storing natural gas and oil products. Research is necessary to make other forms of energy more convenient and efficient to use. Many solar energy products are on the market; maybe you can improve on their efficiency or reduce their cost. Solar energy is received at the rate of one calorie/cm^2/minute; how can a high percentage of it be captured efficiently?

Wood is a replaceable resource that is used for construction, heating, and paper products. How can wood's use be better engineered so less will do more?

Wind energy is studied and has been developed with little commercial success. The payback time for the required investment is too great for most people's thinking, while others think the rotating blades are too dangerous, or that windmills are too noisy. Can consumer fears be overcome while reducing cost and increasing efficiency?

Another problem facing our technological society is the storage and retrieval of **information**. People learn more, know more, write more, and want access to more information. Computers have made some of this job easier. Computers can search through compiled data. How does it all get compiled, and in what fashion so it can easily be found? The stock market is a good example of the use of computer storage and retrieval, but even there the time-honored paper stock certificate is still used.

Good communications systems are required as a result of the mass of information available and the variety of problems and projects people undertake. All forms of communication, oral, written, and pictorial, need to be considered in any system. Short term temporary communication is needed as well as for long term permanent records.

FIGURE 1.4

Industrial air pollution.

Society has come a long way with the telephone, television, radio, printing, telefax, and other methods of communication; but problems of accuracy, timeliness, availability, and permanence still exist.

Transportation still has problems, whether it is a short 0–100 mile trip or a long 1000–5000 mile trip, (see Exercise 7.16). Connections between types of transportation need improving. A one-hour, 500-mile plane ride can end in frustration during the one-hour, 5-mile taxi ride to the final destination. Even connections within a given transportation mode can be bad as anyone who has had to wait at O'Hare airport in Chicago five hours between "connecting" flights can attest.

As airways, highways, and waterways get more crowded, transportation safety becomes more important. Safety involves more than having a buoyant seat cushion available.

Medicine still has unsolved problems. Tools and instruments are needed for the delicate and sensitive work performed on the human body. Existing artificial body parts

FIGURE 1.5

Transportation air pollution.

can always be improved, and additional artificial body parts are needed. Studies on the effects of medicine on humans and animals need to be done. Will a person be safe over a period of years after taking combinations of a variety of medicines? The design and use of laboratory testing equipment are problems for the engineer.

Much work needs to be done to gain access to the products of the **ocean and other waterways**. The waters of the world are more than just a place to go fishing or a place to dump wastes. Fresh water supplies are a great concern to many parts of the world. Potentials of the ocean include mineral extraction and underwater agriculture. The world water supply may hold some of the answers to energy shortages. Figure 1.3 illustrates a growing problem with our water supply, industrial waste dumping.

There are also many problems of **product distribution** faced by manufacturers associated with supplying products to the market place. Products must be reliable and sell at a reasonable cost. Automation can increase the consistency of products manufactured, but the quantity supplied may be more than is needed. Versatility and alternate use of facilities need to be considered.

There are also many problems for humans caused by **natural events** such as earthquakes, floods, spontaneous combustion and fires, tornadoes, hurricanes, volcanic eruptions, soil erosion, natural atomic radiation, avalanches, landslides, or a tsunami. Figure 1.6 is but one example.

There are many problems and too few trained people and funds available to solve them. The solutions require a combination of present and new theory, and new applications of acceptable solutions. This means answers are not found in textbooks; new thoughts and efforts are required.

FIGURE 1.6

Erosion problem.

1.4 References

1. Alm, Rick and Thomas G. Watts. "Critical Design Change Is Linked to Collapse of Hyatt's Skywalk." *Kansas City Star*, July 21, 1981.
2. Bennett, Gary F., Frank S. Feates, and Ira Wilder. *Hazardous Materials Spills Handbook*. New York: McGraw-Hill, 1982.
3. Bolt, B.A. et. al. *Geological Hazards*. New York: Springer-Verlag, 1977.
4. Boraiko, Allen A. "Earthquake in Mexico." *National Geographic,* Vol. 169, no. 5, May 1986, 654-675.

5. Bronikowski, Raymond J. *Managing the Engineering Design Function*. New York: Van Nostrand Reinhold, 1986. Chapter 16 on product liability.
6. Chiles, James R. "The Ships That Broke Hitler's Blockade." *Invention & Technology*. Vol. 3, No. 3, Winter 1988, 26-41.
7. Constance, John. *How to Become a Professional Engineer*. New York: McGraw-Hill, 1978.
8. Cooper, Sol E. *Designing and Structures: Methods and Cases.* Englewood Cliffs, N. J.: Prentice Hall, 1985.
9. Cottrell, Alan. *How Safe Is Nuclear Energy?* London: Heineman, 1981.
10. Cross, Jean and Donald Farrer. *Dust Explosions*. New York: Plenum Press, 1982.
11. Dawson, Gaynor W. & Basil W. Merces. *Hazardous Waste Management*. New York: John Wiley & Sons, 1986.
12. Hopf, Peter S., editor. *Handbook of Building Security.* New York: McGraw-Hill, 1979.
13. Howard, Ross. *Acid Rain: The North America Forecast.* Toronto: The University of Toronto, 1980.
14. Kohl, Larry. "Wall Against the Sea." *National Geographic*, Vol.170, no. 4, October 1986, 526-537.
15. Lindgren, Gary F. *Guide to Managing Industrial Hazardous Waste.* Stoneham, Mass.: Butterworth, 1983.
16. Martin, Edward J. and James H. Johnson Jr. *Hazardous Waste Management Engineering.* New York: Van Nostrand, 1987.
17. Melaragno, Michele. *Wind in Architectural and Environmental Design.* New York: Van Nostrand Reinhold, 1982.
18. Monakhov, V.T. *Methods for Studying the Flammability of Substances.* New Delhi, India: Amerind, 1986.
19. McDowell, Bart. "Eruption in Colombia." *National Geographic*, Vol. 169, No. 5, May 1986, 640-653.
20. Ormerod, Richard. *Nuclear Shelters—A Guide to Design.* London: The Architectural Press, 1983.
21. Petroski, Henry. *To Engineer Is Human.* New York: St Martin's Press, 1985.
22. Petroski, William. "Eastbound Crew Claims Train Left Yard Too Early." *Des Moines Register,* August 3, 1988, 3A.
23. Putnam, Palmer Cosslett. *Putnam's Power from the Wind*, 2nd ed. New York: Van Nostrand Reinhold, 1982
24. Ramson, W.H. *Building Failures—Diagnosis & Avoidance.* New York: E & F.N. Spon, Ltd., 1981.
25. Rogers, Michael. "After Oil; What Next?" *Newsweek*, June 30, 1986.
26. Ross, Steven S. *Construction Disasters: Design Failures, Causes, and Prevention*. New York: McGraw-Hill, 1984.
27. Sykes, Jane, editor. *Designing Against Vandalism.* New York: Van Nostrand, 1980.
28. Wanielista, Martin P. et al. *Engineering and the Environment.* Monterey, Cal.: Brooks/Cole, 1984.
29. Zackrison, Harry B. *Energy Conservation Techniques for Engineers.* New York: Van Nostrand Reinhold, 1984.

1.5 Exercises

1. Find samples from your own life or that of family or friends in which the first solution to a problem did not adequately solve the problem and necessitated additional tries.
2. Research and write a paper on any of the product failures mentioned in the chapter, or on any product failure for which data are available. Include discussion of the failure, what happened, why, the "cost" to society, and how the problem was solved.
3. Research and write a paper on a "successful" design, such as the Brooklyn Bridge, the Mackinac Bridge, the Panama Canal, the Alaska pipe line, or the 1986 Voyager airplane.
4. Research and prepare a paper on the social and environmental implications for one of the following waste disposal projects.
    a) Sanitary land fill
    b) Garbage incineration
    c) Refuse resource recovery
    d) Separation and concentration of hazardous wastes
    e) Biological degradation of wastes
    f) Chemical treatment of hazardous wastes
    g) Waste recycling (paper, metals, glass, plastic)
    h) Radioactive wastes
5. Research and prepare a use proposal on one of the following energy sources. Include reasons for using and reasons for not using the source. Consider environmental effects.
    a) Nuclear
    b) Ocean thermal
    c) Ocean tide
    d) Solar
    e) Geothermal
    f) Wind
    g) Synthetic fuels
    h) Hydroelectric
    i) Coal
    j) Coal gasification
    k) Petroleum
6. Give examples from nature that illustrate the following concepts.
    a) Tubular structural members
    b) Levers and fulcrums
    c) Squeeze pump
    d) Aerodynamic streamlining
    e) Water repellent surfaces
    f) Ultrasonic communications
    g) Light sources
    h) Light sensitivity
7. What is the largest value of $\emptyset$ in degrees such that $\tan \emptyset = \emptyset_R$ to the nearest 0.01? 0.001? 0.0001?
8. Repeat Exercise 1.7 using the sine function. Compare the results.
9. Write an algorithm that determines the $\sin \emptyset$ if the $\tan \emptyset$ is known, but does not determine the value of $\emptyset$.

10. A 10 megawatt power plant burns coal for its energy source. How much ash is released into the atmosphere if the coal used is 8% ash, 65% of the ash reaches the electrostatic ash precipitator, and the precipitator is 96% efficient?

11. Find real life problems that are examples of the 14 points made in Section 1.2 about types of product failure.

12. If the coefficient of linear expansion of brick masonry is 0.0000031 in/in·°F, and that of concrete is 0.0000067 in/in·°F, what concerns must be designed for in normal concrete-masonry buildings? Quantify the problem of a 100 foot length wall and 100°F temperature change.

13. A 3000 x 4000 mm piece of 5 mm thick plate glass is used in the construction of a building.
    a) Consider what will happen when part of the glass is in the shade in the winter, and another part of the glass is in the sun and is warm to the touch, $\Delta T = 60°C$.
    b) Will thicker glass solve any potential problem?
    c) Consider two possibilities:
        (1) The glass is well grouted and free to move in the flexible grout.
        (2) The glass is oversized, the frame is undersized, and there is no grout between them.

14. The design for the center cable on a suspension walk bridge is to have an in-place span of 75 feet and a sag height of 17 feet. The cable is made from a steel alloy, $E = 30 \cdot 10^6$ psi, and supports an assembled load of 500 pounds per foot of walk-bed. The engineer in charge claims a 0.75 inch effective diameter cable, with an allowable tensile strength of 20,000 psi, will be sufficient for the job. Do you agree?

    Explain your conclusion. Reference equation: $T = \{wL/2\}\{\sqrt{1 + (L/4h)^2}\}$, where T = tensile force in the cable at the support, L = distance between supports, w = weight load/unit of length L, and h = sag height.

15. A 95 foot high cylindrical water tower is designed to hold 80,000 gallons.
    a) What is the minimum diameter of the tank?
    b) Is it likely that the diameter is larger than that calculated in part (a)? Why?
    c) What minimum compressive load is placed on the ground by a full tank?
    d) Using the answer from part (a) what amount of uneven settling will cause the tower to lean four feet out of line with vertical?
    e) The condition in part (d) happened to a water tower in Charter Oak, Iowa in 1988. Discuss the likely cause(s) of the leaning tower and why it should not have occurred.

16. In Exercise 1.15, part d, does it may make a difference if the four feet is measured on the surface of the ground or from the corner of the tank that has sunk below the ground? Calculate the amount of sinking using this assumption and compare with Exercise 1.15.

2 Engineering Activities

"Technology is a function of societal selection based on perceived needs—social and economic....Who perceives the needs and shapes the technology? And to whom are they accountable?"
♦ MARTY STRANGE, FAMILY FARMING

The purpose of this book is to investigate those things that are important for an engineer to consider while designing solutions to problems. The professional engineer practices the art of applying scientific theory and principles to the efficient conversion of natural resources for the benefit of humans, to satisfy their perceived needs and desires. Engineers link theory and technology to the needs and desires of the consumer. Part of the engineer's work during the linkage is to convert laboratory-scale operations to the industrial level, from test tubes and beakers to barrels, pipes, and pumps. Problems and restrictions arise during these transitions, and the engineer tries to find optimum solutions to the problems of delivering a product or service to the marketplace. The procedures used to find optimum solutions while solving these problems make up engineering design. Optimum does not mean the utopian best, only the best within the restrictions placed on the problem environment. Optimization techniques will be covered in detail in a later chapter. Engineers are also responsible, either directly or indirectly, for converting natural resources to usable form. Engineers may not dig the iron ore from the ground, but by specifying 1020 HR on a drawing, they are requiring that the digging be done.

2.1 Primary Job

The engineer's primary job is to solve problems relating to the marketing of a product: its creation, manufacture, distribution, safe use, and disposal. The product can be a physical object such as a building or a diesel locomotive, or it can be a system such as a scheduling procedure for using the locomotives, cars, and track of a train network. In the process of getting a product to market, an engineer may be responsible for a product's original design, the proverbial idea person. It is more likely, however, that day-to-day work will be related to many other activities required to get the product to the customer, such as the following:

1. Determine sales forecasts using marketing studies.
2. Determine product specifications as a result of market studies, consumer requests, and competitor product analysis.
3. Determine product cost, required capital, project profitability, and return on investment.
4. Determine project plans including personnel requirements and time schedules.
5. Establish acceptable levels of reliability, maintenance, and repair criteria.
6. Test products to verify levels of performance and reliability.
7. Determine processes and procedures to manufacture the product in a consistent and economical way.
8. Design the machines and processes needed to manufacture the product.
9. Design the systems that will use the product.
10. Write the product maintenance and repair manuals, or conduct the service training school sessions for product dealers and their personnel.
11. Assist with promotion plans and procedures for timely market introduction.
12. Be involved with 101 other necessary details to make a good product and a successful company.

College courses cannot include all the details of the necessary activities to carry out an assignment. Be prepared to learn new things when you begin nonacademic engineering problem solving in business and industry. Learning will include working with practicing engineers, reading current literature, attending seminars and other training programs, taking advanced classes in your specialty, taking classes in new technology, and trying out your own ideas. The designs of the engineer must keep pace with the expansion of technology and the state of the art.

As you read this book, keep in mind that lists, such as the one above, will not necessarily be complete; indeed, I hope you will be able to add to all lists that appear. Secondly, the order of items in a list does not imply an order of importance. The most important items depend on the situation in which you find yourself.

2.2 Necessary Skills

Engineers need many skills and talents to get the required jobs done. Among the talents engineers need are:

1. The ability to express problems clearly, both orally and on paper, using written and graphical representations. Communicating with others is essential, both in the activity of collecting information to solve a problem, and in the dissemination of results. See Section 2.8 for additional discussion on communications.

2. The experience and knowledge to propose ideas for the solution of problems. The range and possible success of ideas is enhanced as your experience and knowledge grow, because you will have more information to apply to the problem.
3. The knowledge and ability to analyze possible solutions for their true merit. This knowledge must include the ability to model problems using mathematics, to test ideas experimentally, to draw proper conclusions from the results, and not be influenced by the prejudice of others.
4. The ability to search out information, study it, and apply it to a new situation. Problem solving is made more efficient, and is directly proportional to the ability to use data sources efficiently.
5. The ability to project ideas from what is known to what is not known. Projecting ideas and state of the art to new applications requires creative thought, one of the most difficult things to learn or teach. See Section 3.2 for additional discussion on creativity.

Some people have a preconceived notion that all engineering work is directed toward the generation of inventions and new products. That is not true, and in general more of the engineer's time and effort is spent making an existing idea work as originally intended than developing new products. The primary job of a research engineer is to develop working prototypes of a problem solution. The primary job of a development engineer is to convert a working prototype of a problem solution into a market-ready, mass producible product. The work of any engineer involves trying new approaches, using old approaches on new problems, testing an idea to see how it works, making changes, trying again, trying a different approach, starting over, revising, reiterating, retesting, and compromising. Only then can a final decision be made which is the best under the given conditions and constraints.

2.3 Engineering Developments And Advancements

Engineers and scientists have built a vast storehouse of data and knowledge to deal with design problems. This knowledge consists of such items as:

1. Theories of the physical universe, many of which have been verified by experiments and observations. Refer to Section 5.2 for a summary of some of the most commonly used physical theories of the physical universe.
2. The data collected from experiments and observations that may lead to new theories and more experimental tests. The appendices of this book contain often used data derived from experiments and theory.
3. The mathematical models that are used to quantify and qualify the results of experiments and observations. Mathematical models are discussed in a later chapter, and are used extensively for quantitative manipulations.
4. The procedures used to control the behavior of products, processes, and people for the desired results. Efficient procedures are necessary to capitalize on theory, data, and models.

5. Rules of thumb, preliminary estimating procedures, orders of magnitude of estimating, and heuristics. A heuristic aids to the direction of solution, but it is not necessarily justifiable.

Engineers are not often thought of as experimenters and while some people think of scientists only in this way, both groups do experiments, however with different emphasis. Engineers are concerned with solving a problem that is of immediate concern, or at least is a known concern. A scientist may be working on an experiment and may not be involved with a current consumer problem. This difference causes an engineer to be under more pressure. Someone—the public, the government, or an employer—is waiting impatiently for a solution.

Engineers and scientists do share concerns. Both are dependent upon information and data that have preceded them. Developments are made, changes are implemented, and the wealth of knowledge expands. What is known today is more than yesterday and less than tomorrow. As knowledge expands, specialities are developed. Engineers specialize to keep up with the knowledge expansion. Civil engineering may be the oldest field of engineering because the first problems to be solved on a need basis were related to water supply, land division, sewage, and roads; today however, nuclear engineering and electrical engineering are disciplines in themselves because the need for these specialties has grown as technology has advanced and consumer needs have expanded. As knowledge and data accumulate, it becomes more apparent that divisions of effort are necessary. No one can remember all the data and knowledge available, so engineering becomes more and more a team effort.

The development of products, processes, and procedures follows the expansion of knowledge. The first self-propelled automobiles were similar in design to the horse-drawn carriages of the time. No one was concerned about suspension, tires, exhaust fumes, or driving in the snow. Nevertheless developments came, sometimes slowly, sometimes dangerously, but always as a result of trying to satisfy the needs of the consumer, or at least the needs the manufacturer thought the consumer had.

During the evolution of the automobile, for example, many developments occurred. Among these were progress in the development of metals and their alloys, plastics, electronics, rubber, glass, manufacturing processes, petroleum refining, paints, fabrics, heat transfer, storage batteries, safety, noise control, pollution control, road construction, bridge construction, and traffic control devices. Figures 2.1, 2.2, and 2.3 are samples illustrating some of the most obvious changes that have occurred in automobile design.

Developments that occur in one industry, such as the automobile industry, carry over to other industries. Like the automotive industry, the aviation industry is concerned with materials; they both need materials that improve their product's performance. The aircraft industry was delighted when a manufacturing process was improved to reduce the thickness tolerance of the airplane skin material to a smaller value. The smaller tolerance made possible planes with a higher guaranteed payload because engineers could calculate more accurately how much the plane would weigh. The automotive and aviation industries both make use of the high octane fuels from the petroleum refining process. Petroleum refining gives products other than gasoline, so other petroleum-based research and products, such as polymers and medicines, were developed. The successful nine-day global nonstop flight of the airplane Voyager in December, 1986 illustrates a technological advance in the use of composite polymer materials.

Many of the advances of modern technology are interrelated. When more iron is

FIGURE 2.1

Early vintage Cadillac.

FIGURE 2.2

1955 Ford.

FIGURE 2.3

1986 Mercedes.

required for ships, cars, bridges, and buildings, steel mills need equipment to produce more steel in less time, and to produce new alloys and new shapes. Mines need machinery to extract more ore in a shorter period of time. The ships that carry the ore to the mills need to be larger and faster. Figures 2.4, 2.5, 2.6, and 2.7 show some of the equipment that has been designed and built as a result of the increased need for iron and steel. Advances bring about needs, needs bring about problems, and problems challenge engineers. The problem range is almost limitless, from product development to efficient production methods.

The engineer spends time developing products, making them easier to use, stronger, more efficient, less expensive, safer, more reliable; or tries to adapt a product to a different application or environment. Consider how automobiles, airplanes, ships, buildings, and bridges have evolved since their first conception. Engineers are always trying to make things better. They push materials, processes, and procedures to the limit, often learning the hard way that a limit has been reached and that a different approach is needed. When the speed of airplanes exceeded the speed of sound, new forces, reactions, and design solutions had to be considered.

FIGURE 2.4

Equipment for underground coal mining.

Sometimes theory comes before application, while in other cases a physical phenomenon can occur and be used long before theories are proposed as to why it works. Electricity is an example of using a phenomenon for practical uses without "seeing" what is happening. When Edison worked on the design of an incandescent light bulb and the electric dynamo, he made improvements with limited knowledge of theory. He just knew his design was getting better, because he compared input and output of previous models to new designs. Although he could not see electricity, he designed to control the effects of it efficiently.

In metallurgy, it is known that when hot steel is cooled rapidly it is harder and can be manufactured into an edge that holds its sharpness. In ancient metallurgy practice, however, this hardening was thought to be a special event that was only successful when a hot sword was thrust through a slave; not a very humane process, but one that worked. Today we know why rapid cooling is effective in hardening steel, and we have developed efficient and practical methods of performing the process.

Product development often follows interesting paths. A major change in chain saw blade design came in the early 1940s when Joe Cox observed the way a timber beetle, *Ergates Spiculatus*, ate through wood. The blade improvement made a chain saw more efficient, and other improvements soon followed. Saw weight, for example, was reduced from 72 to 16 pounds for an equivalent machine between 1938 and 1966. Other

FIGURE 2.5

Equipment for open pit mining.

modifications were also made, such as spark arrestors to prevent brush fires, and design improvements to reduce vibration, noise, and exhaust fumes.

As engineers make technological changes, they must be alert to the side effects of the changes. For example, nonstick surfaces make cleaning of cooking utensils easy, but rob people of a source of iron in their diet. Cast iron cookware supplies trace amounts of iron to the food prepared in them. Today many people, women in particular, suffer from iron deficiencies and need to take iron supplements because they no longer use hard-to-clean cast iron cookware. The next example illustrates how complex project side effects can be.

FIGURE 2.6

Equipment to transport ore to refineries.

EXAMPLE 2.1:

Aswan Dam.

Side effects of the Aswan Dam, built in Egypt at a cost of $1 billion to control the Nile River, may have been overlooked or ignored. The primary purpose of the dam was to provide electric power and water for irrigation purposes. Side effects, which became apparent after the dam was built and placed in operation in 1964, include the following:

1. The dam prevents the annual flooding of the farm lands beside the Nile, which had brought nutrients to the soil, kept ground water levels high, and flushed salts from the soil. The farmers now have to purchase and add fertilizer to maintain the same crop level at a cost of $100 million per year, and the increased salt level reduces the overall productivity of the soil.

2. The dam prevents the nutrient laden silt from reaching the Mediterranean Sea, where it had supplied food for sardines, mackerel, shrimp, and lobster. The harvesting of these "crops" has about vanished from the area.

FIGURE 2.7

Equipment required to process the ore.

3. The clear water that now flows is washing out the river's bed and bridges downstream from the dam. Solutions to the washing out problem include the proposal to build additional dams between the Aswan Dam and the Mediterranean Sea at an estimated cost of $250 million.

4. The lack of silt is causing the delta areas to be eroded by the sea, and farm land is being lost.

5. The higher water levels behind the dam were coming close to washing out the 3000 year old Ramses Temple at Abu Simbel, and the temple was dismantled and moved to save it.

6. The lake and irrigation distribution of water has improved the habitat of snails that carry schistosomiasis, a disease that affects the strength of humans. The disease has spread, and there is no known cure.

7. It was projected that the dam would be full by 1970, but it was only half full in 1985 because of underground seepage and surface evaporation.

Engineers must be aware that the wealth of information available to them cannot be held in constant readiness to be applied the instant it is required. It does not take too long to become aware there is more to know than one person can keep up with; that's why specialties are developed. It is also apparent that good reference sources, such as those listed in Section 4.2, are necessary. Each area of engineering has its own emphasis, its own jargon, its own realm of expertise. Regardless of the area of engineering that becomes your specialty, the primary objective of your work will be to work with others to solve problems.

An engineering problem is often the direct result of trying to adapt to a new situation. Consider the Liberty ships that had been used satisfactorily in some waters, but began breaking up in the North Atlantic during World War II. In this case an existing design solution, a ship that carried men and equipment, had performed satisfactorily; but the new need, to move men and equipment in the cold North Atlantic sea, required a modified design. What was learned from this situation led to the specialty area of fracture mechanics, the study of cracks in metals and of crack growth under load-time-temperature variations. Engineers now know that cracks can grow at low temperatures, and at stresses below ultimate or yield strength levels.

Another problem that arose due to technological advances is the effect of neutron radiation on the materials used in nuclear power plants. Neutron radiation increases the

FIGURE 2.8

Truss-designed Queensboro Bridge, New York City.

temperature at which a metal changes its characteristics from ductile to brittle. Until nuclear reactors were used regularly, the rates of reaction to neutrons and the limits of change were not known.

The aerodynamics of bridges was not considered a design analysis criterion until after the Tacoma Narrows Suspension Bridge failed due to wind-induced oscillations. Bridges built before this time had been designed to withstand aerodynamic effects, but designers were not aware they had accounted for them. The Golden Gate Bridge, for example, was built on design principles similar to those of the Tacoma Bridge, but it had a different depth/width/length ratio and was safe from aerodynamic effects, something the engineers weren't aware of at the time of design. The variety of bridge designs shown in Figures 2.8 through 2.12 illustrates that no one design is always the best, and that good designs must be adaptable.

The development of automobile springs is a case where the best attempts at designing a more reliable spring, even with more expensive steels, did not solve the problem. When engineers decided to revert to less expensive steel, they discovered that the less expensive steels made the springs last longer. Later they learned that the lower grade steels were not susceptible to hydrogen embrittlement. At the time it was just not known what the operating characteristics and reasons for failure were, just that something else worked better. Today we know of other improvements to make springs more

FIGURE 2.9

Draw bridge, St. Joseph, Michigan.

FIGURE 2.10

Arch bridge in Salzburg, Austria.

reliable, such as shot peening the surface which was introduced to commercial production by the Associated Spring Corporation in 1929.

Additions to the information available to engineers may come as slow developments and improvements to existing products, or as breakthroughs in technology. Transistors, for example, made vacuum tubes obsolete, and are one of the biggest contributors to the practicality of small computers. Computers themselves made obso-

FIGURE 2.11

Walt Whitman suspension bridge, Philadelphia, Pennsylvainia.

lete such items as the slide rule and the mechanical calculator. Internal combustion engines made the horse-drawn carriage and buggy whips obsolete, but helped make engine-powered airplanes a possibility. Engines helped advance the conversion of energy forms including the capacity to generate more electricity. Electricity gave us the potential to design the telephone, telegraph, radio, television, and radar. Many technological advances are interrelated this way.

Some engineering designs are less sensational, but still important for our way of life. The design of a fire escape procedure from a skyscraper may never be put to the test, but if the procedure is not designed correctly, a fire can cause the loss of life, and may result in a lawsuit against the design engineer. The life boat shown in Figure 2.13 may never be used, but if needed, the launch system must function as intended. There won't be a second chance.

Designs used today are the result of many trials and failures. Edison tried more than 30 different materials, and in many different forms, in his search for an appropriate filament for the incandescent light bulb before he succeeded in finding one that lasted long enough for practical use. He proceeded with tests by trying alternative forms of the material, different filament geometries, and alternate filament support systems, always trying to improve the efficiency, output, and life of the light bulb.

Sometimes, failure is the result of not having enough knowledge, not making use of existing knowledge, or not having the right equipment. An early experiment performed to determine the speed of light is an example. Two men with lanterns stood on facing hills. The first man uncovered his lantern and started timing. When the man on the other hill saw the light, he uncovered his lantern. The first man stopped timing when he saw the light from the second man's lantern. By using the recorded time and the distance between hills, they thought the speed of light could be calculated. As we are now aware, this experiment was doomed to failure because the apparatus and technique were not capable of doing the job.

FIGURE 2.12

Arch truss bridge, Cape Cod Canal, Massachusetts.

2.4 Responsibilities

As engineers design and problems are solved, responsibilities are undertaken and engineers are involved in a wide variety of tasks. Some engineers work specifically on new product development, which arises when the consumer or a company recognizes a need that is not being met. New products range from a supersonic airplane to an

FIGURE 2.13

Life Boat Launch System.

automatic door opener for a building. The product can be in the public eye, such as the space shuttle, or as inconspicuous as a new coin sorter used at a toll booth.

Engineers probably spend more time improving existing products than doing any other design job, and many reasons account for this. A product may fail before its time; it may not perform as intended; or sales may be dropping due to competition. An improvement may be required to reduce cost, or a purchased part or material may no longer be available so a change in design becomes necessary.

Product modifications may be needed when a larger model or an adaptation of a product to a different environment is required. For example, an existing crane may lift 50,000 pounds (22,680 kilograms) to a height of 100 feet (30.5 meters) and someone wants to lift 70,000 pounds (31,750 kilograms) to a height of 130 feet (39.6 meters). This situation may be solved with an adaptation of a current product, or a new product may be required. The decision depends on whether an existing design can be modified and extended to meet the new objective, or whether a totally new concept is required. If a new concept is required to solve the problem, then the task belongs in the new product category. If changes in material, material size, or part configuration can be made to meet

FIGURE 2.14

Computer input tablet and specifications.

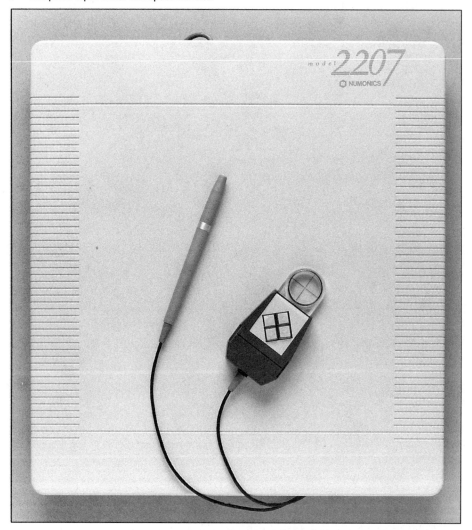

the new conditions, then a product modification or update is appropriate. The advantage of redesigning a current product is that existing parts, components, and design theory can be used. If a new product is necessary, additional time may be required to develop a new technology, capital expenditures may be required for new processes and facilities, and reliability and safety may become questionable.

Before serious design work is started, the design problem must be defined and specifications goals established. Product specifications may include requirements in any category including mechanical, chemical or electrical. A working capacity specification might be: dig 10 feet deep, lift 1000 pounds, travel at 10 miles per hour, withstand 2000 psi of internal pressure, process 1000 transactions an hour, operate at 120° C, consume

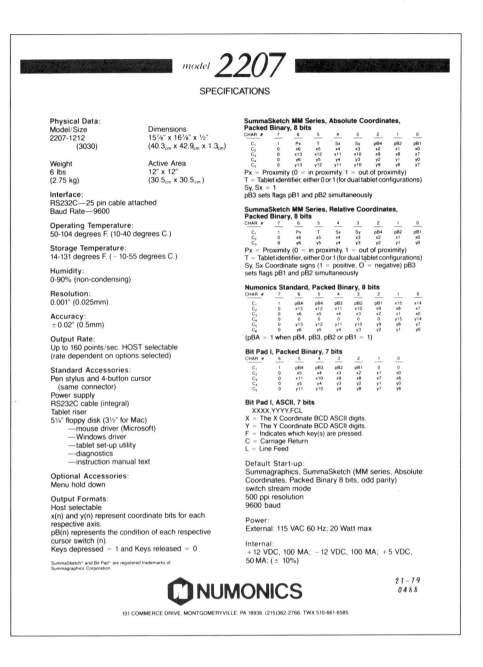

less than 5 kilowatts of power. Or they may involve physical capacities, a maximum or minimum weight or dimensions, color, and geometric shape. Specifications may change as a product is developed, and the final specifications may not be the same as the original. Specification goals are important however, so that development effort is focused in the right direction. Figure 2.14 is an example of a product and its final specification.

Some designs end up with "frill" specifications that do not contribute to the function of the product. Does a faucet have to be gold plated to function? No, but the contact points on an electronic component board might have to be gold plated so they won't corrode in use. Never confuse a well-designed product with an over-featured product. A Cadillac gets a driver to a destination as does a bicycle. The Cadillac just has more features, not the least of which is a roof to keep the occupants dry when it rains. Does a car need a computer to determine an estimated time of arrival, or is it just a nice feature?

Engineers spend much of their time determining part function, ranging from press fit assemblies to wind fit clearances. These decisions in turn determine part tolerances, manufacturing processes, and subsequently the cost. Deciding the function of a part so a product performs properly requires a combination of functional design tolerances, experience, and testing. An inexperienced engineer may specify small tolerances which are safe but expensive, while an experienced engineer will specify larger tolerances, adequate to make a part function, but at a lower cost.

A design cannot be manufactured unless drawings of the parts are made, as illustrated in Figure 2.15. The drawing includes sizes, tolerances, the material to be used, and other data that the engineer feels is necessary to insure a functional part.

The details on part drawings determine how parts are manufactured. Parts should be designed so they can be made with the least expensive processes available. The cost of tooling, gages, and machine tools as well as the amount of time to manufacture can be reduced with thoughtful component design. Parts are put together in assemblies and

FIGURE 2.15

Production part drawing.

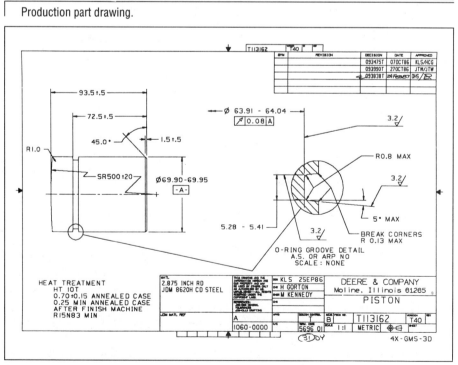

Courtesy of Deere & Company.

documented using a product bill of material. A bill of material includes not only the manufactured parts, but also the purchased parts such as nuts and bolts, nails, glue, washers, paint, and special packaging. Part names and required quantities are included. The bill of material is used for assembly purposes as well as for determining final product cost, and often includes assembly instructions. Figure 2.16 is an example of a subassembly bill of material.

Engineers also are responsible for the life of a product. Establishing life limits of a product is not done on an independent basis; input from the customer, edict from management, concern for the safety of the consumer, and legal restrictions must all be considered. Should the part last a week, a year, 100 years, 1000 cycles, 1,000,000 cycles? These are decisions that must be made, because product design is based on these decisions, and the decisions affect cost.

Engineering decisions affect product cost. The cost of a product includes the cost of the material from which the parts are made, the cost of the manufacturing time required to make them, and the cost of supporting facilities. Information needed to determine cost includes process times; required process tooling (dies, jigs, and fixtures), determined by manufacturing engineers; wage and overhead rates, determined by the accounting

FIGURE 2.16

Sample Bill of Material, Courtesy of Deere & Co.

AT 107691 SLIDING FRAME INSTALLATION

Quantity	Component Part Number	
1	AT 114649	Frame (Sliding with pistons)
1	AT 106584	Sliding Frame with bushings
4	R 81195	Ring, Back-up
4	R 83469	Packing, O-Ring
4	T 113162	Piston
0.001	UN 6012	Gals of oil, JDM J20A Transmission: Apply to O.D. of locking cylinder pistons before installing into sliding frame.
2	JD 7806	Fitting, Lubrication, 90 Degree
1	T 106066	Bar (Sliding Frame Locking Bar)
2	T 106398	Shim
2	T 106399	Shim
0.010	UN 5826	Lbs of JDM J13E3 Lubricating grease: Grease area of sliding frame which contacts lower rail of backhoe frame.
0.010	UN 6012	Gals of oil, JDM J20A Transmission: Apply to O.D. of locking cylinder pistons.
4	19M7369	Hex Screw
4	24M7241	Washer, Hardened

Position sliding frame onto backhoe frame making sure it seats completely down on either the top rail or bottom rail. Place lock bar into position. Install as many shims as possible under the lock bar (need not have equal amounts on each side). Now remove 2 mm worth of shims. Note: Thick shims are 3 mm and thin shims are 1 mm thick. Install cap screws with washers and tighten.

department; shipping costs; required return on investment, determined by first level management; and administrative costs.

An engineer is also responsible for specifying service requirements necessary after the sale. Regular service and recommended maintenance of a product are important so that it will function and last as long as intended. The preparation of instructions and requirements for service activities is as important as any other step in designing a product for the customer. Good service is needed to back up the sale so that the customer remains satisfied. The original design should always take into consideration how a product will be serviced and repaired. There is no value in placing a grease fitting on a moving joint if a maintenance person cannot get to it. Periodic adjustments that are hard to make probably won't be made. Clearly specifying maintenance procedure in owner's manuals is also a legal requirement meant to protect consumers from injury caused by a malfunctioning product.

Engineers must be concerned with safe and efficient ways to get a product to the customer in usable condition, as well as safe use by the consumer. Sharp edges and corners must be avoided, and glass and other fragile parts must be packaged safely and protected during shipment. Surfaces that could rust while being stored must be protected. Openings must be protected from the weather if outside storage is required. Special packaging is required when hoisting or forklift equipment is used for moving the product. If the limit on width of railroad or highway shipping is 12 feet, then the product must be designed within that limit. Some products are designed so they can be easily disassembled for shipment and reassembled upon delivery. Many products cost more to package, protect, and ship than to manufacture.

Safe use requires that the consumer be protected from potential dangers. In the age range of 1 to 44 years, more people are injured or die from accidents than from any other cause. Engineers cannot prevent all accidents from happening, but products must be designed to prevent normal-use accidents from happening. Consumers must be shielded and protected from very hot or cold objects, radioactivity, electric shock, chemicals, high sound levels, and excess vibration. Products must be designed to reduce the level of undesirable side effects. Ample and clear warning labels and instructions are a must.

2.5 Day-To-Day Concerns

Regardless of the role an engineer plays in getting a product from the conception stage to the customer, there are day-to-day concerns that are common to all engineering activities.

An engineer must be willing and able to accept an assigned responsibility, whether working for someone else, or self-employed. An engineer has to accept the challenge of the problem, make decisions, and accept the consequences.

Engineers must advance themselves and the company they work for by delving into areas in which they are not expert, as well as by probing deeper into their own area of expertise. Without taking some risk you and your employer can stagnate. As knowledge grows and new technology is developed, so must you grow and develop.

Engineers are as certain to make mistakes as the sun is to rise in the morning. In this respect engineers are no different from anyone else. To minimize mistakes engineers must check their work, test the possibilities, and check again, but mistakes will happen. An error may result in a catastrophic event, such as the failure of the Tacoma Narrows Bridge; or be as embarrassing as the wash tank that was drawn to half size, and then dimensioned and built as though drawn to full size. Two things are worse than making a mistake. From a personal development standpoint, worse than making a mistake is not learning from it. From a professional and ethical standpoint, worse than making a mistake is knowing it was made and not trying to correct it.

Trying to satisfy other people's needs and desires is the main source of engineering problems. Don't get trapped into thinking you are the only person a design has to satisfy. There isn't much profit to be made if the only customer of a product is the engineer who designed it.

All engineering projects are limited by the amount of money, time, material, or people that are available. A good engineer must make efficient use of available resources. Having resources left over at the end of a project is a sign of good management. Even if an overall budget is approved, it doesn't give the engineer a blank check for all the money so budgeted. Proper ethical decisions must be made. Each expenditure must be justified as supportive of the project objectives.

Engineering ideas need to be tested before the consumer has the opportunity to purchase a product. Testing may show up a problem that was not accounted for during the theoretical design of a product. Testing of a complete product or just a portion of a product may be appropriate. Frequently, testing will be used for design analysis because no combination of theories can be used to estimate an answer; the total design will be developed experimentally.

Engineers are also involved in training people who will use a new product; it may be different enough to require special instruction, or require the learning of new skills. Engineers may also have to train the people who will run product tests and experiments, or who are responsible for assembling a new product.

Engineers make decisions. Decisions are made on such details as material, material form, fabrication methods, and system controls. Final designs rarely satisfy everyone, and tradeoffs (compromises) among such items as cost, function, appearance, reliability, serviceability, and maintenance are the rule. Balance of goals is needed, and good engineers make judgments that lead to optimum, but user safe, solutions. Many designs have to be reviewed, tested, redesigned, and retested several times before satisfactory results are obtained. Don't be frustrated when an activity must be redone; knowledge and progress are taking place and the product is getting better.

All projects require the proper use and scheduling of people, machines, and money. Personnel support must be efficiently used or a project will fall behind schedule. Machines, whether they are testing apparatus or being built to manufacture the parts, must be coordinated so tests are run on time and production parts are manufactured on time. Money must be used efficiently, for if it runs out before the project is complete, an out-of-budget condition will exist, and a good project may be canceled.

Engineers also analyze project proposals so that resources are properly directed. Analysis includes activities that supply data so intelligent decisions can be made. This includes comparing designs, checking test data, running computer simulations, studying company warranty records, calculating critical sizes, and checking modes of possible failure.

Engineering projects and proposals must be communicated to others. Communication is one of the most used and most neglected skills an engineer has. Engineers must communicate to complete a project efficiently. How else can the details of what is needed be known, and how can a response to a design proposal be considered without good communication? Good communication is necessary as an engineer collects data and ideas; reports status; prepares work orders; reports test results; prepares drawings, specifications, and assembly procedures; or sells a completed project. See Section 2.9 for more discussion on communications.

2.6 Constraints and Limited Resources

Design optimization requires finding a compromise that is the best solution within the constraints of the problem. Constraints include anything that in some way controls the outcome of a project including time, money, personnel, facilities, and laws. Some of the constraints imposed on engineering projects are within the engineer's control, others are not.

A schedule of expected events by dates or amount of **time** is almost always a constraint for an engineering project, and usually outside the engineer's control. For example:

1. Build a bridge before the tourist season starts.
2. Build a boat before the ice goes out of the river.
3. Overhaul the boiler system before the winter heating season starts.
4. Finish remodeling the student union building before fall semester begins.

The following example illustrates problem solving under the restraint of time.

EXAMPLE 2.2:
Time constraint.

How many Ping-Pong balls will fit into the room you are in? You can probably think of several methods of obtaining an answer. Now add the restriction that an answer must be given in 30 seconds. Perhaps none of the ideas you have will work in 30 seconds and you will simply have to guess. Any guess seems appropriate under the severe restriction of a 30-second time limit. An answer might be: Greater than 1000, 10 million, or some other random guess. Unless you have remarkable powers, a 30-second answer has to be a guess.

Now suppose 30 minutes are available to determine an answer. In 30 minutes you can measure the room, calculate its volume, and divide by the estimated volume of a Ping-Pong ball. A Ping-Pong room might be close by so you can measure the diameter of the ball to calculate the volume more closely. But the answer is still an estimate, although probably a better one than the 30-second answer, for you still don't know for sure how many Ping-Pong balls will pack into the room.

Suppose 30 days are available to obtain an answer. In 30 days you might buy enough Ping-Pong balls to fill the room, and then count them. You might also be creative and just do part of the room and calculate a total answer by proportion. Either way your answer is more accurate than in the two previous methods.

A **budget** is a necessity for the control of a project. A budget lists the cost estimates necessary to get a job done, and projects a profit based on estimated sales and an estimated sales price. After a company agrees to a proposal, the amount of money committed to it must be controlled so funds are adequate to complete the project. If not, a financial loss can result, and the company may be forced out of business. The solution to the Ping-Pong problem may be restricted by a budget constraint if there is not enough money to buy enough Ping-Pong balls to fill the room.

Another limitation is **support staff.** A project is generally assigned to a limited number of engineers and supporting staff. No project can afford the luxury of unlimited staff.

An engineer is often told what to design based on **consumer needs** and abilities. These needs come from current users of your product, or through market surveys. A product that will be used by an average person must be designed within the limits of human size, reach, and strength.

Available materials are always a constraint within which you must work. The combination of desired material properties and what is commercially available, generally forces a compromise between such factors as weight, strength, corrosion resistance, electrical conductivity, machinability, cost, and delivery dates. This restriction also includes standard purchased parts such as nuts, bolts, springs, keys, light bulbs, and pipe.

Available manufacturing facilities are as critical as available material. If your company only has casting facilities for sand casting iron, and you design an aluminum part to be die cast, a compromise will be necessary. The part needs to be redesigned, purchased from another company, or new casting equipment must be purchased. Purchasing a part rather than manufacturing it may add lead time to the project or idle existing capital equipment.

Products must be designed to be used in **different environments**. A product designed to be used in the salt water of the ocean might be designed differently than a product used on dry land. A structure built in the earthquake zone of California might be designed and built differently than one in Chicago.

Another restriction that engineers must take into account are **laws, codes, and government regulations.** There are many laws that place restrictions on designs. The following laws, codes, and commission regulations are a sample of sources of constraints that may affect your design work. The example following this list illustrates the details that codes force an engineer to consider.

- Local building codes
- Federal Trade Commission
- Flammable Fabrics Act
- Refrigeration Safety Act
- Federal Boat Safety Act
- Highway Safety Act
- Toxic Substances Control Act
- Occupational Safety And Health Act
- Federal Housing Authority regulations
- National Traffic And Motor Safety Act
- Federal Food, Drug, And Cosmetic Act
- Federal Coal Mine Health And Safety Act

- Local fire regulations
- Department Of Labour
- Poison Prevention Packaging Act
- Magnuson-Moss Warranty Act
- Department Of Transportation
- Federal Railroad Safety Act
- Federal Hazardous Substance Act
- Environmental Protection Agency regulations
- Consumer Product Safety Commission
- Radiation Control For Health And Safety Act
- Federal Metal And Nonmetal Mine Safety Act

EXAMPLE 2.3:
Code restrictions.

An existing city building code contains a section on the construction of parking lots that includes the following restrictions and requirements.

1. No building permit or occupancy permit shall be issued, and no construction, grading, or other land development activity listed below may be commenced on property unless a Site Plan has been submitted and approved for such activity as set forth in this code.
2. The Site Plan shall include one or more appropriately scaled maps or drawings of the property clearly and accurately indicating the following:
    a) Complete property dimensions;
    b) The location, grade, and dimensions of all present and/or proposed streets or other paved surfaces and engineering cross-sections of proposed curbs and pavement;
    c) Complete parking and traffic circulation plan, if applicable, showing location and dimensions of parking stalls, dividers, planters or similar permanent improvements; perimeter screening treatment, including landscaping;
    d) Location and full dimensions of all buildings or major structure, both proposed and existing, showing exterior dimensions, number and area of floors, location, number and type of dwelling units, height of buildings;
    e) Existing and proposed contours of the property taken at regular contour intervals not to exceed five feet, or two feet if the City Planner determines that greater contour detail is necessary to satisfactorily make the determinations required by this ordinance;
    f) The general nature, location, and size of all significant existing natural features, including but not limited to sidewalks or paths, tree or bush masses, all individual trees over four inches in diameter, grassed areas, surface rock and/or soil features, and all springs, streams, or other permanent or temporary bodies of water;

g) A locational map or other drawing at appropriate scale showing the general location and relation of the property to surrounding areas, including where relevant, the zoning and land use pattern of adjacent properties, the existing street system in the area and location of nearby public facilities.

3. All Site Plans shall include a report or narrative containing the following:
   a) Legal description and address of the property.
   b) Name, address, and phone number(s) of the property owner(s).
   c) Name, address, and phone number(s) of the developer(s) or contractor(s), if different that the owner(s).
   d) Proposed use(s) for all non-residential buildings or structures.
   e) Data clearly identifying the following: total number and type(s) of dwelling units on the property; number and type of all structures or buildings, whether residential or non-residential; total area of the property; number of dwelling units per acre; total floor area of each building.
   f) Proposed landscaping schedule indicating plant types, number and timing for installation.
   g) Proposed construction schedule of all structures and physical improvements indicating the timing and sequence of each major structure and improvement.
   h) Present zoning classification(s) of the property.
   i) Present and proposed type and number of parking spaces on the property.

4. The Site Plan must show that a reasonable effort has been made to conserve and protect those natural characteristics that are of some lasting benefit to the site, its environs and the community at large.

5. Slopes which exceed ten percent shall be protected by appropriate measures against erosion, run-off, unstable soil, trees and rock. Measures shall be taken to stabilize the land surface from unnecessary disruption. The Soil Conservation Service shall be consulted for soil erosion control practices and allowable soil loss as permitted by the State of Iowa. Said erosion control and soil loss limitations shall be the responsibility of the property owner.

6. The placement of buildings, structures, fences, lighting and fixtures on each site shall not interfere with traffic circulation, safety, appropriate use and enjoyment of adjacent properties.

7. Adequate illumination shall be provided to parking lots, sidewalks and other areas for vehicular and pedestrians circulation. In no case shall illuminating devices be placed above fifteen feet in height in a residential district.

8. All parking spaces shall be clearly marked in accordance with Parking Lot Design Standards. Signs and pavement markings shall be used as appropriate to control traffic access and egress.

9. All areas designed for vehicular use shall be paved with a minimum of either an eight inch rolled stone base and two and one-half inch asphaltic concrete mat, a six inch Portland cement concrete pavement, or other equivalent pavement approved by the city. The paving surface must be so designed and maintained as to allow prompt and effective drainage of natural precipitation. No water drainage across sidewalks shall be allowed.

10. The minimum single car exit or entrance access lane width shall be fifteen feet.
11. The minimum combined exit and entrance access lane width shall be twenty-five feet.
12. Standard car stall depth shall be nineteen feet for 90° parking, twenty-one feet for 60° parking and twenty feet for 45° parking.
13. The minimum isle width for standard cars shall be twenty-four feet for 90° parking, eighteen feet for 60° parking, and thirteen feet for 45° parking.
14. Single-family, two-family dwellings, and parking lots shall maintain a minimum of twenty percent of lot area as a permeable and uncovered surface that contains living material.
15. All parking lots abutting a residential district or public right-of-way shall be screened from grade level to a height not less than three feet.
16. All commercial and industrial uses that abut residential, office, or institutional districts, shall maintain screening not less than six feet along the abutting property lines.
17. Screening will be equivalent to:
    a) Fences with at least fifty percent opaque construction; or
    b) Hedges, shrubs or evergreen tress of at least thirty percent opacity at the time of installation and fifty percent opacity maintained within three years of installation; or
    c) Berms or graded slopes of not less than three feet of mean height. Such berms or graded slopes shall contain at least fifty percent living material.

Product life is also a design constraint. Engineers must design a product to last an acceptable length of time. In some cases, such as a fire extinguisher, the life may be one use. In the case of an automobile valve spring, the life must be greater than 10^6 cycles. A typewriter ribbon may be designed for 10 cycles of use, oil in an engine crankcase for three months use. The life of a product can be measured either by the number of cycles used or a fixed period of time.

Product design is also influenced by the anticipated **disposal** after the product has served its useful life. For example, disposal of spent fuel rods from a nuclear reactor is an important concern. A product designed to be biodegradable might be better received by the consumer, and certainly makes for a better resource-use cycle than one that isn't. An engineer who is not concerned with the disposal of a used product is not concerned with the total solution of a problem. Figure 2.17 illustrates one major product disposal problem in the United States, junk cars.

The projected product **sales price** is also an engineering restriction. In most cases the sales price goal of a product is set before a project is complete, and it is up to the engineer to design a product whose cost meets the established goal. Cost invariably becomes the ultimate restriction, and engineering decisions affect product cost. If the cost is so high that no one will buy a product, it doesn't matter how well the product functions.

Another restriction, though sometimes nonfunctional, is one of **styling and appearance**. This restriction is more common in products that people will look at as well

FIGURE 2.17

Product disposal problem.

as use, as illustrated in Figure 2.18. Houses, cars, furniture, jewelry, and clothes fall into this category, and large companies employ engineers who spend the majority of their time on styling. On the opposite side, where styling and appearance have little consequence, are such products as drainpipes and electronic components that are rarely seen by the customer, and whose appearance is not critical.

An engineer might be told to design an **unnecessary feature** into a product, one that has nothing to do with the function of the product. Many of these features relate to appearance and styling just mentioned; however, they may also be features that are just nice to have, like automatic door locks and power seat adjustments on an automobile, or automatic re-dialing on a telephone. This does not imply that features aren't necessary or very useful for certain people; they just aren't needed by the majority of people. A remote control on a television, for example, is much needed by a bed-ridden invalid desiring to watch television, but it is not really needed by others.

Size and weight limits exist on the **shipping** and delivering of completed products. If a required product exceeds these restrictions, then final construction and assembly must be done at the use site, not at the factory. Prefabricated homes and offshore oil drilling platforms, Figure 2.19, are products which are usually assembled on site.

Engineers must specify all the **maintenance** required so a product remains functional. This includes specifying what must be done, and how often, such as greasing the bearings once a week, checking the fluid levels each day, or changing the oil every 2000 miles of operation.

FIGURE 2.18

Products which appeal because of styling

Company policy may also be restrictive. If the company you work for has a history of using hydraulic mechanisms, your superior might take a dim view of a pneumatic mechanism design.

2.7 Project Planning

Engineers are managers of time and other resources, and so they must plan, schedule, and follow-up on those plans. Plans for project implementation and logical schedules for events are crucial to the success of a project. Project planning starts with a statement of the intended goal. The more specific this statement can be made, the better the plan will be. An acceptable planning statement might be: Design a 100 horsepower, diesel powered bulldozer, and have it ready for sale in 19 months.

After the overall project plan is specified, the project is divided into major project activities. Completion times, personnel requirements, and cost estimates are made for each activity. The time limits and cost for each activity are based on similar completed projects, or from estimates by those who must do the work. Methods of making preliminary time and cost estimates include using a per part, per pound of product, per

FIGURE 2.19

Construction of an off-shore oil drilling platform.

floor of a building, or per cubic yard basis. Major product activities include first design, first build, first test and evaluation, first cost, redesign, retest, plant modification, build production tooling, establish production methods, final cost, and production. The actual list depends on the project (See Exercises 2.37, 2.38, and 2.39 for examples).

The sequence of project activities must be determined so an overall time plan can be developed and personnel, money, and plant resources scheduled. The sequence requires input from each department and work subgroup that has some responsibility for the project. This is an interdepartmental effort and includes product design, test and evaluation, purchasing, manufacturing, production, sales, service publication, plant layout, and accounting.

Most projects allow overlap of activities because design changes, redesign, rebuild, and retest can be done on component parts at the same time that overall product build and test is in progress. Parallel activities such as product cost, plant facility modifications, and sales literature design and promotional activities can be completed while product test and design are going on. Gantt charts are often used to map out the overall project schedule, showing the major activities and their overlap (see Figure 2.20).

Once a preliminary schedule of activities and times is presented, total period costs and personnel requirements can be estimated by passing down the Gantt chart for any interval of time. In Figure 2.20, for example, November, 1991 shows six activities occurring at the same time. This is only possible if the resources used for first build test are not required for second build test, as the two activities overlap. The resources include people as well as equipment. Either different resources must be made available, or the schedule of 19 months must be extended.

The advantage of the activity overlap of the Gantt chart is also one of its shortcomings. The overlap makes a critical time path difficult to determine. The general chronological order is down and to the right, but all of the preceding events do not have to be complete for the next level of activity to begin. In Figure 2.20, for example, six activities are shown to be active during the month of December, 1991. If a project is broken into smaller, more specific increments of work, or shorter elements of time, then progress can be checked more accurately; however, the chart can become large and cumbersome when more activities are added.

A network chart, often called a Critical Path Network (CPN), overcomes the problems of the Gantt chart because of the use of directed activity arrows. In a network chart it is assumed that an arrow activity must be complete before the next activity can begin. The time and effort required to conduct the detailed project analysis and determine activity prerequisites and logical activity order is time well spent. It forces attention on the details of a project and it may reveal a problem that otherwise might go unnoticed. Key personnel or critical test equipment may not be available at the time required to meet schedule. Plan ahead for special events such as holidays and vacations; schedule problems don't go away by neglect.

To use the network chart, you must break down activities into segments small enough to show their interdependence. Because of the sub-activity detail required, network charts quickly become complicated, and the use of a computer becomes helpful. Subdivisions of network charts are often based on subassemblies, material types, manufacturing method, or major functional groups such as engine, transmission, sheet metal, seats and upholstery for an automotive project. The extreme subdivision for a project development network is to list every part separately. This becomes necessary if specific parts require the use of new untried material or manufacturing processes.

Figure 2.21 shows an activity network and the interdependence and nondependence of the activities. Nondependent activities can be worked on at the same time. Vertical analysis of resources can be done on a network chart more efficiently if the activity arrows are drawn in proportion to their time estimates. If parallel activities do not require the same resources, then they can be completed anytime within the confines of the overall plan. Perhaps the greatest advantage of the network chart is that the critical

FIGURE 2.20

Sample Gantt Chart.

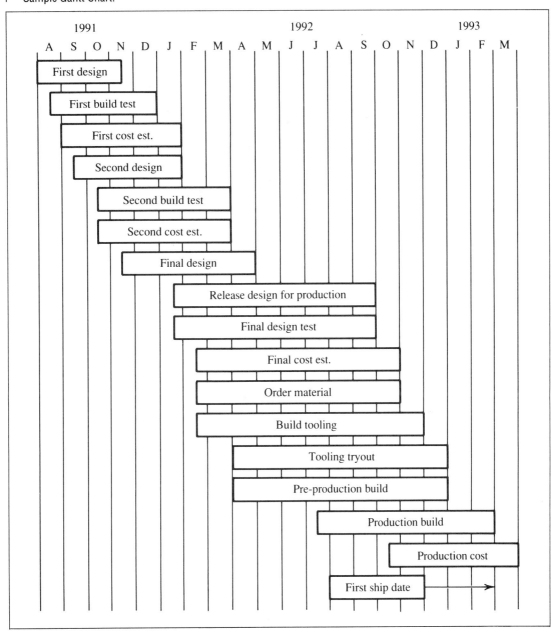

FIGURE 2.21

Sample Activity Chart.

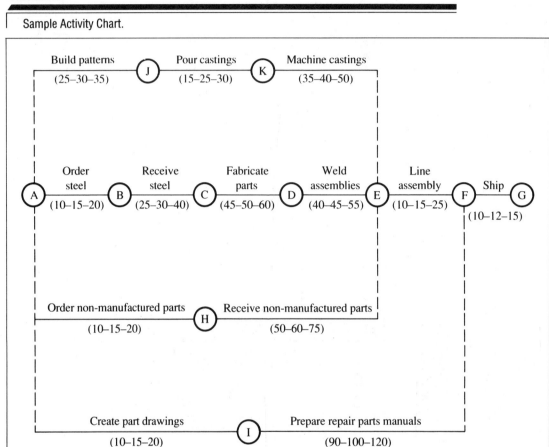

Numbers represent working days. (Optimistic Estimate–Normal Estimate–Pessimistic Estimate)

path can be determined. The critical path is the chain of dependent events that encompass the entire project, and whose total time is a maximum. Major management and work force emphasis should be placed on the current critical path.

In Figure 2.21 using the normal time estimates, path A-B-C-D-E-F-G is the critical path and projects a project length of 167 days. Path A-J-K-E and A-H-E cannot be the critical sub-path because their most pessimistic times of 115 and 95 days are less than the most optimistic time of 120 days for the parallel path A-B-C-D-E. Path A-I-F could become critical because its pessimistic time of 140 days is greater than the most optimistic time of 130 days for path A-B-C-D-E-F.

A Program Evaluation and Review Technique (PERT) chart is a network chart showing earliest and latest completion dates for each activity based on the no-slack critical path of activities. PERT requires constant update when activities are completed, and the results require many calculations that are usually implemented using available computer software. PERT on computer was first implemented in 1958 by Admiral William F. Raborn on the Navy Polaris missile project.

The main problem with using activity networks and PERT is establishing the various sub-activities, their interdependence, and an estimate of their completion time. A well documented PERT chart will be complemented with a table listing the earliest start date, latest start date, and any slack time for each event. The chart must be updated on a regular basis so project status is always current. The earliest start date for an activity is the first date the activity can begin based on the earliest completion date of all preceding events. The latest start date for an activity is the latest date an activity can be started so all succeeding activities can be completed by the project due date. An activity has slack if there is a difference between the earliest start date and the latest start date. Slack time might be negative if a preceding event is late in being completed. Negative slack requires a change in plan, such as changing the due date or reducing future activity time. Activity time can be reduced by assigning more people to the task or working overtime.

2.8 Communication

Communication is vitally important to an engineer working to solve a problem. An engineer receives problem statements from people who may have essential details specified incorrectly or incompletely; indeed they may not even understand the situation. Many people are contacted to establish the proper project groundwork so that money, time, and talent are properly directed.

Communication is also important in gathering support information. An engineer spends time gathering data from books, periodicals, other engineers, production workers, and maintenance personnel. During the life of an engineering project, no resource can be excluded as a source of information. In the search for problem understanding or problem solution, there is no substitute for good communication. The majority of engineering tasks are complex and related to other fields and disciplines. The ability to work with the ideas of others, and the results of their work, is essential to the efficient completion of any project. Communications usually involve both written and oral activities, and both are important as the following example illustrates.

EXAMPLE 2.4:
Typical communication situation.

A meeting had two project proposals on the agenda that were to be presented, discussed, and voted on. The person reporting on the first project distributed a half-page proposal summary. After ten minutes of discussion and a few clarifying remarks, the proposal passed and work started on the project the next day.

The second proposal was no more complicated, but was done orally only. After 20 minutes of confused discussion concerning implementation, a vague motion for acceptance was made which included the phrase "...and will distribute a copy of the final proposal." The motion passed but it wasn't clear if anyone really knew what they were voting for, nor whether the final form of the proposal would be well received. Work on the project didn't start for several weeks.

2.8.1 Written Communication

The need for good report writing cannot be overstressed because from time to time all engineering activities require written communication. Reports are written in response to questions such as: What is the problem with the new control system? What ideas do you have to fix the problem with the control system? What is the status of the warranty replacement program on the control system? What is the progress of the design of the new computer control system? Reports are used to communicate the results of a feasibility study which lays the groundwork for a new project. Periodic reports are also required to communicate the status of projects that cover extended periods of time. Status reports are required at regular time intervals or at major scheduled completion dates, and are critical for determining whether or not schedules and budgets are on target and if adjusting is required. Major scheduled completion dates include those times when budget requests are made, when an important product development step is complete, or when approval is needed for the next phase of a project. Report writing should be thought of as a talent to improve and polish just like any other skill. Regardless of how good an idea is, a report is needed to get approval for its implementation. Reports are also required to maintain contact with management, and to insure project support from other operating departments.

Engineers must be ready to communicate on the many occasions when reports are required. Your management or your client will require a report for a contract bid proposal, a project status report, or a report on a meeting. A request may come to prepare instructions on a new procedure, a new product specification, or a new product release notice. Reports are required as a result of research and development, product testing, or customer reactions to new products. Reports are also needed when trips are made to other companies, when inspection or product manufacturing capability studies are made, and for personnel evaluations. In addition, if you work for a large company the only way other people in the organization will get to know you is through reports you write.

Data and related information required for reports may come from laboratory experiments, library resources, product vendors, field tests, warranty claim records, letters from customers, market surveys, cost accounting records, or product scrap records. Some of the ways information is presented in reports include charts, graphs, photographs, lists, quotes from other people, copies of official documents, or a diary of chronological events (of great value during patent negotiations).

The first phase of an engineering project is usually a **feasibility report** written to consider whether or not a project should be pursued. This report is written to clarify an existing situation, to suggest solutions that are practical, to consider the proper technology to use, and to estimate costs and schedules. This report is made after the problem is identified, ideas are suggested, and preliminary analysis has been made, but before money is allocated to begin long term design, model building, and product testing. Space in the report is allocated to an accurate and complete discussion of the problem, discussion of solution ideas suggested, reasons why certain ideas were rejected, and reasons why specific ideas are considered for further study. A feasibility report may include multiple possibilities, the most acceptable being selected only after extensive research, testing, and development. The report will also contain a discussion of related problems such as manufacturing capacity, material availability, personnel training requirements, and the impact on other products presently being produced. See Section 5.5 for more on feasibility studies.

A **project proposal report** is made to request implementation of a project that has come through the problem solving and feasibility acceptance procedure. This report includes the final recommendation and details that led to the decision to recommend pursuing the project. It is not necessary to discuss all discarded ideas or give them equal time in the report, but it is important to include those that were in the final running, and the reasons why they were discarded. Give enough background to support the recommended decision.

Depending on the complexity and magnitude of a proposed project, and the assignment originally undertaken, the following items are included in a proposal report:

1. An abstract of the problem and the proposed solution, sometimes included in a cover letter.
2. Details of the problem situation as known, its background, needs, and scope.
3. Details of the proposed solution including economic feasibility arguments, the solution which at the time appears to be the best, budget needs, personnel needs, and a time schedule.
4. An appendix with materials that may be important for backup, but not required to follow the logic of the proposal.

There are occasions when the format of a report is preset. This usually happens when a proposal or report is being made to a government agency, another company, or an associated support foundation. Reports of this nature must follow the established guidelines. If they are not, a request for project funding may be denied just because the proper format was not followed; a bid for a job may be rejected even though the bid price is the best; or legal action may be instigated because terms of a contract are not followed. If a report is required before next-phase funding is released, and the information is not complete or not in the proper format, delays in funding may occur, projects will get off schedule, and personnel may have to be released.

A **project status report** is not as lengthy as a project proposal report because only activity related to the original schedule and budget is reported. The main points to include are:

1. Work completed since the last report and any special results.
2. Schedule status, including time and money.
3. Work to be done next, and any anticipated problems.

Extra information need be included only when it is necessary to emphasize particularly good results, or problems that may require extra budget, testing, personnel, or a schedule change. If serious problems occur, then the status report will be supplemented with a new proposal, including much the same information as the original proposal, but extended to rationalize the necessity for a change. In the discussion of the reasons for a change other people or departments are not to be attacked. More important are facts supported by test reports from the field or other reliable information, letters from vendors of material or parts, letters and other documentation from marketing people, or information on changes in the state of the art of the technology being used.

Technical writing is simple and to the point. For clarity, sentences are short and words used are common to engineering and understandable by the reader. Be sure that verb tense is consistent throughout the report. Double check pronoun antecedents; if in doubt repeat the proper name rather than use a pronoun. Do not use the pronouns he, his, him, she, or her unless in reference to a specific person. Avoid opinion when presenting data. Do not use ambiguous words and phrases such as large, lightweight, mostly, higher priced than, and more significant than.

EXAMPLE 2.5:

Communicating with specifics.

1. Rather than saying the mixture has significant amounts of water in it, say the mixture has 35 percent water in it and the next highest ingredient, latex, is 30 percent of the mixture.

2. Rather than saying the group was made up mostly of dealer representatives, say in the group of 50, 35 were dealer representatives.

3. Rather than saying the part is worn out, say the bearing is worn so that the shaft rotates 0.10 inch out of line.

In many ways writing a report is just like any other problem that needs solving, and the steps of the problem solving procedure can be of value. The following discussion offers suggestions to help you write better reports. You need to develop your own style and at the same time cover the necessities. Engineers develop an individual style of writing as well as a report draft procedure that works best for them. There are, however, elements common to many types of writing.

Begin a report by stating what you want the report to do; make a proposal, report a status, ask for action from someone else, request money, or change a schedule. This statement may not be in the final report but it will serve as your objective as you write the report. Write a draft or outline of your thoughts without specific regard for sentence and paragraph structure, spelling, or consistent tense. The purpose of the first draft is to get ideas down, as in the design problem solving procedure. The first draft can include phrases, incomplete sentences, or a list of words. Some writers, for example, outline all the points that need to be covered, place them in preferred order, and then go back and fill in sentence details under each point. Still others start with the conclusion and work backward through the logical developments that precede the conclusion. As with geometry proofs, the end conclusion is known, and the thought process works backward to the beginning assumptions.

Read the first draft over, looking for the best logical sequence of the elements of the message you have to deliver. Rephrase and combine similar thoughts in the same paragraph. Eliminate expressions such as more than, better than, higher than, significant amounts, made up mostly of, superior to, and safer than, and replace them with quantitative data such as $350 compared to $400, no failures in 10,000 hours of test, or lasted 1000 cycles compared to 600 cycles. Eliminate unnecessary words and condense sentences to basic information.

Rewrite your marked up draft. Start polishing by checking for proper word usage, proper spelling, consistent tense, and correct punctuation. Use a dictionary, thesaurus, or a guide to grammar and punctuation for reference. Generally other changes will come to light as this second draft is being worked on, perhaps a change of order of topics, or the addition of more recent test results. Reworking your own draft helps refine thoughts and their logical sequence. Some engineers think that typing should only be done by secretaries, and their own time is too valuable to be spent on redrafting. With the advent of computer word processing, however, writing and editing is done efficiently, and with the ease of making changes, writing quality should improve. This redraft-reread cycle may occur several times; it is similar to refinement in the problem solving procedure.

When the final draft is complete, allow some time to pass, such as overnight, or from morning to afternoon if time is short. Rereading a draft after a time break will permit better judgment on the completeness and coherence of the writing. Additional thoughts and developmental ideas may come to mind and can be included. Request a co-worker to read the draft to help you decide if the message is clear. If a co-worker has to ask a question for clarification, it is a sure bet that other people reading the report will have questions too.

As you reread the draft yourself try to read as the receiver of the information rather than the sender. Be certain you know the background of your reader. Make it clear if you want action taken, want a response of acceptance or rejection, want information, or want an activity started. Specify dates, dollar amounts, and people's names. A good report often has more influence on the final approval for implementation than does a good idea.

2.8.2 Oral Communication

Daily engineering activities and project work require communication, and information is regularly passed between people orally, in a casual manner. Oral communication is usually used when instructing someone on a new procedure, explaining how to use a new piece of equipment, clarifying a new operating policy, or informing someone about an event of mutual interest. Oral communication, used more than written communication, has a problem that written communication does not: oral communication cannot be studied by the receiver for full and complete meaning. For this reason effective oral communications and presentations require some approaches that are different from those for written communication.

Oral communications require information to be passed in smaller increments; long complicated sentences that are written and can be reread, studied, and interpreted are likely to be confusing when spoken. A report that is read out loud requires too much work on the part of the listener, and is usually ineffective because too much information is coming in too short a period of time. Most listeners do not like to be read to; it makes them seem uninvolved and unimportant.

Oral communication is more effective if important statements are made shorter than when writing. Listeners, even if they take notes, need information in small pieces so it can be digested before more information can be received and processed. Since a good oral presentation will be easy to outline by the listener, prepare an oral presentation by outlining the points you wish to make. Use one note card or one piece of paper for each major point. Write larger than normal and leave space between lines so you can read easily.

The main advantage of oral communication is that it permits an exchange of ideas and information over a short period of time; clarifying questions can be asked and clarifying statements made. To be effective, good oral communication requires some feedback so the giver of the information knows if the information has been received correctly. This can be accomplished by asking questions of the receiver or by having the receiver write the information as it is heard, interpreted, and will be used. General discussion between the giver and receiver, along with examples, assist the oral transfer of information.

There are many times, however, that individual discussion and question and answer feedback are not practical, such as a presentation before a large group, or when the presentation is recorded for future use. For occasions like those it is better to be overprepared for a presentation than underprepared. Careful and complete preparation for a presentation, informal or formal, will result in a smoother and more effective transfer of information.

Some oral report preparation is the same as for a written report. The background and technological level of the audience must be known so the presentation can be given at the correct level; terminology must be appropriate, the common interest of the group must be pursued, and the purpose of the presentation must be clear. Is the presentation being made to inform the group of a new project proposal? A present project status? A request for additional time and money? Different purposes require different approaches.

Concern for the length of an oral presentation parallels that of a written report. A presentation that is too long will lose listeners just as a report that is too long will lose readers. Be well organized; get to the point, state the facts, summarize, and stop. A report over 20–30 minutes may be trying to cover too much information, but if more time is required be sure the presentation is straightforward and logical, and not a rambling sermon.

A strong introduction is important to attract and hold an audience's attention. When delivering a technical report do not use the typical after dinner speaker's opening joke. State concisely and precisely the purpose of the meeting. The following openings may be appropriate:

1. To office workers who may have to alter their parking procedures. "I want to explain to you what will happen next week when construction begins on the building addition."
2. To department members prior to the vacation season. "I am going to review the vacation policy because last summer there were several problems that occurred."
3. To engineers who will be designing parts and to engineers who determine production manufacturing procedure. "Manufacturing has installed a new metal spraying process. It is important we all understand the advantages and limitations of this process."

Oral presentations can usually benefit from visual aids, slides, videos, poster-size graphs and charts, and overhead transparencies. These should be prepared ahead of time and reviewed by practicing the presentation. The proper timing of visual aids is important. Don't show a slide and then say "I'll get to this in a few minutes." The visuals must keep pace with the presentation.

If overhead transparencies are used, do not turn your back on the audience to read the information to them. Stand near the projector and use a sharp pencil to point to key items as you discuss them. Limit the amount of information on each overhead to one idea or an outline of main points; too much information presented at one time will not be easy to absorb and understand. Be sure your overheads can be read from the back of the room you are using for the presentation. Practice ahead of time, and check to be sure a spare projector bulb is available.

The use of slides can also enhance a presentation. The point of each slide must be appropriate, and as with overheads, they must be visible from the back of the room. Try to avoid complete darkening of the room to project visuals, as many people like to take notes and a dark room will prevent this.

The use of video is often appropriate because it can show motion which is not possible with slides or overhead transparencies. Be sure the point of the video is relevant. The shortcomings of videos include:

1. Poor visibility because the television screen is too small to be seen from a distance.
2. Improper sound levels, too loud for those up front and too quiet for those in the back.
3. Inattention because watching television is usually considered entertainment, not information sharing.

Do not talk while a sound supported video is showing. If comments are required use a remote control and stop the action, or tell the audience ahead of time what to watch for. Video displays should not monopolize the meeting, only support it.

If charts and graphs are to be displayed without using slides or overheads, prepare them in advance of the meeting. Drawing graphs and charts during a presentation is not a skill most people have, and it takes time away from the meeting. Make the charts large enough to show necessary detail and to be seen clearly from the back of the room.

When appropriate, use real samples of the object under discussion. Samples must be large enough to be seen clearly or small enough to pass around. Scale models are also appropriate if the group is small, as at a meeting for a project proposal presentation.

If the information to present is highly technical, or there are many details, consider distributing a handout that contains the major points. Leave space on the handout for additional notes. The last page of a handout can be a questionnaire if you require meeting feedback.

Always practice an important oral presentation ahead of time, preferably to someone, or to a small group of people, who will give you honest feedback. Practice will give you a chance to check the length of the presentation and the appropriateness of visuals, as well as logical order and completeness. Practice in the room the meeting will be held in if possible. This will let you check the effect of visuals, and let you determine the proper voice volume to use. This is particularly important if a sound system is not available.

If a presentation is made as a result of a project feasibility study, or is a project proposal review in conjunction with a written report, be flexible; it is not uncommon for managers who have not thought about a particular problem for a while to change the boundary conditions and restrictions at the last minute. Try to anticipate the consequences of different conditions by considering alternate courses of action before the presentation.

Pace the presentation to a level you would use if talking one-on-one with a co-worker. An audience will tune out quickly if you talk too fast. Use a pace that allows you to emphasize the key points and give the audience a chance to absorb them. Do not fill in normal conversational pauses with "you know," "right," "um," or "uh." A momentary pause is normal; every second does not have to be filled with words. Maintain eye contact with the audience. A speaker who never looks up from notes quickly loses the communication link. Stand on both feet with your weight balanced; don't put your hands in your pockets and jingle coins or keys. Keep your hands away from your mouth so you do not mumble; articulate clearly and complete the endings on words; working, not workin'.

At the end of the presentation summarize the main points. If there is time ask for questions. If questions are asked and you are in a large room, repeat the question before answering so you keep the audience together. If the presentation is limited by time and time is up, stop. End with a "thank you."

2.9 | Personal Characteristics and Abilities

Regardless of the problem situation an engineer is in, to get all the necessary jobs done and solve problems efficiently, many talents, favorable personal characteristics, and abilities are needed. The most important ones include:

1. *Technical competence.* Engineers must know the details of their specialty area and how their area interacts and works with other areas.

2. *An understanding of nature.* Nature's resources, laws, and governing principles must be used to advantage when an engineer designs an optimum solution to a problem. An engineer has to know how nature works. Section 5.2 lists some of the laws of nature to consider when solving engineering problems.

3. *Empathy for the requirements of other people.* An engineer who can stand in the shoes of another person can better reach an optimum solution to that person's problem.

4. *Active curiosity.* Curiosity may have killed the cat, but curiosity allows an engineer to seek out conditions that may influence the results of design work. Seeking answers to "What if?" questions is a sign of curiosity.

5. *Ability to observe without judgment.* Prejudging a condition restricts thinking and prevents an optimum solution from being chosen. Observation and recording of data and details in an impartial way lays the best groundwork for an efficient and economical solution.

6. *An innovative mind.* An innovative engineer will continually think of ideas and solutions that are better than average, easier to implement, and less expensive. Chapter 5 reviews methods which assist innovative idea generation.

7. *Motivation to design just for the pleasure of accomplishment.* This is just a fancy way of saying that an engineer must enjoy the work of design. Self-motivation is an important ingredient to job satisfaction. An engineer must want to get a job done, and get it done well.

8. *Confidence.* Engineers must be confident in their talents, education, and skills. They should feel confident to take on any task in their realm of expertise, even if it is beyond their present status.

9. *Integrity.* Engineers must be honest with themselves and those with whom they work. Talents must be used to support the goals of those who are paying the way. Decisions must be the best possible under the company restrictions. The chapter on ethics includes some concerns that an engineer must consider when making decisions.

10. *Willingness to take risks and assume responsibility.* Some risk is always present; nothing can be certain. At many times in the design process decisions will be required. Engineers have to take risks based on the data available, and be ready to accept the responsibility of their actions.

11. *Capacity to synthesize.* One of the most common methods of problem solving is to synthesize, that is, to combine principles and ideas that have worked before in a manner that will solve a new problem. Many problems are solved by using old solutions in new ways or in new combinations. Creativity is important when synthesizing; see Section 3.2 for hints on being more creative.

12. *Persistence and a sense of purpose.* Some problems require many attempts before a satisfactory solution is found. Keep trying. Iterations, retests, and rebuilds are common procedures in problem solving. Engineers must keep goals clearly before them so work is concentrated and focused toward goals.

13. *Patience with products and people.* Engineers who are patient can keep their mind clear to think of better ideas and are able to think of more possibilities. By being patient, work will be careful and meticulous. This will help prevent errors of judgment as well as errors in computation and data collection. Patience will also help when working with people. The result will be more cooperation and more useful communication.

2.10 References

1. Acret, James. *Architects and Engineers: Their Professional Responsibilities.* New York: McGraw-Hill, 1977.
2. Burstall, Aubrey. *A History of Mechanical Engineering.* Cambridge, Mass.: M.I.T. Press, 1965.
3. Chiras, Daniel D. *Environmental Science.* Menlo Park, Cal.: The Benjamin/Cummings Publishing Co., 1985.
4. Diesch, Kurt H. *Analytical Methods in Project Management.* Ames, Iowa: Iowa State University, 1987.
5. Drescher, Nuala McGann. *Engineers for the Public Good.* Buffalo: U.S. Army Corps of Engineers, 1982.
6. Florman, Samuel C. *The Civilized Engineer.* New York: St. Martin's Press, 1987.
7. Friedel, Robert and Paul Israel. *Edison's Electric Light.* New Brunswick, N.J.: Rutgers University Press, 1986.
8. Gaither, Norman. *Production and Operations Management.* Hinsdale, Ill.: The Dryden Press, 1987. Chapter 10, Resource Requirements Planning, and Chapter 15, Planning and Controlling Projects.

9. Garvey, William D. *Communication: The Essence of Science.* Elmsford, New York: Pergamon Press, 1979.
10. Gastel, Barbara, M.D. *Presenting Science to the Public.* Philadelphia: ISI Press, 1983.
11. Glorioso, Robert M. and Francis S. Hill Jr. *Introduction to Engineering.* Englewood Cliffs, N.J.: Prentice Hall, 1975.
12. Halcomb, James. *Project Manager's PERT/CPM Handbook.* Sunnyvale, Cal.: Halcomb Associates, 1966.
13. Hajek, Victo. *Management of Engineering Projects.* New York: McGraw-Hill, 1984.
14. Howell, James F. and Dean Memoring. *Brief Handbook for Writers*, 2nd ed. Englewood Cliffs, N.J.: Prentice Hall, 1989.
15. Kamm, Lawrence J. "On the Road to Success." *Design News*, March 27, 1989, 158-166.
16. Katz, Michael J. *Elements of the Scientific Paper.* 3rd ed. New Haven, Conn.: Yale University Press, 1985.
17. Kemper, John Dustin. *Engineers & Their Professions*, 3rd ed. New York: Holt Rinehart & Winston, 1982.
18. Kemper, John Dustin. *Introduction to the Engineering Profession.* New York: Holt Rinehart & Winston, 1984.
19. Koen, Billy. *Definition of the Engineering Method.* Washington, D.C.: American Society For Engineering Education, 1985.
20. Krick, Edward V. *An Introduction to Engineering and Engineering Design*, 2nd ed. New York: John Wiley & Sons, 1969.
21. Lannon, John M. *Technical Writing*, 3rd ed. Boston: Little, Brown & Company, 1985. Many good examples.
22. Lesikar, Raymond V. *How to Write a Report Your Boss Will Read and Remember.* Homewood, Ill.: Dow Jones-Irwin, 1974.
23. Lesly, Philip. *How We Discommunicate.* New York: AMACOM, 1979.
24. Markel, Michael H. *Technical Writing,* 2nd ed. New York: St. Martin's Press, 1988.
25. Martina, R.L. *Project Management.* Philadelphia: Management Development Institute, 1968.
26. Mayr, Otto, ed. *History of Science.* New York: Science History Publications, 1976.
27. Meredith, Dale D. et al. *Design and Planning of Engineering Systems.* Englewood Cliffs, N.J.: Prentice Hall, 1973.
28. Miller, Tyler G. Jr. *Living in the Environment*, 5th ed. Belmont, Calif.: Wadsworth Publishing Co., 1988.
29. Petroski, Henry. *To Engineer Is Human.* New York: St. Martin's Press, 1985. Easy reading.
30. Red, W. E. and Benjamin Mooring. *Engineering Fundamentals of Problem Solving.* Boston: PWS Engineering, 1983.
31. Ryckman, W.G. *What Do You Mean by That?* Homewood, Ill.: Dow Jones-Irwin, 1980.
32. Sandler, Ben-Zion. *Creative Machine Design.* Jamaica, N.Y.: Paragon Press, 1985.

33. Sherwin, Keith. *Engineering Design for Performance*. New York: John Wiley & Sons, 1982.
34. Thomas, Elain, et. al. *Safe Chain Saw Design*. Durham, N.C.: Institute for Product Safety, 1983.
35. Van Fleet, James K. *The 22 Biggest Mistakes Managers Make and How to Correct Them*. New York: Parker, 1973.
36. Vidosic, Joseph P. *Elements of Design Engineering*. New York: John Wiley & Sons, 1969.
37. West, Jerome D. and Ferdinand U. Levy. *A Management Guide to PERT/CPM*. Englewood Cliffs, N.J.: Prentice Hall, 1969.

2.11 Exercises

1. Consider each of the following items and outline a sequence of product development. Include technical advances and any products made less important or obsolete in the process.
    a) Television
    b) Computers
    c) Ships at sea
    d) Travel on snow
    e) Bridges
    f) Firearms
    g) Writing with ink
    h) Roads
    i) Airplanes
    j) Electric lights

2. Give examples of products for each of the following restrictions.
    a) Is to travel no faster than 25 mph.
    b) Must be rectangular in shape.
    c) Has a given maximum weight; specify the weight.
    d) Has a minimum weight; specify the weight.
    e) Has a required color; specify the color.
    f) Has a minimum width, length, or height; specify the dimensions.

3. Give an example of a new product on which an engineer might work.

4. Give an example of a product you are familiar with that has a functional problem, and for which a redesign or modification seems appropriate.

5. Give an example of an existing product that might be modified to fill a related but different need.

6. Find a product specification sheet for a product such as an automobile, backhoe, stereo, pump, or electric motor, and determine if the specifications are easy to interpret and whether they are usable to make comparison judgments.

7. Find specification sheets on two or more competitive products and chart a comparison of the items listed.
8. Prepare a bill of materials for a small product such as a chair, table, or pencil sharpener. Include part name, material, basic function, and whether you think the company manufactured or purchased the part.
9. Obtain and read a service, repair, or installation instruction. Evaluate how well the description is written. Is there a chance for misinterpretation or serious error?
10. Study how some product is packaged and evaluate its effectiveness.
11. Give examples of products that sell primarily because of their necessary features.
12. Give examples of products that sell primarily because of the features that are not necessary.
13. Analyze some different products and distinguish between necessary functions and unnecessary, though nice, features.
14. Select a product for analysis and judge which components were purchased by the manufacturer and which were not. Use a product such as a desk lamp, pencil sharpener, stapler or some other small product that can be easily taken apart or for which all parts can be seen without disassembly.
15. Make a list of possible filament materials Edison may have tried while he was developing the incandescent electric light bulb.
16. If an airplane has 5000 square feet of surface area covered with 0.100 inch thick aluminum, by how much could the advertised payload change if the process tolerance used to manufacture the 0.100 aluminum was reduced from ± 0.010 to ± 0.005?
17. Find physical examples of parts that are:
    a) Wind fits, where no physical object comes closer than the wind.
    b) Assembly clearance fits, where parts must clear each other so assembly is possible.
    c) Press fits so parts stay assembled.
    d) Close tolerance no-interference fits.
18. Give examples of products or parts that must work only once before failure.
19. Give examples of products or parts that must work millions of times before failure.
20. Give examples of products that have time life limits rather than cycles-used life limits.
21. Give examples of products that have cycles-used life limits rather than time life limits.
22. Give examples of parts that have performed their designed job if they do fail.
23. List some products that can be used equally well by right and left handed people.
24. List some products that can't be used equally well by right and left handed people and could use right and left hand versions. Suggest design changes on these products so they would be easier for both right and left handed people to use.
25. Discuss reasons why car manufacturers test cars by driving them into walls. Why is that test considered less expensive than other design analysis methods?

26. How would the task of finding how many Ping-Pong balls would fit into an automobile differ from that of finding how many would fit into the room you are in?

27. From a market data search, determine the cost of sugar in the following forms and estimate the cost of packaging from your data.
    a) In as large a container size as possible.
    b) In a five-pound package.
    c) In a two-pound package
    d) In cube form.
    e) In one-teaspoon package form.

28. Why do many foods have "Use before..." warnings on their packaging?

29. Consider and discuss some products, or main features of products that are very human-limit dependent. List the product and the human function limit involved. Example: The use of a car jack requires a certain level of strength in a person's arm muscles.

30. List some products that have been redesigned over the years to reduce maintenance. Specify the maintenance that used to be performed and what the design feature was that eliminated the need for maintenance.

31. List some products that still require maintenance and speculate as to why the maintenance needs have not been designed out of the product. Example: Lawn mower blades still require sharpening. The cost of sharpening is less than the cost of a new blade.

32. List some products that are well designed for after-use disposal.

33. List some products that are not well designed for after-use disposal. Suggest changes so they are.

34. Suggest ways that chain saw weight and exhaust fume levels have been reduced.

35. Compare food and beverage containers made from metal, glass, plastic, coated paper, and uncoated paper on the basis of raw material use, methods of fabrication, disposal by efficient means, reuse, litter life potential, litter cleanup ease, by-product effect on the environment and humans, ease of use by the consumer, and shelf life of the product.

36. For each of the codes, laws, and regulatory commissions listed in the chapter, specify a product you could be working on that would be affected by each one.

37. Draw a network chart for the project whose activities are listed in Table 2.1.
    a) Can this project be completed in 27 weeks?
    b) Determine which events have slack time, and how much.
    c) Assuming each activity is a two-person job, determine an efficient schedule of required personnel for the project.

38. Table 2.2 is a list of some of the preliminary steps used during a new product introduction. Determine a sequence of activities and an estimated minimum time to product introduction. *Note*: This may not represent any actual project you may work on.

TABLE 2.1
Project activities for Exercise 2.37.

Activity	Description	Time, weeks	Preceding Event
A	Remove old boiler and controls	4	—
B	Install new external steam lines	5	—
C	Install new internal steam lines	4	A
D	Construct new footings	5	A
E	Install new thermostats	6	A
F	Install new boiler	4	D, C
G	Install new boiler controls	7	F, B
H	Test boiler and lines for leaks	4	G, E
I	Check thermostat	3	H
J	Install system insulation	2	I
K	Test final system	1	J

TABLE 2.2
New product activities for Exercise 2.38.

Activity	Description	Time, weeks
A	Determine customer needs	4
B	Evaluate competition	3
C	Perform market survey	3
D	Prepare product specs	3
E	Project sales forecast	2
F	Evaluate marketing methods	2
G	Design product	5
H	Build test product	4
I	Test product	5
J	Plan marketing activities	4
K	Mount advertising campaign	4
L	Distribute sales literature	1
M	Print sales literature	4
N	Establish sales price	2
O	Build customer product	3

TABLE 2.3

Activities for building a small building, Exercise 2.39.

Activity	Description	Time, hours	Personnel
A	Draw plans	24	1 designer
B	Obtain building permit	3 (days)	1 designer
C	Clear site	12	2 workers
D	Purchase and deliver materials	6	1 worker
E	Excavate for footings	4	2 workers
F	Form and pour footings	8	2 workers
G	Allow footing cure time	overnight	0 workers
H	Form and pour frost walls	16	2 workers
I	Allow frost walls cure time	36	0 workers
J	Pour floor	8	2 workers
K	Allow floor cure time	36	0 workers
L	Erect walls	16	2 workers
M	Erect roof trusses	16	3 workers
N	Install siding	24	2 workers
O	Lay plywood roof	8	2 workers
P	Shingle roof	16	2 workers
Q	Install windows	6	1 worker
R	Paint siding, 2 coats	16	1 worker
S	Form and pour walkways	16	2 workers
T	Allow walkway cure time	3 (days)	0 workers
U	Install Doors	8	1 worker
V	Clean up site	6	2 workers

39. Prepare a project network chart for the construction of a small storage building using the activities listed in Table 2.3. Add any activities you think are missing
    a) Determine a critical path for an optimistic time schedule and the number of days required. Assume an 8-hour work day.
    b) Determine a pessimistic time schedule assuming rain every other day.
    c) Determine the required number of workers for each day.

40. An executive from a ski manufacturing company said, "Computer analysis defines the ski in theory, but determining the best combinations of characteristics requires field testing." Explain the meaning of this statement.

41. Explain the difference between the meaning of these two statements:
    a) Parts are also difficult to obtain.
    b) Also, parts are difficult to obtain.

42. Explain the difference in the meaning in the following three phrases:
    a) high priced method
    b) higher priced method
    c) highest priced method

43. Explain why each of the following statements is not well defined.
    a) Coors is the best beer on the market.
    b) Henry studied very hard for the exam.
    c) The car stopped running.
    d) The workers in the assembly department are overworked.
    e) This gear box is too heavy.
    f) The backlog of orders is unbearable.
    g) This is a good milling machine.
    h) The warehouse is crowded.
    i) This flame cutting operation is a dangerous operation.
    j) The finish on this car is beautiful.
    k) The new transmission tested better than the old one.

44. For each of the statements in Exercise 2.43 explain what detail or details you would want to clarify to improve on the accuracy and precision of the statement.

45. The following two sentences imply an understanding of economics but are poorly worded. Reword the sentences and list other information needed to clarify the statements even further.

    "When solar heating devices are installed, customers usually consider overall cost. This cost is offset by fuel savings and maintenance expenditures."

46. What is wrong with the logic of this statement? "The pipes are inaccessible, making repairs costly and time consuming."

3 Problem Solving

"One of the best things you can do when you have a problem is to talk about it."
♦ *MOSHE F. RUBINSTEIN & KENNETH PFEIFFER*

Engineering design is a process of using compromise (mutual concession and agreement) and iteration (repetition for improvement) to solve problems relating to the creation of a product or service to satisfy a real or perceived human need. The variety of problems encountered requires that an engineering design procedure be flexible and adaptive. Procedures must be flexible because problems rarely have the same conditions and restrictions, and will not be solved using the same activities. The engineer must be creative in the approaches used, and sensitive to the restrictions and the environment surrounding the problem, and must decide how to proceed with a problem by judging its complexity and considering various courses of action. A complex problem, such as designing a space station, will require the efforts of a team of engineers, each contributing knowledge and solutions. Experienced engineers recognize that there are many items that need to be considered when solving a problem, and they understand that design is not just following a grocery list of individual tasks. Design is a series of interrelated activities that lead to optimum use of available resources under the restrictions imposed. Many activities are done more than once. New ideas and possibilities may occur at any time, and feedback and reevaluation are important. Iteration and refinement are required to narrow the gap between what is, and what is desired.

3.1 Sources Of Problems

Where do the engineering problems come from? Whether you work for a large corporation or have your own consulting company, problems will come to you from many different sources.

One primary source of problems is existing products. An existing product or procedure may have an **observable deficiency**. Deficiencies may exist because a product was designed using older state of the art analysis, or because it is being used differently than originally intended, or because it was poorly designed or manufactured. Product deficiencies are encountered in many different places, and include such items as:

1. A signpost that bends over in the wind.
2. A conveyor that stalls when too many packages are placed on it.
3. A plastic tool handle that bends or shatters when the tool is used.
4. A copier paper feeder that feeds multiple sheets when only one is required.

Engineers also work on **unmet needs** and design new products or services such as:

1. A storm sewer system for a subdivision to control water runoff.
2. Desks and proper lighting so students can study efficiently in the library.
3. A new style drilling rig for offshore drilling in deep water.
4. A process to detect explosives in airline baggage.

An engineer may be assigned to search for a **product improvement**. Perhaps the present cost is too high and the job will be to reduce cost. A competitor's product may be selling better and a redesign is needed to meet competition. Examples include:

1. Redesign a furnace to increase its efficiency by 10%.
2. Redesign a computer printer to print twice as fast and to be 25% quieter while it is printing.
3. Redesign a lathe so that it will hold tolerances of ± 0.0003 on turned diameters.
4. Redesign a temperature sensing system so it will send a signal to the security office if the temperature is greater than 120°F.

A company's product or service may be influenced by a **change in a law.** Law changes occur in building codes, fire regulations, environmental constraints, or may be caused by newly discovered product side effects or increased liability for a particular existing defect. Some recent law changes require:

1. An average automobile mileage of 25 miles per gallon of gasoline.
2. An engine that runs efficiently on unleaded gasoline.
3. An automatic passive restraining system for airplane passengers.
4. An automatic blade brake on a power mower that activates when the mower handle is released.
5. A limit on the use and disposal of asbestos.

You may have a **natural curiosity** to solve a problem whether the problem has an immediate application or not. This is a common trait of engineers, for new and unique situations arouse interest and stimulate creativity. Curiosity can also create valuable experience to be used later when solving other problems as well as improvements or inventions that are patentable. The following example illustrates a natural curiosity problem.

EXAMPLE 3.1:
Nine dot problem.

Connect the nine dots shown in Figure 3.1 with four straight, sequentially connected lines.

Problems like Example 3.1 provide excellent opportunities to practice problem solving, and often make people aware of barriers that can interfere with creative problem solving. The difficulty in finding a solution is the restriction placed on the problem: four sequentially connected straight lines. The basic problem of connecting all the dots is no challenge; there are an infinite number of possibilities. And so it is with most engineering problems; the restrictions, whether they be weight, volume, cost, material, or time, are the real challenges you will face. See reference (1) for interesting solutions to Example 3.1, including some that use only one line.

3.2 Creativity

A productive, creative engineer has the capacity to propose innovative, workable solutions to problems. Creative people can visualize alternatives without being inhibited by outside factors, and have the ability to mull over in their mind present input and past experiences while searching for a logical problem structure that removes solution conflicts. Most engineers want to be creative, but often daily chores get in the way of creative work. Working on unstructured or open-ended problems is good practice in creativity. An unstructured problem is one in which any solution that is not precluded by the problem statement is acceptable. Finding a door-holder-opener is an unstructured problem. Locating a wedge to hold the door open is not, unless material selection is the unstructured search.

Not all people can be brought to the same level of creative problem solving, but practice will help, the more practice the better. One of the purposes of practicing creative problem solving is to change the way people look at the world around them. Engineers who become more aware of surroundings will be better problem solvers. Everyone can be creative, but some people are better prepared for creative thought than others. Various studies indicate that creative people tend to be adventurous and willing to take risks to

FIGURE 3.1

Connect the nine dots with 4 sequentially connected straight lines.

```
. . .
. . .
. . .
```

achieve high gain; are perceptive in observation rather than judgmental; and are emotionally sensitive, self-confident, and open-minded.

There are a number of personal traits that will help you in your problem solving activities:

1. Have an inquiring and questioning attitude. Ask questions to determine what is expected and what restrictions and limitations are present and to discuss options. Ask "What if?" questions to gain understanding and to open the door to new ideas. See Example 4.3 and Section 5.1.7 for sample questions to ask about a problem to get the creative juices flowing.

2. Develop communication skills. Information and ideas transfer from person to person most efficiently when communications are good. This applies to both spoken and written communications. Many people communicate orally with success because misunderstood points can be repeated. Not so with written words. Practice writing to sharpen your communication abilities. Reading other people's work is one aid in learning to write, because recognizing examples of clear writing will help you judge your own work.

    Improve communications with yourself. If an idea or question crosses your mind, make a note of it, as it may never again surface in your thoughts. Make lists of personal priority goals and objectives, about projects to undertake, and investigations to make. Know what you are thinking and how your thoughts relate to the problem at hand.

3. Be physically and emotionally healthy. Physical well being improves concentration ability and stamina to carry out experiments and studies. Emotional health allows you to put a lid on distracting problems such as sorrow, fear, and worry. Energy spent on emotional problems is energy that can't be used on productive problem solving.

4. Accept reality and be self critical. Accept the fact that conflict and tension will occur from time to time, but don't let frustration result. Frustration blocks the normal thought process, saps needed energy, and limits creative thinking.

5. Be willing to consider new ideas and not worry about your ego. Even if an idea conflicts with your present thoughts or past experience it must not automatically be discarded. An idea that didn't work before may work now. The circumstances may be different, or the acceptance levels of others may have changed. Don't ever wave away an idea by saying "We tried that once and it didn't work."

6. Be curious and observant. Notice what goes on around you. Take note of things you see that work well and also those that don't.

7. Allow time for subconscious thinking. Some people consider their problem-solving efforts a failure and become frustrated if a great idea doesn't occur to them after a brief period of time. Meditation and thinking time are important when a solution is elusive. Some suggestions that may improve your concentration and help ideas to flow are:
    - Take a walk away from disturbances.
    - Play music to force concentration on a problem.
    - "Sleep on it" and take a fresh approach the next day.
    - Read a book.
    - Work on another problem.
    - Do something physical: cut the grass; wash the car.

Everyone can improve innovativeness and creativity, but it requires practice. While personal intelligence may set an upper limit on how creative a person may become, high IQ does not guarantee high creativity.

Highly creative people get great personal satisfaction from solving problems in new and unusual ways. Their satisfaction comes from a sense of pride, loyalty, pleasure, joy, thrill, and competition. The chance of doing something creative favors the prepared person. Creativity will not come to you by revelation. For additional dialogue on creativity see references (1), (5), (6), (17), and (18).

3.3 Roadblocks To Creative Problem Solving

Roadblocks to creative problem solving can come from external sources, as well as from sources within ourselves.

Past experiences, and habits that result from them, may lead an engineer to think there is only one way something can be done. Habit may cause a problem solver to have difficulties such as:

1. A limited range of basic knowledge search. Many people feel that once they have something that works, there is no other choice: "This is the way it has always been done." Or if a shy person rarely offers an idea, some might think "Don't ask him, he never has a good idea."

2. Failure to consider all the details of a problem. Relevant details may be ignored due to preconception of the answer. Obvious details must always be considered in balance with details found through deep investigation. Example: The witness saw the smoking gun and the bullet hole in the chest, but the coroner found poison in the stomach.

3. Failure to distinguish between cause and effect. This can lead to wasted effort and time pursuing something that is not the real problem. Example: If I frequently run my car out of gas, and the engine stops running on my way to work, I will probably conclude I ran out of gas again. But there are many other reasons why an engine will stop running; no spark, loss of engine coolant and the engine "freezes," or a plugged fuel line.

4. Failure to see what is around us. We look, but we don't see. For example, how many people can correctly describe the face of a telephone, including the locations of the numbers and the letters? Creative people see more than noncreative people see. They also visualize images more easily and consider alternatives just by thinking about them. To increase your creative ability try practicing your visual imagery. For practice, try visualizing each of the following; a bullfight, a baseball double play (How many different ways can it happen?), an airplane flying upside down, a bird flying backward, hitting a golf ball made from a marshmallow, poured water flowing up rather than down, drinking cold coffee, drinking hot lemonade. Did you have more trouble with images that you have never seen before? Most people do and often comment "Why should I think of things I haven't seen?" Exactly the point. Creativity and innovativeness assist you to think of something new, or a new application of an old idea, not just something you have seen or done before.

Our **culture and society** pose roadblocks. The culture in which a person has been raised can have effects that limit creative freedom. Certain ideas or methods will not be a valid choice because of upbringing mores or social customs. Examples would include the different philosophies and life styles of the Mormons, Amish, Quakers, Eskimos, Japanese, Chinese, Hindus, or Africans.

Cultural restrictions prevent innovations and make large changes in day-to-day living difficult. A change may be ready before its time. Examples include the iron plow, which early opponents claimed would poison the ground, birth control for societies that consider wealth a function of family size, or nutritious food that is considered fit only to feed cattle.

Personal emotions can also be barriers that must be overcome. Emotional barriers include sadness, worry, and perhaps the biggest emotional barrier, fear. This includes fear of failure, and personal embarrassment; fear of criticism from those we respect; fear of ridicule, both public and private; fear of embarrassment; fear of loss of job; and fear of appearing uninformed or ignorant. Today's society places emphasis (sometimes too much) on knowledge. Students are pressured to know facts and to be able to regurgitate details. As a result, to ask a question (an attribute of a creative person) is to cast doubt on one's intelligence. However, once asking questions becomes a part of your way of life, a must for gathering data, the fear will disappear.

Associated with the fear of appearing stupid is the mental block of prejudging an idea before it is suggested or tried out. Prejudging, and not offering a suggestion prevents open discussion and often excludes the best idea from being considered. Even an idea that will not work may lead to an idea that will. Refer to Section 5.1.5 for more on the danger of prejudging ideas.

One method of overcoming potential embarrassing situations is to write down all the possible results of your actions you can think of. What is the worst thing that can happen if you try your idea or ask your question? Now write down the good things that can happen if you try your idea or ask your question. What is the best thing that can happen? If you have more to gain than to lose, then do it.

Roadblocks can also be imposed by supervisors and co-workers. If a co-worker responds to a suggestion for change with, "We tried that last year and it didn't work," or "That method won't work in our company," or "We don't have time to try that," creative effort can easily be stymied. The following case study illustrates how the problem definition can limit creativity.

CASE STUDY #1:
Door-Holder-Opener

Consider the day the Physics Department was setting up for a conference. The person in charge of the conference wanted the self-closing double doors that separate the lobby from the main hallway open, so guests would feel free to wander about. The person in charge said to me, "Find a wedge to hold these doors open."

A wedge is a logical door-holder-opener, but so are many other things: a brick, a chair, a table, or a person. So as I walked about looking for an efficient solution, I considered several alternatives. Although a student worker was nearby, I didn't think he would like to spend the day holding the door open, so I pursued other alternatives.

My solution to the problem was a rectangular block of wood 1.50 inch x 2.00 inch x 0.75 inch. I opened the door and placed the block between the hinge side of the door and the doorjamb. It held the door open all day, didn't get in the way of people, and saved a student from a most monotonous job. A wedge is a habit solution to the problem. I was sent after a wedge when I should have been sent after a door-holder-opener. If time had not been short, other more permanent and versatile solutions could have been considered. The name given to a problem can itself be a roadblock to a creative solution. The device could be called a door anti-closer, doorstop, or door-block to allow for more creative thought.

3.4 Problem Solving Activities

Chapters 4 through 7 contain specific details of problem solving activities, but an overview now will provide the framework for the details in those chapters. An engineer engages in many activities in the quest for the solution to a problem. No list of activities is complete, nor is a given list the best for every type of problem. Problems are rarely the same, either in terms of resources and talent available, or in potential reward. So the following activity outline is not intended to be a fail-safe or complete list, merely a reminder of general responsibilities and procedures.

3.4.1 Define the Problem

Before a problem can be solved, it must be determined what the problem really is. If the activity of defining, clarifying, and understanding the problem is properly done, subsequent work is more efficiently performed. Nothing is as frustrating as to work on a project for some time and then realize you're not working on the real problem. The problem statement usually includes two main categories:

1. The demand specifications, criteria that are essential.
2. Criteria that are nice to have if they are affordable and do not jeopardize the performance of the demand specifications.

The criteria specified must not be so specific that they determine the solution to the problem (use a maximum of 500 watts of *power* rather than use 500 watts of *electricity*). The focus of problem definition is directed toward *what* has to be achieved, and on the problem limits and boundary conditions. If the problem of Case Study #1 had been described as "We need a wedge," I might not have found one in the building and a student might have had a long tiresome day holding the door open. But by stating the problem as "We need something to hold the door open," many possibilities can be considered. Methods to assist in defining problems are discussed in Chapter 4.

3.4.2 Consider Alternative Solutions

After a problem is adequately defined, possible solutions must be presented and the focus shifts to *how* the demand specification criteria can be accomplished. As the number of alternative solutions is increased, chances for an optimum solution increases.

If time is short, this activity can be brief, and the first idea that comes to mind may be used. For example, if a door is swinging toward you, and your arms are full of groceries, the idea of sticking your foot out to stop the door may be the only idea you can come up with in the brief time you have before the door hits you. However, if time is available, considering only one possibility may lead to expensive and inefficient solutions. Considering alternate solutions can also be a complicated activity requiring research and investigation lasting several years. Major projects, such as designing a space station, will consider many alternatives following lengthy research.

As ideas are generated, the scope of the project may be modified. This requires redefining or refining the problem statement. For example, the original problem statement may have been to create a computer program to generate payroll checks. After deliberation, the problem may be changed to include the generation of monthly, quarterly, and annual IRS and workman's compensation reports. If the scope of a project is changed, be sure to adjust the resources for the job, such as staff support, money, and due dates. Suggestions for generating ideas are presented in Chapter 5.

3.4.3 Refine Ideas

One objective of idea refinement is to narrow the selection of possible ideas and weed out ideas that have little chance of succeeding. Remaining effort, time, and money are then used to develop those ideas with the best chance of being workable and profitable. Some ideas may be discarded simply because the technology is not available for implementation. In other cases high preliminary cost estimates will cause an idea to be dropped early in the project. An equally important objective of refinement activities is to concentrate effort on improving how a problem will be solved. Good ideas are made better, and sub-problem solutions are combined for overall performance improvement.

Refinement activities may also require additional project scope changes. A refinement in the problem statement can result in new ideas to consider. An original problem statement "Design a ten-horsepower, front-tined rotary cultivator," may be modified to "Design an eight-horsepower, front-tined rotary cultivator with forward and reverse power drive."

Additional ideas for solution can be generated from a problem statement refinement made during idea refinement activities. Any activity may result in iteration back to a previous activity. No activity is performed once and then forgotten; always consider the whole picture. Suggestions for refining ideas are presented in Chapter 6.

3.4.4 Analyze Ideas

Analysis includes recording and studying all the efforts being made to verify that an idea is workable and will function. The questions that need answering are:

"Will the idea work as desired under the restrictions imposed?"

"Can a weak idea be altered so that it is viable?"

"Can this idea be implemented under the cost and time constraints imposed?"

Verification of function can be done in many ways, such as mathematical modeling, computer modeling, or actual building and testing. Additional refinement is usually made, and more ideas may be generated as the results of the analysis. Keep in mind that most projects require iteration through the idea, refinement, and analysis sequence of activities. On some projects, such as the development of a new material, process, or product, the iteration process may last for several years. Consider superconductors, for example, whose electrical resistivity properties are favorable, but which have low power carrying capacity and are difficult to manufacture. Their development will continue for a long time.

Models or prototypes are built and tested to verify function, and can be a portion of the product or the complete unit. For example, during the development and analysis of large computer systems, a program is often broken down into subprograms that can be tested independently of the total system. This is possible by using subroutines and a common data base. Some models and prototypes will be experimental in nature and will use experimental optimization procedures.

Product testing can be done using complete assemblies, or components. A shaft in a transmission can be tested outside of the complete transmission assembly, or it can be assembled into the transmission and then tested. The transmission can be tested by itself or installed in the completed vehicle and tested. Suggestions for the analysis of ideas is combined with refinement in Chapter 6.

3.4.5 Decision

The time comes when a decision on the final direction of a project is required. This may happen during the problem identification or idea generation phase of a project, but usually it comes after some amount of refinement and analysis has been done. In the worst case, none of the ideas being refined will function to an acceptable level, and the whole project must be started over or discontinued. It may be that the problem definition is too restrictive or demanding, or that time and money are inadequate to continue. If the decision is made to cancel the project, then salvage the work done to date, record accurately what has been done, and move on to a new project.

In the ideal case, one idea will clearly be the best, will be the least expensive, have a shorter lead time to production, and will suit the fancy of 1,000,000 customers.

In the usual case, no idea will work as well as desired, and a compromise between such items as function, cost, sales potential, and time to production will be made.

Perhaps the problem will have to be narrowed in scope. The reverse feature on the tiller may be discarded to reduce cost; or the four-horsepower engine already in production will be used to reduce development time and lower sales price; or about this time the Sales Department will report that only rear-tined cultivators will sell. Any number of unexpected changes may be required. Methods to assist in decision making are discussed in Chapter 7. In today's high technology environment of computers and Computer-Aided-Design/Computer-Aided-Manufacture (CAD/CAM) systems, perhaps the expression "Back to the drawing board," should be updated to "Back to the computer terminal."

3.4.6 Implementation

Once an acceptable solution is chosen, and assuming profit potential still exists, the final drawings, production methods, construction of buildings, hiring of people, ordering material and other supplies, and installation of required processes and procedures can proceed. This does not mean that new ideas and additional refinement are halted. New ideas and refinement should never be terminated; if changes are extensive, however, they may not be implemented on the first production model. They may have to wait for subsequent models. Implementation activities are included in Chapter 7 with decision making.

3.5 References

1. Alger, John R. M. and Carl V. Hays. *Creative Synthesis in Design*. Englewood Cliffs, N.J.: Prentice Hall, 1964.
2. Andrews, James G. "In Search of the Engineering Method." *Engineering Education.* Oct. 1987, 29-30, 55-57.
3. Bailey, Robert L. *Disciplined Creativity for Engineers.* Ann Arbor, Mich.: Ann Arbor Science, 1978.
4. Beakley, George C. and Ernest G. Chilton. *Design, Serving the Needs of Man.* New York: Macmillan, 1974.
5. Beakley, George C., et al. *Engineering, An Introduction to a Creative Profession,* 5th ed. New York: Macmillan, 1986.
6. Brightman, Harvey J. *Problem Solving: A Logical and Creative Approach.* Atlanta: Georgia State University, 1980.
7. Cross, Nigel. *Engineering Design Methods.* Essex, England: John Wiley & Sons Ltd., 1989.
8. George, F.H. *Problem Solving*. London: Duckworth & Co., 1980.
9. Harrisberger, Lee. *Engineersmanship*: The Doing of Engineering Design, 2nd ed. Monterey, Cal.: Brooks/Cole, 1982; 143 pages of easy, but stimulating, reading.
10. Kaku, Michio and Jennifer Trainer. *Nuclear Power: Both Sides*. New York: W.W. Norton & Co., 1983.
11. Love, Sydney F. *Planning and Creating Successful Engineering Designs: Managing the Design Process.* Los Angeles, Cal.: Advanced Professional Development Inc., 1986.
12. Meadow, Charles T. *Applied Data Management*. New York: John Wiley & Sons, 1976.
13. Osborn, Alex F. *Applied Imagination,* 3rd ed. New York: Scribner, 1979.
14. Pearl, Judea. *Heuristics*: Reading, Mass.: Addison Wesley, 1983.
15. Polya, G. *How to Solve It*, 2nd ed. New York: Doubleday Anchor Books, Doubleday & Co, 1957.
16. Rubinstein, Moshe F. *Patterns in Problem Solving*. Englewood Cliffs, N.J.: Prentice Hall, 1975.
17. Rubinstein, Moshe F. and Kenneth Pfeiffer. *Concepts in Problem Solving*. Englewood Cliffs, N.J.: Prentice Hall, 1980.
18. Ruggiers, Vincent Ryan. *The Art of Thinking*, 2nd ed. New York: Harper & Row, 1988.
19. Smith, Karl J. *The Nature of Mathematics*, 5th ed. Monterey, Cal.: Brooks/Cole, 1987.
20. Spotts, M.F. *Design Engineering Projects*. Englewood Cliffs, N.J.: Prentice Hall, 1968. Chapter 1 on creativity.

3.6 Exercises

1. List as many ideas as you can that could answer the question, "How can I get from my dorm to a movie theater."

2. List at least one advantage and one disadvantage of each idea listed in Exercise 3.1.

3. Rephrase each of the following statements to allow more room for creative solutions:
   a) Get a towel to dry your hands on.
   b) Take the car to go get a loaf of bread.
   c) Find a pen to write down a message while on the phone.
   d) Find a table to set our lunch on.
   e) Get a boat so we can row across this river.

4. Discuss this definition of design: Design is the combination of basic principles and available resources to solve a problem.

5. Discuss the implied meaning of the phrase "Trial and error," often heard in relation to the design process.

6. Give an example of a constraint so severe that an acceptable answer would be impossible.

7. When the Chernobyl nuclear accident occurred in Russia in April, 1986, the "China Syndrome" question was proposed: Where in the world is directly opposite Kiev/Chernobyl?
   a) Estimate an answer in 60 seconds.
   b) Discuss how you would go about making a more accurate answer.
   c) Follow your procedure in part (b) and determine an accurate answer.

8. Two ropes hang from the ceiling of a normal room, but they are too far apart for one person to reach both at the same time while they are hanging straight down.. Discuss ways to get the ropes together so the ends can be tied. Try out your ideas.

9. What is the next letter in the following sequence of letters? OTTFFSSENTETTF Explain your reasoning.

10. In the morning going to work, a man rode an elevator alone to the tenth floor and then walked up five flights of stairs to his office. At night when he left work he rode the elevator down all 15 floors. List possible explanations for this behavior.

11. Discuss what are often called "contingency plans." How would they be used in setting up an original project schedule and budget?

12. Make a list of methods for holding a door open. Include pictures or descriptions of commercially available products. Visit hardware stores and building supply houses for ideas.

13. List design shortcomings of each of the following products. User interviews may be required.
    a) Home vacuum cleaner
    b) Home clothes drier
    c) Electric stove

d) Lawn mower
e) Gas stove
f) Snow blower
g) Dishwasher
h) Popcorn popper
i) Refrigerator
j) Coffee maker
k) Clothes washer
l) Toaster

14. Give some examples of development projects that you feel could not be speeded up by adding more money and personnel compared to some that could. These need not be absolute examples, only relative to each other.

15. List some possible design deficiencies with the signpost that bends over in the wind.

16. List some possible design deficiencies with the conveyor that stalls when too many packages are placed on it.

17. List some basic criteria to consider in the need for a storm sewer system.

18. List some basic criteria to consider in the need for study desks and proper lighting.

19. List some basic criteria to consider in the need for a new offshore drilling platform.

20. What can be measured to determine the success or failure of the redesigned furnace?

21. What approaches would you consider in the solution to the problem of designing an automobile engine that averages 25 miles per gallon?

22. Discuss possible sources and reasons why the problem statement was changed from "Design a five-horsepower, front-tined rotary cultivator," to "Design a five-horsepower, front-tined rotary cultivator with forward and reverse drive."

23. List advantages and disadvantages of the differences in testing a transmission shaft by itself, in an assembled transmission, and in a transmission in an automobile.

24. Make a list of five to ten machines in order of their versatility. Example: sewing machine, low versatility; computer, high versatility.

25. Make a list of five machines that run themselves compared to five machines that must have an operator to control them. Example: An electric clock runs itself, a bicycle requires an operator.

26. For each machine listed in Exercise 3.25 that does not need an operator, describe a design or equivalent situation that would do the same function with an operator. Example: A clock that needs an operator would be an hourglass.

27. For each machine listed in Exercise 3.25 that needs an operator, describe a design or equivalent situation that would do the same function without an operator. Example: A "bicycle" without an operator is used on overhead cables to carry loads from one location to another.

28. Discuss this definition of problem solving: "To reach a least-cost method of achieving a transformation from one state of affairs to another."

29. A passenger in an airplane, properly seated and buckled in for takeoff noticed a sign

86 Problem Solving

on the back of the seat in front of him. The sign said "This outboard seat back does not recline." Which seat doesn't recline, his or the one in front of him? Discuss the sign labeling.

30. In 1976, Kenneth Appel and Wolfgang Haken, professors at the University of Illinois, Urbana-Champaign, announced they had proven that any map could be colored with only four colors, and no border would have the same color on both sides. They used the computer to check "all possible combinations." Any exception, of course, disproves their hypothesis. Does the map in Figure 3.2 disprove their hypothesis because no color is available for the inside circle? Explain.

31. Select a simple product, such as a crank pencil sharpener, and list the verb-noun functions of all the parts. For example one part of the pencil sharpener must have the function of "catch shavings."

32. Analyze a procedure, such as installing a chain link fence, changing a car tire, or building a book shelf, by listing the steps and the necessary tools for each step. Consider alternate sequences of activities to do the same job.

33. Three switches control the lights in a particular hallway. Two are 3-way switches, one at each end of the hall to control the even numbered lights. The third is a single switch controlling the odd numbered lights. The single switch is up if the lights are on, and down if the lights are off. A 3-way switch can be up or down when the lights are on depending on the position of the other 3-way switch. One 3-way switch is mounted next to the single switch at one end of the hall. For each of the following cases only one set of lights is on, either the even ones or the odd ones, but you don't

FIGURE 3.2

Map for Exercise 3.30.

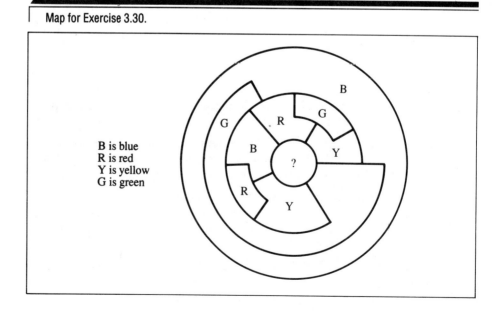

B is blue
R is red
Y is yellow
G is green

know which. Your problem is to change the position of *one* of the two switches and, based on the results, regardless of what the results are, determine which switch is the 3-way switch and which is the single switch.

Case 1: S_1 down and S_2 down

Case 2: S_1 down and S_2 up

Case 3: S_1 up and S_2 down

Case 4: S_1 up and S_2 up

34. Same situation as Exercise 3.33 except both the even and odd lights are on for each of the cases of switch settings.

35. Same situation as Exercise 3.33 except no lights are on for each of the cases of switch settings.

36. If four cars are stopped at each corner of a four-way stop, how many ways can the cars proceed, all going at the same time, without colliding with any of the other cars? Describe them.

4 | Problem Definition

"The engineer's first problem in any design situation is to discover what the real problem is."
♦ GEORGE C. BEAKLEY

Problem definitions can range from a simple exclamation, "Help! My boat is headed for the waterfall!" to a book of performance specifications such as those for a proposed power dam. A preliminary problem statement may also be quite vague. A customer may say "I sure could use a good door-closer." A problem statement with that little detail is not clear enough for an engineer to proceed with a design. Does the person want a product that is light, colorful, rustproof, installs with a screwdriver, one that closes and locks the door, closes the door in less than two seconds, or costs less than $50? Is it possible that what the customer wants is already available? The purpose of the problem definition is to clarify the situation so the true problem or need can be worked on. Communication with the people affected by the problem is essential. Make contact with as many people as practical, and discuss the situation with them. Go to the problem site to observe and collect information. Seeing the problem firsthand will clarify potential questions, and the physical restrictions will be observable and measurable. Take good notes.

4.1 | Methods Of Problem Definition

A carefully thought-out problem definition includes as much information as possible relating to the problem. There are five general categories in which to accumulate project information. Case Study #2, concerning the movement of material between production departments, will be used to illustrate the steps of a typical problem solving situation.

The case study will be interrupted several times in this chapter with general problem definition discussion, and will also be continued in Chapters 5, 6, and 7.

Begin with a **basic statement of need**. A basic statement of need or want may have vagueness and generality. This is acceptable in an opening statement and allows more creative thought. The opening statement should allow general discussion and further clarification without assuming a solution. The basic statement of need should not include a preconceived solution.

CASE STUDY #2:
Parts moving problem; basic problem statement.

The statement "We need a four-wheel dolly to move this load of parts to the next department" includes the solution and restricts creative problem solving. A better statement, one that allows more creative solution ideas, is a question "How can we most easily and efficiently move these parts to the user department?"

Start problem clarification by listing **basic requirements**. As the basic problem statement is considered and discussed, certain desired outcomes and performance criteria should become apparent. These specifications should be listed or clarified with detailed sketches. They eventually become the technical goals for the project. Figure 2.14 is an example of product specifications resulting from a completed project.

CASE STUDY #2 CONTINUED:
Requirements.

1. Only one person should be involved in the movement of parts.
2. The movement of parts should take five minutes or less.
3. The movement of parts must be safe for the person doing the moving as well as for other people working in the area.

Basic limitations are as important to consider as basic requirements. As discussion and research on the problem continue, features may emerge that are considered undesirable, things you don't want the product to do. These limitations should be clarified and recorded with detailed sketches and/or word descriptions. There may also be constraints over which you have no control, such as relevant laws or codes. They are real constraints, and should be identified early so generated ideas can take them into account. Ideally, of course, the fewer restrictions the better, though all real problems have some constraints. There are also indirect constraints with engineering projects. They are the constraints of the effect of a change on the present status. All design projects will affect some component of a present system or of a related system. For example: If a four-lane bridge is built to replace ferry service, queuing is eliminated, approach roads may be inadequate, a company may be forced out of business, and tall-masted sailing boats may not be able to pass by when the water level is high. These side effects may cause the scope of the project to broaden (rebuild the access roads), may cause special features to be added (make the bridge a draw, swing, or lift bridge), or require a business relocation subsidy. Recognizing and including the consequences of project implementation early in the planning process will result in wider acceptance of engineering projects.

CASE STUDY #2 CONTINUED:
Limitations.

1. The parts must not come in contact with each other while moving because the surfaces have been ground to a 16-microinch surface finish.
2. A solution must be developed within ten days.
3. The budget for the project is limited to $10,000.
4. The union contract limits the weight one person may carry to 50 pounds (22.7 kg).
5. There is a building wall doorway between departments that is 8 feet wide and 8 feet high (2.44 × 2.44 m).

Everything that comes up during discussion about the project should be noted. Anything not clearly a requirement or limitation is listed as **other data**. This includes information about company policy, styling considerations, or available plant facilities. Law or code restrictions might also be specified, as well as other restrictions beyond the control of engineers. An example of an uncontrollable restriction is the contour map of a lot being considered for a shopping center.

It doesn't really matter whether a condition is listed as a requirement, a limitation, or as other data. The main objective is to allow room for a variety of considerations so a good framework and basic understanding of the problem can be established.

CASE STUDY #2 CONTINUED:
Other data.

1. Part orientation while moving is not critical, parts do not have an up or down side.
2. A variety of parts will be moved. Their weight ranges from 5 to 25 pounds (2.27 to 11.34 kg).
3. All of the parts are shafts that range in length from 6 to 16 inches (152 to 406 mm), and in diameter from 0.75 to 1.50 inches (19.05 to 38.10 mm).
4. Quantities of any given part to be moved will range from 10 to 100.
5. The distance to the next department is 150 feet (45.7 m), on the same floor.
6. That department uses the shafts for assembly into gear cases.

Most of the time when a new problem or project is started, **questions** are raised that cannot be answered immediately. Make note of those things that need to be answered during the course of solving the problem. Questions about performance specification, unknown restrictions, or related products commonly occur. As the answers to the questions are found, the scope of the problem may expand or contract.

CASE STUDY #2 CONTINUED:
Questions and answers.

1. How much floor space is available in the sending department and the receiving department? This will affect how often during the day parts must be moved.
2. Is part inspection required before moving or at the second department when received? This influences the time involved to prepare for moving, or the time for processing after the parts are received, as well as floor space required.
3. Can holes be put in the wall separating the departments? There might be a shorter way to the second department.
4. Is the wall a bearing or a partition wall? This is a logical follow-up question after question #3.

Answers to questions.

1. Parts are presently placed on flat pallets in a move area, and are moved whenever the move area gets crowded or when parts are needed in the next department. Parts are moved an average of four times a day. The pallets are moved with a hand operated pallet mover by the department parts expediter. The volume of parts produced has been increasing because two new grinders have been added to the department, making a total of eight machines.
2. Part inspection is done before the parts leave the machines so the parts are ready to move when they reach the move area.
3. *and 4.* Holes can be put in the walls, but they are partition fire walls, and there must be a heat activated automatic closer installed if an opening is cut.

If a new product is being considered, questions that need answering include the following:

1. Does this product replace an existing product, or does it add to the product line?
2. Is the product an extension of an existing series of models so that outer styling must be matched?
3. Are there parts already manufactured or purchased that should be considered for use rather than starting from scratch?
4. Are there existing plant facilities for manufacture, paint, assembly, and storage?
5. What size limitations for manufacturing exist relative to material handling, tonnage of presses, and bed sizes of machine tools?
6. What basic manufacturing methods are available and do the work stations have time capacity remaining to make additional parts?

Keep in mind that any statements made at this point in the project are not final. Many conditions can, and most likely will, be altered as the project proceeds. Keep in mind that a specific requirement on one project: "Power shall be a five-horsepower engine," might be a question on another project, "What horsepower is required?" In the first case, five-horsepower is a preliminary specification that can be changed, whereas in the second case, not enough information is available to make a preliminary decision. Perhaps the market survey isn't complete, or the power requirements of a required cultivation width aren't known. Don't let a project stagnate because some pieces of data are not available. Proceed and let data catch up. Work on the data available and keep the project moving.

The following example illustrates the five areas of problem definition for a proposed new product.

EXAMPLE 4.1:

New product preliminary problem definition.

1. Basic Statement—Design a garden rotary cultivator.
2. Preliminary Requirements
    - Power will be a five-horsepower gasoline engine.
    - The tines will be in the front.
    - Forward and reverse power operation will be available.
    - All controls will be reachable from the normal standing operating position.
    - Cultivating width will be 18 inches (457 mm).
    - Movement under power must stop if the controls are released.
3. Preliminary Limitations
    - Total weight will be 110 pounds (50 kg) or less.
    - Maximum storage area will be eight square feet ($0.74\ m^2$).
    - Selling price will be $350 or less.
    - Maximum sales will be 1000 units per year, 70 percent in the spring, 30 percent the rest of the year.
    - The cultivator must be market ready for April selling in two years.
4. Other Data
    - This product is an addition to the current line of home garden equipment; no other cultivators are manufactured.
    - Exterior appearance should be compatible with existing products.
    - The manufactured components of a garden tractor now in production are larger than the expected size of the parts of the cultivator, so the size of existing machine tools and other manufacturing equipment is probably large enough.
    - Design the cultivator with features found on other products in the line so common parts, such as power train components and electrical controls can be used, and existing methods of manufacture will be adequate.
5. Questions
    - Will five-horsepower be adequate for an 18-inch (457 mm) cultivating width?
    - Should the drive mechanism be belt, gear, chain, or something else?
    - Should the tines come in segments, so narrow cultivation is possible?
    - Should the cultivator be self-propelled?

- Where should the starting controls be located?
- What parts should be painted? What color?
- Should any parts be plated?
- Should multiple use be considered so that the power unit can also be used for lawn mowing or snow blowing?
- Should the engine be two-cycle or four-cycle gasoline, or electric?
- Should regular maintenance be necessary or should the unit be maintenance free?

Remember that the lists are not necessarily complete or exhaustive. You will be able to add to all of them, particularly the questions. The more questions asked, the more thorough your thought process, and the greater the potential for creative solutions. An important part of problem identification is to make a record of details as they occur. A good beginning is important so that during the other stages of problem solving activities no aspect of the design will be neglected or overlooked.

When a problem is well defined, it is identified with a statement that includes expected results. Problem identification requires an understanding of the situation, specifying known relevant data, knowing what additional data is required, establishing limiting assumptions, and making a decision on priorities and personal values. Problem identification statements must be done carefully so they are not too restrictive or too vague to be approachable.

The following example illustrates how a problem statement can change based on different personal and environmental circumstances.

EXAMPLE 4.2:
Alternate problem statements.

Consider the different problem statements, all coming from the same situation.

1. My car engine has stopped, how do I get to work?
2. My car engine has stopped, how do I get it started?
3. My car engine has stopped, how do I stop safely?
4. My car engine has stopped, what do I do with the car while I go for help?
5. My car engine has stopped, how do I get the car out of traffic?
6. My car engine has stopped, what repairs will have to be done?

Problem statement #1 indicates my biggest concern is getting to work. Choices can include hitchhiking, calling a taxi, walking, or taking a bus.

Problem statement #2 indicates a desire to get the car restarted and continue on. Choices can include checking to be sure there is enough fuel, trying to get a push, jump-starting if the battery seems low, or calling emergency road service for help.

Problem statement #3 indicates I was traveling at some speed, and if the engine isn't running, I will not have power steering or power brakes. Various choices include slamming on the brakes, coasting to a stop on the shoulder (if there is one), or shifting to a lower gear to slow the car down gradually.

Problem statement #4 suggests I have already stopped, but perhaps in the middle of traffic or where there is no safe place to park. I could just leave the car where it is and hope for the best, or get help to push it out of the way.

Problem statement #5 definitely means I am in the middle of things and want to get out of the way. Getting a push or calling road service are among my choices.

Problem statement #6 might indicate I have been having trouble with the car and have recently spent money getting it fixed. Having more trouble now, and worrying about repairs might suggest I am ready to trade in the car and don't want to spend much money getting it repaired.

The next example illustrates how questions can be used to develop and clarify a problem when the initial statement has no restrictions or guidelines listed. These kinds of problems are often referred to as open-ended problems, as no one knows for sure where they will lead.

EXAMPLE 4.3:

Question approach to problem identification.

Consider the general problem statement "Design a desk."

Getting more information is a key to properly defining this problem and making clear what the parameters are. Among the questions that could be asked about designing a desk are:

1. What is the purpose of the desk? Study? Writing? Drawing? Storing material? Typing? Interviewing? Collecting money?
2. Who will use the desk? An adult? A child? A handicapped person?
3. How will the desk be used? Sitting down? Standing up? Straight-back chair? Chair on wheels? Wheelchair?
4. Will the desk need drawers? What for? Pencils and small articles? Papers? File folders? Books? Products to sell?
5. If drawers are needed should they be lockable?
6. Should drawers have bottom or side guides?
7. Should drawer guides be slides or rollers?
8. What material should be used for the desk? Solid wood? Plywood? Particle board? Metal? Plastic? Some combination?
9. What style is desired? Modern? Early American?
10. Is there a maximum permissible size? Does the desk have to fit a particular space?
11. Is there a maximum or minimum weight?
12. Should the desk be movable by one person?
13. Should the desk be stackable for storage?
14. Is color a concern for matching a particular decor?
15. Should there be room for permanent use of a typewriter, word processor, or computer?

16. Is a roll top appropriate?
17. What cost range is appropriate?
18. How many desks will be made?
19. What is the time schedule for completing the project?
20. How many people will be working on the project?
21. Are any differences required for right- or left-handed people?
22. Is a privacy screen desirable?
23. What height is comfortable?
24. Is a swinging door for a storage area appropriate?
25. Should the desk be made to be assembled by the customer?
26. What desks are presently available that come close to satisfying the current need?

Getting answers to questions like these will get the project off to a good start. Be sure to document answers in writing, including the source. One possible set of preliminary specifications and project guidelines coming from the questions are:

1. The desk will be used by a computer programmer.
2. The user will be a average sized adult. Refer to anthropometric charts in reference (11) and (12) listed in Section 6.4 for dimensions of average adults. For example, 99 percent of adult males have a sitting hip width less than 16.2 inches (411.5 mm), while the maximum is 18.11 inches (460.0 mm).
3. The programmer will be sitting down, so the desk needs space for leg clearance.
4. The desk needs one drawer for stand-up storage of $8\frac{1}{2} \times 11$-inch file folders, and one for pencils, pens, paper clips, and a stapler. A shelf or other storage area is needed for 24 inches (610 mm) of reference books, 12 inches (305 mm) maximum height.
5. Security is a problem so each drawer must be lockable. Same key for all drawers.
6. Drawers must slide and not cause static electricity or loud noises, but guide location is arbitrary.
7. No preference to drawer slides or rollers unless static electricity generation is affected.
8. Wood is preferred material, it has less potential for stray magnetic fields.
9. No preference on style, function is main guideline.
10. The maximum size allowed is 32 inches (813 mm) deep and 60 inches (1524 mm) wide.
11. No weight limitation.
12. Desk does not have to be movable by one person.
13. Stacking storage not required.
14. Natural wood color preferred to blend with the general office decor.
15. Room for a computer terminal required; a terminal and movable keyboard require 20 inches (508 mm) width and 25 inches (635 mm) depth, minimum.

16. No roll top.
17. Budget has been set at $400 maximum per desk.
18. Department will have ten people, five now, five more in about one year.
19. Five desks must be available in 90 days.
20. One person is assigned to the design of the desk, construction may be contracted out.
21. No right-hand or left-hand problems anticipated.
22. Privacy screen required.
23. See reference in item 2.
24. Do not eliminate swinging doors from possibilities.
25. Desks should come ready to use.
26. Designer will have to check catalogs, and local furniture and office equipment stores.

4.2 Related Activities

There are many activities in which an engineer might engage while preparing a problem statement. No single source will always be the best, so a wide range of possibilities must be kept in mind.

Conduct market surveys to establish customer wants and needs, price limits, and sales forecasts. A market survey can include personal interviews, telephone surveys, or mail questionnaires. Important sources of statistical business information include:

- *Census of Manufacturers,* U.S. Department of Commerce
- *Statistical Abstract of the United States,* U.S. Bureau of the Census
- *Metal Statistics,* American Metals Market, New York
- *Minerals Yearbook, Vol. 1, Metals and Minerals,* U.S. Bureau of Mines

Check products and material available for purchase on the market, and current standards to see how well the need is already being met. Use references such as:

- Annual catalog, American National Standards Institute.
- *Annual Book of ASTM Standards.* The 1981 Book of Standards is in 48 volumes and contains over 6400 standards.
- Chemical engineering catalog: *The Process Industries' Catalog*, Van Nostrand Reinhold.
- *Index of Specifications and Standards*, U.S. Department of Defense.
- *Index of Federal Specifications and Standards*, U.S. General Services Administration.
- *MacRae's Blue Book* (Hinsdale, IL), a multivolume collection of manufacturers' catalogs.
- National Association of Manufacturers.

- *Sweet's Catalog File* (McGraw-Hill, New York), a compilation of manufacturers' catalogs with emphasis on machine tools, plant engineering, and building materials.
- *Thomas Register of American Manufacturers* (New York), a comprehensive directory of addresses of manufacturers arranged by product.
- *VSMF Design Engineering System,* a compilation of vendors' catalogs and product specifications for mechanical and electronic parts; on microfilm, updated monthly.

Check patent records to determine the extent of workable ideas and protected ideas, as well as to determine a progression of technology. Use the weekly publication, the *Official Gazette* of the U.S. Patent Office, for the latest patents. For reference by class and subclass, you must visit the Public Patent Search Facilities of the Patent and Trademark Office, 2021 Jefferson Davis Highway, Arlington, VA. Thirty major libraries in the United States also have patent records arranged in numerical order.

Read books written on a specific subject to determine its history and development. This will provide data and insight into what has been tried and was successful, and also what has failed and why. This activity might also uncover legal concerns that should be considered, such as previous lawsuits, court rulings, governing laws, and code restrictions.

1. Dictionaries
    - Ballentyne, D.W.G. and D.R. Lovett, eds. *A Dictionary of Named Effects and Laws in Chemistry, Physics and Mathematics*, 3rd ed. London: Chapman and Hall, 1970
    - Bennett, H., ed. *Concise Chemical and Technical Dictionary*, 4th ed. London: Arnold, 1986
    - McGraw-Hill editors. *Dictionary of Engineering.* New York: McGraw-Hill, 1985
    - McGraw-Hill editors. *Dictionary of Mechanical & Engineering Design*. New York: McGraw-Hill, 1985
2. Thesauruses
    - Engineer's Joint Council editors. *Thesaurus of Engineering and Scientific Terms*. New York: Engineers Joint Council (AAES), 1969
    - Roget, Peter M. *Roget's Thesaurus of English Words.* New York: St. Martin's Press, 1978
    - *Thesaurus of Metallurgical Terms*, 4th ed. American Society for Metals. Metals Park, Ohio: American Society For Metals, 1980
3. Encyclopedias
    - Besancon, R.M., ed. *Encyclopedia of Physics*, 3rd ed. New York: Van Nostrand Reinhold, 1985
    - Considine, Douglas M., ed. *Encyclopedia of Chemistry*, 4th ed. New York: Van Nostrand Reinhold, 1984.
    - *McGraw-Hill Encyclopedia of Science and Technology*, 5th ed. New York: McGraw-Hill, 1982
4. Handbooks
    - Baumeister, T., ed. *Marks' Standard Handbook for Mechanical Engineering*, 9th ed. New York: McGraw-Hill, 1987

- Beaton, C. F. and G. F. Hewitt, eds. *Physical Property Data for the Design Engineer.* New York: Hemisphere Publishing, 1988
- White, John A., ed. *Production Handbook*, 3rd ed. New York: John Wiley & Sons, 1986
- Crocker, S., and R.C. King, eds. *Piping Handbook,* 5th ed. New York: McGraw-Hill, 1967
- Cummins, A.C., and I.A. Given, eds. *SME Mining Engineering Handbook.* New York: Society of Mining Engineers, Inc., 1973
- Fink, Donald G., and H. Beaty. *Standard Handbook for Electrical Engineers,* 12th ed. New York: McGraw-Hill, 1986
- Fink, D.G., and D. Christiansen, eds. *Electronics Engineers' Handbook,* 3rd ed. New York: McGraw-Hill, 1988
- Harper, Charles A., ed. *Handbook of Electronic Systems Design.* New York: McGraw-Hill, 1979
- Hewitt, Geoffrey F., ed. *Hemisphere Handbook of Heat Exchanger Design.* New York: Hemisphere Publishing Corp., Harper & Row, Inc., 1989
- Higgins, L.R., and L.C. Morrow: *Maintenance Engineering Handbook,* 44th ed. New York: McGraw-Hill, 1987
- Karassik, I.J., et al. *Pump Handbook.,* 2nd ed. New York: McGraw-Hill, 1985
- Maynard, H.B., ed. *Industrial Engineering Handbook*, 3rd ed. New York: McGraw-Hill, 1971
- Merritt, Frederick S., ed. *Standard Handbook for Civil Engineers,* 3rd ed. New York: McGraw-Hill, 1983
- Parmley, R.O. *Standard Handbook of Fastening and Joining,* 2nd ed. New York: McGraw-Hill, 1989
- Slote, Lawrence, ed. *Handbook of Occupational Safety and Health.* New York: John Wiley & Sons, 1987
- Tapley, Byron D. *Handbook of Engineering Fundamentals,* 4th ed. New York: John Wiley & Sons, 1989

5. Reference Books
    - Blake, L.S., ed. *Civil Engineer's Reference Book,* 4th ed. Boston: Butterworth, 1989.
    - Siemens Teams of Authors. *Electrical Engineering Handbook* . New York: John Wiley & Sons, 1982
    - Parrish, A., ed. *Mechanical Engineer's Reference Book*, 11th ed. Boston: Butterworth, 1973

Read technical journals to learn about applications and developments that are current. No attempt will be made here to select key journals from a list that includes over 50,000 titles. Reference to them, however, is guided by abstracting and indexing services such as:

- *Applied Mechanics Reviews*
- *Applied Science and Technology Index* (formerly *Industrial Arts Index*)
- *Building Science Abstracts*
- *Ceramic Abstracts*

Related Activities

- *Chemical Abstracts*
- *Computing Reviews*
- *Corrosion Control Abstracts*
- *Energy Index*
- *Engineering Index*
- *Environmental Index*
- *Fuel Abstracts*
- *Highway Research Abstracts*
- *Instrument Abstracts*
- *International Aerospace Abstracts*
- *Metals Abstracts*. Combines *ASM Review of Metals Literature* (U.S.) and *Metallurgical Abstracts* (British)
- *Nuclear Science Abstracts*
- *Public Health Engineering Abstracts*
- *Science Abstracts A: Physics*
- *Science Abstracts B: Electrical and Electronics Abstracts*
- *Solid State Abstracts*

Computerized data bases for abstracting and indexing are listed in Table 4.1.

Attend trade shows to see current products, collect literature, and ask questions.

Check your own company records for any research work that has already been done on your project. Include discussions with those who might have worked on the project before you.

Contact component product suppliers. They are generally eager to talk about their products and will contribute specification sheets and offer product demonstrations.

Contact government agencies. There may be policies, codes, or laws that will affect your project. This would be particularly important if your project involved waste by-products. The Government Printing Office is the source of all federal government publications. A Monthly Catalog of U.S. Government Publications lists all available publications. The following list is a brief guide to government technical literature:

TABLE 4.1

Abstracting and indexing databases

Data Base	Supplier	Number of abstracts
Metadex	American Society for Metals in cooperation with The Metals Society of London	555,000
CA Search	American Chemical Society	4,000,000
Compendex	Engineering Index, Inc.	766,000
NTIS	National Technical Information Service, U.S. Dept. of Commerce	725,000

- Directory of Information Resources in U.S., Vol. 1, Physical Sciences Engineering, 1971. National Referral Center, Sciences & Technology Division, Library of Congress
- Monthly Catalog of U.S. Government Publications, Government Printing Office
- Government Reports Announcements and Index (semi-monthly), (prior to 1971 was U.S. Government R&D Reports), National Technical Information Service, U.S. Dept. of Commerce. References R&D reports for government-sponsored reports plus translations of foreign material. Copies of reports may be purchased from NTIS (formerly DDIC and ASTIA).
- Scientific and Technical Aerospace Reports, (STAR), Abstracts compiled by NASA, Purchased from Government Printing Office
- Nuclear Science Abstracts, Abstracts compiled by Department of Energy, purchased from Government Printing Office

The government has also established specific information centers, which are necessary for the large amount of specific technical data available, including:

- Metals and Ceramics Information Center, Battelle Columbus Laboratories
- Mechanical Properties Data Center, Battelle Columbus Laboratories
- Plastics Technical Evaluation Center
- Shock and Vibration Information Center
- Machinability Data Center, Metcut Research
- Thermophysical and Electronic Properties Information Analysis Center
- Thermodynamics Information Center
- Atomic and Molecular Processes Center
- Metal Matrix Composites Information Analysis Center
- The Smithsonian Science Information Exchange (SSIE). The SSIE is the only source of information on all federally funded research in progress. The data are in machine-readable form and can be searched by computer.

Specific government agencies may also be of help. Some of the better known agencies are:

- U.S. Chamber of Commerce
- Interstate Commerce Commission
- Federal Communications Commission
- National Aeronautics and Space Administration
- U.S. Department of Agriculture
- U.S. Department of Commerce
- U.S. Department of Health, Education and Welfare

Contact trade associations for literature. Trade associations produce and review voluntary standards, and these may affect your project. Those that have produced a substantial number of standards include:

- American Petroleum Institute
- Association of American Railroads
- Electronics Industries Association
- Manufacturing Chemists Association
- National Electrical Manufacturers Association

Contact product testing organizations for information concerning specific products. These organizations include:

- National Fire Protection Association (NFPA)
- Underwriters Laboratories, Inc. (UL)
- American Gas Association
- American Society for Testing and Materials
- American Society of Heating, Refrigerating and Air Conditioning Engineers

Contact consulting firms that deal with the general area of your project. There are many specialty companies whose experiences would be most helpful. The telephone yellow pages may list companies to contact.

Contact finance and insurance companies. They may have price index, market money rates, bond discount data, product liability data, and other financial data that would prove helpful.

Contact professional and technical societies. They have also made important contributions through standards activities. Those societies that have contributed to standards development include the following. See reference (4) for addresses and details.

- Aerospace Industries Association
- American Concrete Institute
- American Institute of Steel Construction
- American Society of Agricultural Engineers
- American Society of Mechanical Engineers
- American Welding Society
- Building Officials and Code Administrators International
- Electronic Industries Association
- Institute for Interconnecting and Packaging Electronic Circuits
- Institute of Electrical and Electronics Engineers
- Instrument Society of America
- National Electrical Manufacturer's Association
- National Safety Council
- Prestressed Concrete Institute
- Society of Automotive Engineers
- Technical Association of the Pulp and Paper Industry

4.3 Cautions

Be aware of several things at this point. First, realize that for a given project, you will not use all the sources mentioned; in fact, you may not use any. The project could be handed to you with most of the preliminary work done, or you may already be familiar with the situation and will add details and solutions from experience. If you know the river's depth from previous encounters, a quick solution can be, "Jump out of the boat and wade to shore; the water is only two feet deep." The possible activities are just that, possible. Use them if you feel unsure, if you need more data, or if you simply want to gain additional insight into the problem.

Second, realize that you should not wait until a problem is handed to you to search out information. Whether your company deals primarily in office equipment, electronic equipment, computer services, conveyor equipment, or space shuttles, you should study the product and the related industry on a regular basis. Read books and periodicals, attend seminars, go to product shows, visit with customers and dealers. Keep informed and up to date; technology and the times change and you must change and develop with them.

Third, realize that projects in most companies are team projects, and expertise will come to the project from other people and other departments. There is too much for one person to know, so make good use of the knowledge of your co-workers. Check with others even if they are not assigned to your project; no source of information should be neglected.

Fourth, realize the difference between a statement of fact and a problem statement. Some people get themselves trapped by issuing a statement of fact and not defining a problem, as the following example illustrates.

EXAMPLE 4.4:
Circumstances and problem statements.

1. "My pen is broken," is not a problem definition; it is a statement of fact. The problem may be to find a service center capable of fixing the pen, or to find something with which to write.
2. "My car won't start," is a statement of fact. The problem to be solved could be "How do I start my car?" or "How do I get to work?" or "How do I get the car out from in front of the garage so I can use the car inside the garage?"
3. "The furnace won't run," is a statement of fact. The problem to be solved could be "How do I get the furnace running again?" or "How do I keep the pipes from freezing?"

Your first introduction to a problem may be preliminary specifications established by your company or your client. Also included might be basic restrictions such as sales price, sales volume forecasts, and a time schedule of development and market introduction. If this preliminary information is not adequate, or you feel there are too many unknowns, ask questions. The earlier clarifications are made, the earlier serious work can proceed. To do a good job of solving a problem, you must first do a good job of identifying the problem. This requires knowledge and understanding of the background of the project.

During the problem identification activity, ideas for solution generally come up as requirements and limitations are discussed. This is normal and should not be suppressed. In a separate record, start listing any ideas that come to mind during the identification process. Do not dwell on solution ideas, and do not spend time now trying to think of the solve-all problem solution. By concentrating on the identification process, you will get the boundary conditions clarified and be in a better position to generate ideas later. The main thing to do now is to think of the problem and record all the items that can possibly have an effect on subsequent activities. Getting the basis of the project clear will make the other steps of the procedure easier. None of the activities in the design process should be done without due regard for all the other steps.

4.4 System Problems

System problems can also be handled with the problem solving methods described. The main difference is that the questions asked will have to deal with discovering the present inputs, procedures, and expected outputs. System problems take a lot of detail work to find out what the users of the proposed system want. Communication skills are most important and can make or break a project. You must ask the right questions, and at the same time not offend people about the way they do their job. Systems involving people have habit built into them. People get comfortable with what they do, and resist change. For this reason it is often hard to find out what a system is either supposed to do, or how it does it. People may feel protective and will not answer completely or truthfully, or a particular event may not happen regularly and is easily forgotten. Later when this event occurs the engineer might be asked why it wasn't considered in the system design. Always maintain contact with the users during a system design. They should have numerous ideas, operational data, good judgment, and they can help evaluate your ideas. The following example illustrates the five areas of problem identification for a system problem.

EXAMPLE 4.5:
Preliminary problem definition for a system.

1. Basic Statement—Design a system to monitor and report on production material inventory.
2. Basic preliminary requirements
    - System must provide a weekly updates printout.
    - Updates must include breakdown of purchases, returns, factory use, scrap, and include costs associated with each category of material.
    - System must be on the central computer so any office can inquire about material status.

3. Basic limitations
   - System must be installed and operating in 6 months.
   - Computer terminals are not available in all departments that handle material.
4. Other data
   - Categories of material include steel sheets, steel plates, iron castings, electrical components, and rubber products.
   - The present system is manual and updated only once per quarter.
   - Shipments of material are received every day.
   - The last complete inventory was done three months ago.
   - The next complete inventory is scheduled to be done in nine months.
5. Questions
   - What support is available from the computer programming staff?
   - How much do computer terminals cost? What is their availability?
   - Does the present computer have connector ports available for additional terminals?
   - Does the present computer have enough storage capacity for added programs and data files?
   - Does the system have to be on line, or are daily update entries acceptable?
   - Is there room to locate computer terminals in all departments.
   - Will data entry be made by people experienced with computers or will training sessions be required?
   - Are any of the data already on the computer in accessible files: material deliveries, schedules, uses, or cost?

System designs are no less important to a profitable and well-run company than product design. Keeping track of the records of a company allows management to make timely and well informed decisions concerning purchases, investments, loans, new product introductions, personnel requirements, new markets (including foreign countries), and product pricing.

4.5 References

1. Caplan, Ralph. *By Design*. New York: St.Martin's Press, 1982.
2. Evans, Powell, Talbot, ed. *Changing Design*. New York: John Wiley & Sons, 1982.
3. Faupel, Joseph and Franklin E. Fisher. *Engineering Design: A Synthesis of Stress Analysis & Materials Engineering,* 2nd ed. New York: John Wiley & Sons, 1981.
4. Koek, Karin E. and Susan Boyles Martin, eds. *Encyclopedia of Associations*, 22nd ed. Detroit, Mich.: Gale Research Company, 1988. A guide to over 25,000 National and International Organizations.
5. Middendorf, William. *Engineering Design*. Boston: Allyn & Bacon, 1969.
6. Robinson, Judith Schiek. *Subject Guide To U.S. Government Reference Sources*. Littleton, Col.: Libraries Unlimited, Inc., 1985.

7. Ross, Steven S. "Rules' Role." *Graduating Engineer*, March 1985.
8. Smith, Ronald. *How To Plan, Design & Implement a Bad System.* New York: Petrocelli Books, 1981.
9. Whyte, R.R., ed. *Engineering Progress Through Trouble*. London: The Institute of Mechanical Engineers, 1975.

4.6 Exercises

Items 1–10 are potential problem statements. List at least five questions that you would ask to clarify each situation so that the real problem can be identified and solved. Some of the statements may result in more than one problem.

1. As I was driving to work, the hood of my car popped open.
2. As I was cleaning fish, the knife slipped and I cut myself.
3. This door won't stay open while I carry packages into the house.
4. When I graduate, where will I go to look for a job?
5. Next spring I want to plant a garden.
6. Next summer I want to add a bedroom to my house.
7. What will I do with my income tax refund?
8. If I fail my math class, will I be able to continue my engineering studies?
9. Where is that smoke coming from?
10. I can build a better lawn mower than that one.
11. Consider the following problem identification statement. Do you feel the statement is satisfactory? Why or why not? Outline the next series of activities to be followed on the project. Be specific.

Design a writing desk that has at least one drawer for pencils and small office supplies, one drawer for letter size file folders, and one drawer to store necessary sales books in. Additionally, the desk will be used by a person sitting down. There should be ample room on top of the desk for a lamp, a telephone, an open loose leaf binder, and room to write out quotation sheets. A typewriter will not be used. Customers will be seated opposite the desk, so a privacy screen is required, and the desk must be attractive so the customer will feel at ease and comfortable.

12. Compare the advantages and disadvantages of the following three methods of conducting a market survey.
    a) Personal interview
    b) Telephone interview
    c) Mail questionnaire

Often a statement of fact will generate many different possible problems and courses of action. For Exercises 13–15, list different possible problems that could be the result of the statement. After listing possibilities, rank them in order of most probable. Give reasons for your ranking.

13. I flipped the light switch and the light did not come on.
14. I tried to unlock the car door and I could not.
15. I pulled the rope on the lawn mower starter and the engine did not start.
16. Give a more complete problem definition for each of the following:
    a) Design a lamp
    b) Design a bike rack
    c) Design an apartment building
    d) Design a computer
    e) Design a mechanical heart
17. Which of the problems listed in Exercise 4.16 would you feel most confident in undertaking? Why?
18. A camping trailer, a powered vehicle to pull it, and a trailer hitch constitute a system problem because the components can be purchased, but they must match each other for safe and efficient use. List some considerations that would be necessary for this system to work effectively.
19. Make a list of as many components or subsystems you can think of for a bicycle. Each item listed may have a unique or multiple function.
20. As an example of code restrictions on design, consider floor loading levels required in buildings. Residential homes can be designed to 40 lbf/ft^2, but public buildings require 100 lbf/ft^2 design. Discuss the reasons for the differences.
21. A TV situation comedy was being acted out in a building during a rain storm, and someone entered the scene and said, "There is five feet of water on the roof." Even if the roof had a five foot high edge to contain the water, would you expect a building still to be standing if the water depth on the roof did reach five feet? Quantify your answer.
22. Many design problems have conflicting control conditions, that is, restrictions that cannot be met simultaneously. For example, suppose electric lights had to produce light and not reach a temperature higher than normal body temperature. Those would be conflicting restrictions, because the increase in temperature is a result of the light producing phenomenon. Give other examples of conflicting restrictions in design.
23. List one problem situation that each of the following objects helps solve if used properly. Then give an alternate solution to the problem. Compare both solutions.
    a) Airplane
    b) Electric motor
    c) Television
    d) Elevator
    e) Radar
    f) Computer
    g) Linotype machine
    h) Electric heat
    i) Phonograph
    j) Table lamp
    k) Automobile
    l) Car bridge

m) Telephone
   n) River dam
   o) Electric water pump
   p) Traffic light
   q) Rifle
   r) Car signal lights
   s) Non-stick cooking utensils
24. A participant at a conference in Washington D.C. looked foreign and was wearing a turban. Another participant commented, "I didn't realize foreigners could attend this seminar. After all, it is paid for by the U.S. Government." Discuss any implied or stated restrictions, assumptions, generalizations, boundary conditions, limitations, or controls that were made by the person making the comment.
25. How were metals welded together before the invention of arc and acetylene welding? Are any methods similar to the first method of welding still being used? Describe them.
26. Many times people try to use machines beyond the range of use they were designed for. Example: A sewing machine designed for sewing cloth is used to sew leather, and the needle frequently breaks. Give other examples of attempts at using machines beyond their design limits and the possible consequences of that use.
27. Studying only the current solution idea is often neglecting the real problem. This frequently occurs and the situation is called "treating the symptom instead of the cause." In each of the following cases discuss what the real problem might be.
    a) A student was often late for class and blamed rush hour traffic, saying, "I tried different routes, but they're all crowded."
    b) A bicycle rider continually got the leg of his pants caught between the chain and the sprocket and blamed the chain guard design.
    c) A student always had errors in computer programs and used poor typing skills as an excuse.
    d) A bracket on a machine was breaking with high frequency under warranty. The engineer in charge wanted to increase the cross-sectional area of the bracket.
28. Complete the problem identification steps for one of the following design problems or select one of the problems from those listed in Section 7.6.
    a) A reserved seat system for a passenger train
    b) A taxi service system
    c) Retractable window awnings
    d) An automatic manufacturing and assembly line for picture frames
    e) An automatic brick manufacturing process
    f) An automatic ball bearing sorting machine
    g) A device to control ambient temperature for a light bulb life experiment
29. List and discuss some related concerns that are indirectly controlled or influenced by each of the following engineering projects.
    a) A dam on a river
    b) A subway transportation system
    c) A skyscraper office building
    d) A downtown high-rise apartment complex

5 Problem Solution Idea Generation

"The more ambitious plan may have more chances of success."
♦ G. POLYA

The next activity on the road to successful design and problem solving, after adequately defining the problem, is to generate solution ideas. Generating ideas will exercise your creativity. Creativity is always important, but when you are starting with a blank page or a blank computer screen, originality and inventiveness are essential. Generating ideas does not cost much at this point, just the thinking time and a few pieces of paper or some computer time. Unencumbered thinking allows you to approach the problem with diversity and ingenuity.

The more ideas you formulate, the greater the chance you have to arrive at a great solution instead of just a mediocre one. Concentrate your effort on the generation of an ample quantity of ideas, as well as quality ideas. The first idea that comes along should not be judged the best solution. Accepting the first solution idea that comes along is one of the most frequent causes of failure for beginning engineers. Of course if time is very short this guideline becomes less critical. As your boat nears the waterfall you aren't going to get out a pad of paper and start listing alternative solutions to your predicament. In most engineering projects, however, time is needed to contemplate various approaches to a problem; sometimes several days, weeks, or months are required.

If you have done a thorough job of problem identification, you already have a start on the generation of ideas. The activities used to research the background of a project during problem identification in Section 4.2 are also sources of ideas.

Another source of ideas is the series of questions that were asked during problem identification. Solutions are certainly being suggested when someone asks if a mechanism should be driven by a belt, gears, or a chain. Recall that there are no sharp dividing lines between the activities used to solve problems. The generation of ideas is discussed as a separate activity in this chapter, but all the activities in the problem solving procedure must be meshed to achieve optimum results.

Search existing products for solution ideas to avoid making the same errors that others have made. Remember that a shortcoming in an existing product design is one source of problems, and one of your objectives is to avoid design shortcomings. Whether examining a competitor's product or your own, you can determine if there are still problems to be solved, or if the same job can be done less expensively or more safely. There is no single answer to a design problem. Consider, for example, all the styles of lawn mowers. They all answer the same basic problem question: "How can we cut the grass?"

A second reason to search existing products is to find out the current state of the art. Don't reinvent the wheel conducting research and development in an area that has already been adequately studied and developed. Review the basic concept, the theory, and the details of an existing technology if the need exists to apply it differently, but try to use the work of those who have come before you in an efficient and effective manner.

Engineers must also be able to call upon their own experience and creativity for ideas. Experience comes with time, and each engineering project you work on will increase your exposure to new ideas and techniques. Living through an experience will deepen your understanding, broaden your expertise, and enable you to draw on personal experience as you undertake more advanced problems.

5.1 | Specific Considerations

Many people have trouble at the idea stage of a project; the paper is blank and they have difficulty getting started. Drawing the first line, or thinking of the first idea is often a roadblock to problem solving. Generating ideas requires more creativity than other activities performed during the problem solving procedure. Review the roadblocks that were discussed in Section 3.3.

Ideas don't cost money, and the more ideas that can be generated at the beginning of a project, the better will be the chance of finding an acceptable solution. A good idea generated too late may be costly in time and money spent developing a poorer idea. An engineer can improve the success rate for generating good ideas if some basic guidelines are followed.

Review the original problem statement, including all the questions, to refresh your mind about the specific task at hand. This is particularly important if you are working on more than one problem at a time, if some time has elapsed since you have worked on the problem, or if it just seems right to take a fresh look at the situation. Review is a reminder of the restrictions, limitations, and the boundary conditions within which you must work. Reviewing the unanswered questions will remind you of data that need to be gathered.

5.1.1 Break Problem Into Parts

Consider breaking a large problem into smaller problems. This will not be necessary if your problem is simple, such as "Find something to put in front of the tire so the car doesn't roll downhill." In the case of a project of great complexity, however, breaking the problem into sub-problems often makes idea generation easier. One way to break a problem into sub-problems is to address all functional inputs and outputs that can be identified. List specifically what the design result is to accomplish. One important sub-problem of all physical solution problems is material selection. Review materials and their properties to promote idea generation. The following two examples illustrate possible sub-problems of a larger problem.

EXAMPLE 5.1:

Design sub-problems.

Consider a primary problem of developing a surface vehicle for moon exploration. Sub-problems to consider include:

1. Power source; possibilities include solar, fossil fuel, or batteries.
2. Steering system; possibilities include manual, hydraulic, gears, or electric motor.
3. Motion control; possibilities include manual, mechanical linkage, radio control, umbilical cord from a control station, or on-board computer control.
4. Terrain contact; possibilities include rubber tires, driving tracks, moving "feet," rolling balls, or rolling cylinders.
5. Suspension system; possibilities include air cushion, hydraulic, springs, or direct with no damper.

EXAMPLE 5.2:

Design sub-problems.

Consider the design of a commercial building. Sub-problems would include site preparation, water supply and distribution, electric supply and distribution, the foundation, the exterior walls, the interior walls, windows and doors, people flow patterns, elevators, escalators, furniture, interior decorating, landscaping, heating, cooling, the roof, vehicle access, and parking. Each of these sub-problems generates its own set of concerns. On large scale projects, a different group of engineers might work on each sub-problem. Each group is required to coordinate its work with that of the other groups to be sure the solution to each sub-problem fits together with the solutions to all the other sub-problems. Before the final solution is determined, all the sub-answers must be meshed together to create the best overall solution, requiring more compromise.

5.1.2 Value Analysis

Value analysis is a technique that analyses a product or solution idea by breaking the problem into sub-problems using verb-noun functional expressions. Each verb-noun sub-problem expression is considered independently and ideas are generated for each one. Then the sub-problem ideas are joined for a total solution. This method assists in determining the essential functions and also the nonessential functions. Ordinarily, value analysis is used to evaluate an existing design, but the procedure is equally valid for helping to describe projected functional results.

The basic emphasis of value analysis is to consider the cost of providing a given function. Functional evaluation is aided by describing requirements in simple verb-noun statements. Phrases such as "support weight," "contain fluid," and "punch hole" are used. An overall project function might be more general, such as "sort mail," compared to a specific component function such as "weigh envelope" or "check stamp."

Each product, part, system, or procedure can be broken down into cost/function relationships. Individual functions must support the overall project goal. If a part has no function relative to the overall goal, it should be eliminated. There are no tables of minimum or maximum cost/function ratio values. The values are used only for comparison between ideas. Judgments about the maximum allowable cost for a function will be made, and the objective is to improve on that value. Most companies have standard cost procedures that use the resources of process engineering, plant layout, tool design, purchasing, and the accounting department to determine the necessary costs.

Value analysis places emphasis on creativity to determine other methods of performing the same function at less cost. Higher value is placed on a solution that accomplishes the same function with less cost. Guidelines such as cost per horsepower, cost per mile/hour, cost per pound, cost per inch-pound of torque, cost per unit of area covered, or cost per unit volume stored are often used. Understand that to make for less cost does not mean to make less reliable. The required task must be performed satisfactorily or the solution is not considered acceptable. The following example illustrates the verb-noun approach of value analysis to generate ideas, even though some of the identifying phrases are longer than two words.

EXAMPLE 5.3:
Environmental problem.

Consider the environmental problem of contaminated ground water and some broad objective functional sub-problems.

1. Raise ground water level.
2. Reduce natural contaminants (salts, sulfur).
3. Reduce human caused contaminants (pesticides, fertilizer residue, runoff from livestock yards).
4. Control use of water.

When these broad objectives are broken down in more detail, sub-problems and sub-problem solution ideas are more apt to come to mind.

How can ground water levels be raised? Increase precipitation. Reduce water runoff. Reduce water use. Reduce population. Increase water absorbing vegetation. Reduce evaporation.

How can natural contaminants be reduced? Reduce irrigation because it promotes salt accumulation in the ground. Pump water only from "clean water" areas.

How can human-made contaminants be reduced? Eliminate the use of septic tanks. Reduce the use of agricultural chemical fertilizers. Reduce the use of pesticides. Encapsulate all wastes to avoid leakage. Reduce the size and number of livestock yards. Collect animal and human wastes and process them for natural fertilizers. Eliminate land fills. Stop the manufacture of dioxins (chlorinated hydrocarbon compounds), organic solvents, and PCBs (polychlorinated biphenyls). Send all waste to outer space.

How can water use be controlled? Control flooding and store water for dry periods. Use pump-sprinkler irrigation rather than flood irrigation. Build more dams. Melt icebergs for a fresh water supply. Place automatic volume control shutoffs on home and business water supplies.

Notice that some of the ideas will cause additional problems to be considered and the scope of the problem to become larger. Many engineering problems have implications that prevent an arbitrary boundary cutoff. The related implications cannot be ignored. For the interested reader, reference (7) has an excellent discussion of the ramifications of water use control.

5.1.3 Combine Ideas

Another method to use, when generating ideas, is to try different combinations of the separate sub-function ideas, and then think of ways to put the parts together. Placing sub-function ideas together usually leads to new ideas. The number of possible combinations increases rapidly as the number of sub-functions and the sub-function solution ideas increase. If three sub-functions each has five ideas then there are $(5)(5)(5) = 5^3 = 125$ combinations to consider. The next example illustrates the combination of sub-problem procedure.

EXAMPLE 5.4:

Sub-problem combinations.

Reconsider the problem of designing a rotary cultivator. Some of the sub-problems and related ideas are:

1. Power source ideas
    a) Electric motor with cord plug in
    b) Electric motor with battery
    c) Gasoline engine, four-cycle, vertical shaft
    d) Gasoline engine, four-cycle, horizontal shaft
    e) Gasoline engine, two-cycle, vertical shaft
    f) Gasoline engine, two-cycle, horizontal shaft
    g) Gas turbine engine
    h) Diesel engine
    i) Air motor with compressed air from a compressor
    j) Hydraulic motor
2. Control ideas
    a) Push-pull controls similar to bicycle brake cables
    b) Solid rods, push or pull

c) Turning crank
   d) Electric solenoid
   e) Hydraulic, similar to actuating automobile brakes
3. Ideas to power the drive and power tine rotation
   a) Belt drive, single or multiple
   b) Chain drive similar to bike chain, single or multiple strands
   c) Gear drive similar to motor bike
   d) Air motor
   e) Centrifugal clutch
   f) Flexible shaft
   g) Hydraulic motor
4. Ideas for a support frame
   a) Weldment from bent and formed steel.
   b) Cast iron or cast aluminum
   c) Molded plastic
   d) Bolted hardwood (oak, hickory)

The total number of combinations for the four sub-functions = (10)(5)(7)(4) = 1400. One of the combinations is a gasoline engine, two-cycle, vertical shaft, using push-pull cable control, chain drive on a molded plastic frame. Additional ideas must now be generated to put these pieces together. If tines rotate about a horizontal axis, and the power shaft is vertical, then the connection between them must change the direction of motion. Possibilities include a gear box to use with the chain drive, a friction wheel disk drive, or belts to change direction and chain connect before or after the change in direction.

5.1.4 Research

Research available sources of information for ideas, many of which are listed in Section 4.2. The more sources you use, the more ideas you should be able to generate. Don't let false pride (another roadblock to creativity) get in the way of obtaining ideas from others. It's what you do with the ideas that is important. Of course you will not steal a trade secret or infringe on patent or copyright protected information. If these sources yield the best workable ideas, then proper license and payment contracts can be worked out; obtain adequate legal counsel.

The use of a dictionary or thesaurus should not be overlooked as a source of ideas. Alternate meanings and methods of expressing thoughts will assist the idea generation activity as the following example illustrates.

EXAMPLE 5.5:
Alternate word idea generation.

Supposing one of the verb-noun generalized objectives generated using value analysis guidelines was "protect surfaces." Ideas are always needed for protecting products. In Roget's *Thesaurus*, protect leads to preserve, which has the following list of possibilities: embalm, deep-freeze, cold pack, refrigerate, boil, dry, dehydrate, ensilage, can, tin, prophylaxcize, moth-ball, pickle, salt, life belt, gas-mask, chair cover, bottle, marinate, smoke, paint, coat, cover, whitewash, kyanize, waterproof, and shore up. Granted that the thesaurus does not include electroplate, varnish, or anodize, but the point is to use a source to get a fresh perspective for ideas.

5.1.5 Record Ideas

Make a record of all ideas. The form of the ideas is not important. Use words, sketches, photos, xerographic copies from books or journals, or supplier catalogs. The recording of ideas cannot be over-stressed. It is from recorded data that future work will be done. A good idea, poorly recorded, or not recorded at all, may be lost due to ambiguity, oversight, or poor memory. When personnel are moved from project to project, as often happens in larger companies, good records are essential for project continuity. Many times a project will be handled by several people, or someone on a project may be reassigned. If the work done by one person is not properly documented, then the next person on the job, a co-worker or replacement employee, may have to repeat work already done. The new person may not think of as many ideas or as many good ideas, and time and effort will be lost.

Proper documentation, including dates and witnesses, is also critical for patent applications and legal protection. Recording of all project activity is a good habit to get into, whether a patent is your goal or not. If a product later causes property damage, or injury to humans, a well-documented trail of design activities and decisions is essential for a defense at court trials.

One of the largest roadblocks to creative design occurs during the idea generation activity. This roadblock is the mental mistake of judging an idea before it is recorded. Too many times an idea is discarded before it is recorded because it is judged to be unworkable. Do not fall into this trap. No idea is perfect, there is always something wrong with every idea. The point of generating ideas without judging them is to get as many ideas as possible. During subsequent activities, adaptations and combinations can be made to improve the ideas. *Do not judge ideas at this time*.

5.1.6 Answer Questions

During the problem identification activity, questions were asked. During the current activity of generating ideas some of the questions should be answered, and the product specifications clarified and quantified. Must all questions be answered? No. In fact, new ones may be added. If air or hydraulic power is considered as a possible solution, for example, then it must be determined whether such power is available at the site of use, or what the cost is to add it.

Some questions may require lengthy model or lab testing to answer. Some questions cannot be answered yet, but a variety of ideas should be considered to account for variations that may be found. Plan ahead for alternative possibilities.

If material selection has not been made, then the strength and availability of alternate choices must be checked out. This also may not be possible until after many tests on prototypes are complete. Tests will determine stresses on components, but preliminary material choices need to be considered.

An answered question may also change the scope of the project. If it is decided to make the rotary cultivator power unit also usable for snow blowing and grass cutting, ideas will have to be generated for connect and disconnect, and power takeoff concerns. It is not at all unusual to loop back to problem identification at this point in the procedure and repeat portions of activities previously done. Remember that no activity in the problem solving procedure is ever done and forgotten.

5.1.7 Ask Questions

Occasionally you will be stymied and unable to begin generating ideas. It is often helpful to ask questions to get over the starting hurdle and to start the ideas flowing. Even if ideas have been generated, answers to additional questions can broaden the range of ideas. The following questions are suggested to assist in the generation of ideas.

1. What materials could be used? Consider ferrous and nonferrous alloys, polymers, wood, and ceramics.
2. What safety features should be incorporated? Consider the possibility of fire, explosion, electrical shock, chemical burns, and cuts from sharp edges.
3. What methods of support can be considered? Consider support from underneath, overhead, or from an adjacent member. Consider welded, bolted, glued, friction forces, or riveted construction.
4. What sizes and weights seem appropriate? Consider human as well as equipment limitations.
5. Are there features that might only be used some of the time? Should there be more than one model? Is there more than one way to use a given model?
6. Are there features that should be attachments rather than part of the main structure? Some features can be added on for special cases, such as reach extenders, extra load brackets, or heavy duty shock absorbers on cars used for pulling trailers.
7. Can an idea or feature be enlarged or reduced? Large fuel tanks allow more mileage but add weight, while smaller electronic computer components allow faster processing time.
8. Can something else be substituted? Use a snap ring in place of a cotter pin; a plastic part for a metal one; a self tapping screw in place of a bolt and nut.
9. Can two or more ideas be combined? Design one part to do the job of two.
10. What are the advantages and disadvantages of other designs that have been considered? Consider features that make an existing product good or bad.
11. How can a particular disadvantage be overcome? Focus on a small detail and make improvements.
12. How can an advantage be improved? Make something good even better; faster, lighter, smaller, or stronger.
13. What is the scientific basis for the ideas to date? What other basis could be considered? Consider making gravity, centrifugal force, or electromotive force work for you.
14. How can reliability be improved? Work to increase the life of a part.

15. How can appearance be improved? Good appearance won't hide a bad product, but it can enhance a good one. Consider painting, plating, anodizing, or special fabrics.
16. How can a specification be improved? Make your product dig deeper, run with less power, have a greater payload, or operate faster.
17. Can an idea be put to another use? A different application of an idea may prove beneficial.
18. Can an idea be modified? Studying one idea carefully may lead to other related ideas.
19. Can components be rearranged? Rearrangement often allows combinations and alternatives not considered before.
20. Will reversing, or using the opposite work? Inside out? Upside down? First last? Last first?
21. Can the shape of a component be changed to fit better, to be made less expensively, or to do multiple tasks?
22. What happens if a part is made longer, shorter, thicker, or thinner?
23. Can some other power source be used? Electric, hydraulic, pneumatic, chemical, solar, wind, or nuclear?
24. Can a part be left out? Does it actually have a function related to the problem to be solved or is it an add on?
25. Can extra value be added at low cost? Better looking, easier to use, or more reliable?
26. What happens if a concept is taken to an extreme? Very short, very long, very light, very heavy?
27. Can the product be made more compact? Portability or storage requirements improved?
28. Can motion be added or subtracted to make an operation more efficient?
29. Can a feature be located at either end? In the middle?
30. Will a moving part work better sliding or rotating?
31. How can assembly be made easier? Nuts & bolts, rivets, welding, cotter pins, keys, splines, glue, Velcro™, or staples?
32. How can an operation be made safer and either faster or slower? A coin-operated toll gate may need speeding up whereas an automatic door closer may need slowing down.

You should also review your own feelings about the problem by asking questions about your basic assumptions and the boundary limits you have established. Questions should include:

1. What am I assuming?
2. What experiences and facts support my assumptions?
3. Is wishful thinking getting in the way of critical thinking?
4. Am I confusing chance and probability with cause and effect?
5. Will my assumptions pass a rigorous logic test?

5.2 Physical Laws

Another activity to consider when trying to generate ideas is to review the physical laws of the universe and determine if any are applicable to the problem at hand. The properties of materials and the principles of the physical universe have been studied and developed into a coherent set of relationships by the work of many people. The following list includes a brief description of many of these common laws and principles. Additional self-study is warranted depending on the problem under investigation. For example, the Piezoelectric effect in crystals might be a substitute for strain gages in a pressure transducer problem.

Ampere's law: Force relationship on a current-carrying conductor in a magnetic field.

Archimedes' principle: The buoyancy force on a floating object is equal to the weight of the displaced fluid.

Avogadro's law: There are $6 \cdot 10^{23}$ molecules per mole of gas at standard temperature and pressure.

Bernoulli's theorem: Conservation of energy in fluids, relates changes in normal pressure with axial velocity.

Biot-Savart law: Magnetic field produced by an electric current.

Biot's law: Linear conductivity of heat in a material.

Blagdeno law: Depression of freezing temperature due to impurities.

Boyle's law: Pressure times volume is a constant for a given gas at constant temperature.

Bragg's law: X-ray beam reinforcement as a function of wave length.

Brewster's law: The tangent of the polarized angle for a substance is equal to the index of refraction.

Brownian movement: Random motion of solids in a colloidal solution.

Charles' law: Pressure times temperature equals a constant for a given gas at a constant volume.

Christiansen effect: Refraction of powders in a liquid.

Corbino effect: Considers current flow induced in a metallic disc while rotating in a magnetic field.

Coulomb's law: Relates forces between electric charges.

Curie-Weiss law: Considers the temperature for transition from ferromagnetic to paramagnetic behavior.

Dalton principle: In gas mixtures, the total pressure is the sum of the partial pressures.

Debye frequency effect: Relates the increase of conductance of an electrolyte with frequency.

Doppler effect: Shift of received frequency due to motion of sound source.

Edison effect: Thermionic emission from heated metal in a vacuum.

Eötvos effect: Relates change in body weight with east-west motion at the surface of the earth.

Faraday electrolysis: Chemical change is proportional to electricity flow across an interface.

Faraday induction: Generation of electromotive force by change of magnetic linkages.

Gauss effect: Considers the increase of resistance of magnetized conductors.

Gibbs phase rule: Considers the equilibrium conditions in mixed systems.

Gladstone-Dale law: Relates change in index of refraction with density.

Graham's law: Osmotic flow of gases and solutions through a porous wall.

Hall effect: Voltage is generated perpendicular to the current in a magnetic field.

Henry's law: Behavior of a dissolved gas in a liquid.

Hertz effect: Effect of ultraviolet light on spark discharge.

Johnsen-Rahbek effect: Increase of friction of viscosity with applied voltage across interfaces between conductor, semi-conductor, and insulator.

Joule's law: Heat produced in an electrical resistor by current flow and time.

Kepler's laws: Planetary orbits are ellipses, equal areas are swept by radius vector in equal times.

Kerr effect: Optical polarization due to electric field applied to a substance.

Kirchoff's laws: Sum of currents into a circuit node is zero, algebraic sum of voltages around a circuit is zero.

Kohlrausch's law: Limiting conductance of an electrolytic.

Lambert's cosine law: Illumination of a surface varies with the cosine of the angle of incidence.

Le Chatelier's law: Equilibrium adjustments of a system under stress.

Lenz's law: Current and force induced so as to oppose motion of a conductor in a magnetic field.

Leidenfrost phenomenon: Non-wetting of a hot surface by a dancing liquid drop.

Meissner effect: The absence of a magnetic field inside a superconductor.

Nernst effect: Electromotive force generated by heat flow across force lines of a magnetic field.

Newton's laws: Acceleration is proportional to an applied force, actions are opposed by equal reactions.

Ohm's laws: Current through an impedance or resistance is proportional to the voltage drop.

Paschen's law: Sparking voltage of gas between electrodes.

Peltier effect: Absorption or liberation of heat by electric current flowing between different materials.

Piezoelectric effect: Dimension changes of electrified crystals, generation of electricity by stressing crystals.

Pinch effect: Reduction of cross section of liquid conductor carrying an electric current.

Pyroelectric effect: Charges developed between opposite crystal faces when the temperature is changed.

Purkinje effect: Color sensitivity change under dim light in the human eye.

Raoult's dimension effect: Change in electrical resistance with a change in length.

Snell's law: Relations between incident, refracted, and reflected rays at an optic interface.

Stark effect: The split of spectral lines when source is in a strong electric field.

Stefan-Boltzmann law: The heat radiated from a black body is proportional to the fourth power of its temperature.

Stokes' law: The wavelength of light emitted from a fluorescent material is always longer than that absorbed.

Tribo-electricity: Electromotive force produced at the interface of two dissimilar metals by relative motion and friction.

Weber-Fechner law: The minimum change of stimulus necessary for a response is proportional to the level of the stimulus already existing.

Weidemenn-Franz law: The ratio of thermal to electrical conductivity is proportional to absolute temperature.

Wien effect: The increase in the conductance of an electrolyte with applied voltage.

Wien's displacement law: As the temperature of a body is increased through incandescence, the color of its emitted spectrum shifts toward the blue.

5.3 Brainstorming

Another method used for the generation of ideas is brainstorming. Brainstorming is an activity in which a group of people get together specifically to think of ideas. A group of people can usually generate more and better ideas in a shorter period of time than can a single individual. When people hear one idea it often stimulates another idea: an offshoot of the idea just mentioned, a similar but different idea, a combination of ideas, or just a thought made conscious by the group's verbal exchange.

The following guidelines help to make a brainstorming session work effectively.

1. Arrange for a group of 4–12 people with various backgrounds, but relatively equal status levels. Try to get a few who have brainstormed before so the session will be easier to start. People in a group that is too small may not interact well, and people in a group that is too large may be difficult to control, and some may not participate.

2. Communicate with the people selected about a week in advance so they can prepare for the activity. Ask for their help, list the conditions of the planned session, and give them the topic to be brainstormed.

3. Select a leader who will:
    - Keep good notes and records of ideas generated. A tape recorder or stenographer may be useful. This is important for evaluation and analysis at a later time, usually done by a different group of people.

- Prevent judgment of the ideas presented so that ideas will flow naturally. Judging must not be permitted. If any idea or comment is judged, embarrassment and/or turf protection may result and idea generation will be reduced. The purpose of the session is to generate ideas, not to evaluate them. Non-judgmental questions are permitted. Often a clarification of words will trigger additional ideas.
- Keep participants on the topic so efforts will be concentrated on the specific problem presented. Side stories and other problems must not interfere with the specific problem being brainstormed.
- Encourage participants and help them feel free to contribute ideas.
- Have questions ready to ask to stimulate ideas in case participants' output lags. A leading question may set the group off on more ideas. Refer to the list of 32 questions in Section 5.1.7.
- Keep participants' responses short. Ideas are the main concern, not flamboyant rhetoric.
- Limit time so the session ends on a positive note. A session that drags on is not productive and may leave a bad impression on participants.
- Follow up with a written summary to the participants, thanking them for their help and requesting additional ideas that may have come up since the session. Include a list of the ideas that were generated for their reference.

Brainstorming can be used for a wide range of problem statements. A specific topic such as, "How can parking congestion on campus be alleviated?" or a general topic such as, "What can be done to make our campus more attractive?" can both be approached with brainstorming.

If a brainstorming session becomes nonproductive, whether it be by intimidation or other cause, consider the written brainstorming method. Give each person a sheet of paper and ask each one to write down three ideas, but not to include their name. Each sheet is then passed to another person; each person reads the ideas written, and adds three ideas. This continues until each person has read the ideas from each of the others and has added three ideas to each sheet of paper. Each sheet of paper should end up with three ideas for each person in the group; five people should conclude with 75 ideas, though some ideas may be duplicates because each person has only seen each piece of paper once. If this method stirs interest, the leader can read ideas from the sheets and participants can add ideas orally.

CASE STUDY #2 CONTINUED:
Ideas to consider for moving parts from one department to another.

1. Conveyor belts
2. Roller conveyor
3. Overhead conveyor with hanging hooks
4. Gravity chute
5. Fluid in which to float parts
6. Pneumatic pressure tube

7. Hand carry
8. Tote pans
9. A wheelbarrow
10. A sled
11. Two wheel hand dolly (like those used for moving refrigerators)
12. Three or more wheel push-by-hand carts
13. A forklift
14. A robot train on wheels
15. Catapult (similar to a hay bale handler)

The next case study presents a problem and shows the actual examples that resulted from the idea generating procedure.

CASE STUDY #3:
Idea Generation

A group of student engineers was given the problem, shown in Figure 5.1, of moving point C 3 inches (76.2 mm) vertically by moving a 24 inch (609.6 mm) hand lever through a maximum angle of 80°. Points A and B are available as pivot points. The resistance force at point C is known to be 50 lbf (222.4 N).

The students were prolific with idea proposals and properly recorded their ideas with words and sketches. Some of the ideas are shown in Figures 5.2–5.5. Figure 5.2 shows a preliminary list of five categories of ideas and a sketch of a specific idea for a sliding bar linkage. Figures 5.3 and 5.4 show sketches of ideas considering gears and cams. Figure 5.5 shows sketches of other ideas, including a consideration of the competitors. The ideas shown are not meant to be exhaustive, only a sample of ideas and the format used for recording them. The sketches also include written comments made during subsequent idea refinement and analysis.

FIGURE 5.1

Lever problem, Case Study #3.

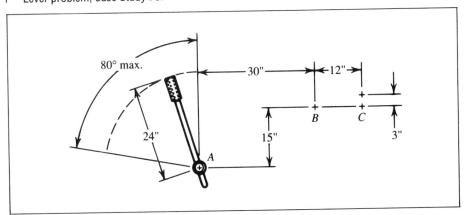

FIGURE 5.2

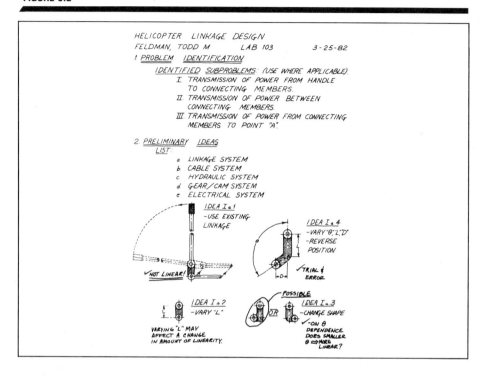

FIGURE 5.3

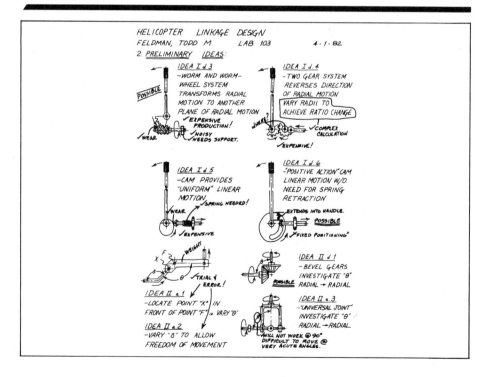

FIGURE 5.4

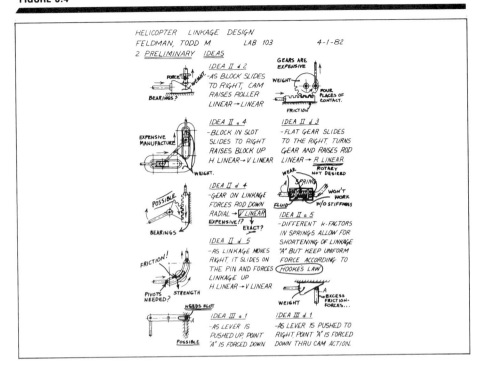

FIGURE 5.5

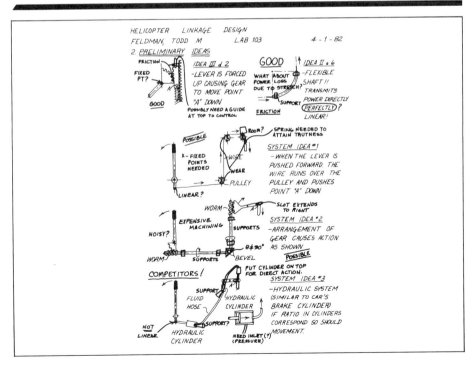

5.4 Work Simplification

Work simplification is a specialty study used to create ideas to eliminate wasted motion and effort while doing a particular job. The objective is to eliminate unnecessary work effort while getting a particular task performed. The usual method of using the work simplification approach is to break a procedure or process into its smallest component operations and determine the best sequence of operations, as well as the best way to perform each component operation. A product assembly line is a good example of a process that has been broken down into component operations, and each one made efficient relative to the whole. The product design itself must take the assembly procedures into consideration, as a design that cannot be efficiently and reliably assembled is not a good design.

Work conditions must be arranged to get the most done with the minimum effort. This includes positioning work so both hands can be used, or designing and supplying special tools or holding fixtures. Questions are asked about the sequence of operations being done and the tools and techniques used. Sometimes just changing the sequence of events will make a job more efficient. Work simplification analysis is most often used on a job already being performed, but the principles should also be used during the planning stages of a project. Efficient operation includes having the right tools available and in good condition, a work area large enough and at the right height, adequate lighting, heating, ventilation, and noise control.

When analyzing the work to be performed in any system it is important to know and understand the differences between the abilities of people and of machines. Humans have a greater capacity for sensory functions such as smell, taste, and sight, but in the areas of hearing and feel there are some tasks that machines can do better than humans. Fixing and matching sound frequencies, and sensing surface microfinishes are but two examples. Humans are very efficient at comprehending a complex situation, perceiving conditions, and deciding on a course of action. This is particularly true if the situation cannot be planned for in advance. A planned situation, such as a rocket flight, or an engine fuel mixture, can be controlled faster and with greater reliability by computers. In this same general area, humans are usually more alert for what is coming and can act on their hunches based not only on what is immediate, but what is anticipated, such as selecting a part from a load of jumbled parts. This makes humans more flexible than machines. A machine can be created to perform a single task and anything out of its range will be a total failure. Humans can at least suggest what to do if their ability range is exceeded. This also gives humans the edge when it comes to reasoning and judgments. Will machines ever be trial judges?

Machines are valuable for many reasons. They can operate faster and more consistently than a human. Most machines perform tasks that might otherwise be slow, dangerous, and difficult, if not impossible for a human to do. Machines perform at a constant pace and do not become bored, lose interest, and make errors because of lack of concentration. Robots are used extensively for applications where repetition of a task dominates a procedure, such as the handling of parts from work station to work station, as well as for applications that are dangerous to humans such as the handling of hot, cold, or radioactive objects. Computers are used extensively to make repeated calculations because they can make them faster and more reliably than humans. It does, however, take humans to control and interpret the results of the calculations. Computers can store and retrieve large amounts of data faster from memory, without error, than can a human. Well designed machines, including the new generation of parallel processor computers, can also perform tasks simultaneously.

5.5 Feasibility Study

The term feasibility study often causes confusion, because a feasibility study creates a special case of generating problem solution ideas for the engineer. Too often people use the term to imply that the project will be implemented as soon as the study is complete, and that can cause misunderstanding and difficulties for the investigating engineer. Often a feasibility study is the link that connects idea generation with analysis because only the feasible ideas, in terms of such concerns as time, money, and state of the art, should be pursued.

Feasibility studies usually cover the situation when someone has an idea and wants to know whether to proceed. The person may be capable of making the study, but will call an engineer in as a consultant to get a fresh outlook and new ideas. Be wary of the person who asks you to "Gather data so I can show the board that this project should be done." You may end up being used as a scapegoat. Always prepare a feasibility study in an honest and fair manner. A biased study may look only at the advantages and not consider the disadvantages, or vice versa. Taking sides will make a study biased, and that would be unethical.

A feasibility study should define the problem, identify factors that limit the scope of the project, and evaluate anticipated difficulties. The study may validate or invalidate the stated need, determine initial direction of a solution investigation, produce an initial set of possible solutions, and suggest a preliminary cost objective and schedule of activities. A feasibility study will use the same problem identification and idea generation procedures as have been discussed. The purpose of the feasibility study is to determine if a particular idea or scheme has enough merit to justify further study and development. But this applies to any engineering project. All projects must remain feasible in order to continue. Never get so attached to a project that you cannot muster the courage to call it to a halt if an insurmountable stumbling block is encountered.

During a feasibility study ask the usual questions about consumer needs and the current solution to the problem. A market survey may be required. Costs of implementation will be estimated at the macroscopic level rather than the microscopic level. The number of component parts required will be estimated, as well as their cost. Any items that can be specifically identified, such as an electric motor or other standard items, can be costed accurately. If the product under study is similar to one already manufactured, a comparison cost may be possible if the current state of the art will work in the new application. Some large equipment cost estimates are made on a per pound basis. In the case of engines or motors a cost per horsepower basis may be used. Perhaps just a cost per part and the estimated number of parts will suffice. Any basis for estimating cost is permissible, but documentation is necessary so follow-up and continued study can be logically done.

Items to consider when thinking of ideas for a feasibility study should include:

1. Is present material handling equipment adequate? This is particularly important if the product being considered is larger than existing models. Overhead hoists,

door openings, paint booths, material storage, and available machine tools all have size limits that may be exceeded with a new product.

2. Can the new product be shipped as current products are shipped?
3. Will the increased work load in the factory replace some current work load, or will additional people be required to meet production schedules?
4. If the new product will involve a new technology, are present employees capable of implementing the technology, or will special training or new hirees be required?
5. Will the present computer have adequate storage and terminal access for the new system?
6. How will the implementation of this project affect other projects and related operations?
7. What other ideas will satisfy the need as defined?
8. Is the public ready to accept the proposed project?
9. Are there unusual environmental issues involved? The Environmental Protection Agency requires an impact study for projects such as roads, dams, bridges, or dam-created lakes.
10. Are there unusual safety issues involved?

During a feasibility study, creativity in component design analysis is not the primary skill necessary. What is needed is the ability to determine whether or not it is possible to do the task being studied and if so, develop a plan for doing it. Consider all ramifications, and determine if it is economical to proceed.

A proper feasibility study will result in a list of objectives and goals for the project, and includes such items as alternative ideas for implementation, target manufacturing costs, time schedules, market penetration, personnel requirements, capital requirement costs, production goals, projected return on investment, and a list of questions not answerable without additional study. These items become the framework within which the detailed design activities must operate. Estimates of all the items listed must be made, and this reinforces the need for a group effort. The sales department can make a market survey for information on consumer needs and products presently available, and propose market penetration and selling price goals. Manufacturing can estimate the cost and availability of tooling and machine tool requirements. Industrial engineers can estimate direct labor cost and assist plant engineers with estimating facilities requirements. The report will give standards to use as a basis for follow-up progress reports. A well-managed project will promote an atmosphere in which all departments will contribute information so the stated objectives will be met.

EXAMPLE 5.6:
Feasibility study.

Consider a feasibility study to investigate adding storm windows to a building. The analysis of expenditures and savings is fairly straightforward. Single pane glass storm windows change the R factor from one to two. The average number of heating degree days can be obtained from the weather bureau or a fuel distribution company. The cost of fuel and the heating system efficiency can be obtained from records or estimated, and fuel savings can be compared to the installation cost.

Other factors should also be considered. More uniform heating may make people inside the building work more efficiently. If the heating system does not have to work as hard, it may last longer before repair or replacement. If the storm windows come with screens, then summer air circulation will be possible without the annoyance of insects. Summer cleaning cost will increase due to the additional glass area to wash. The change in looks of the building may attract more customers. A feasibility study should answer all the questions possible and list the ones that need further study and analysis.

If by now you are asking, "How does a feasibility study differ from the activities of regular problem solving, they both seem the same?" you understand well. There is no difference in activities, perhaps only in the depth and degree of completeness, and the amount of time used. Questions that may lead to a feasibility study include:

1. Would this be a good corner on which to build a parking lot?
2. Should I sell trailers to complement my already established tractor sales?
3. Can our company build and market a sheep's foot soil compactor attachment for highway building contractors?
4. Would it be economical to purchase and install sensors that will turn off the lights in this classroom when no one is in it?
5. Should our free parking lot be converted to a pay-as-you-use lot with toll gates or some other type of fee collection?

5.6 Summary

Methods proposed in this chapter are designed to help generate ideas to solve a given problem. Different problems will suggest different methods of idea generation. This makes the method of idea generation a problem in itself; so why not come up with your own ideas on how to generate ideas. Try to think of a method to generate ideas that was not discussed in this chapter.

The following method of generating ideas, though perhaps not original, came to me as I was reading the section on brainstorming in reference (1). Write down one solution idea to the problem at hand. List one advantage and one disadvantage of the idea. Think of an idea to overcome the disadvantage. List one advantage and one disadvantage of that idea. Think of an idea to overcome the disadvantage. List one advantage and one disadvantage of that idea. Continue the process of thinking of advantages, disadvantages, and ideas to overcome the disadvantages until a good overall solution to the problem takes shape. Not only should you have an acceptable solution, but you will know its advantages and how it will handle potential disadvantages. This procedure can also be done in a small group, but it is not brainstorming in the usual sense because judgments are made at each step of the procedure.

5.7 References

1. Adams, James L. *Conceptual Blockbusting: A Guide to Better Ideas.* New York: Norton Publishing Co., 1980. Includes a 9-page discussion of guidance to other readings on creativity.
2. Beakley, George C., et al. *Engineering, An Introduction to a Creative Profession,* 5th ed. New York: Macmillan, 1986. Chapter 12.
3. Freedman, George. *The Pursuit of Innovation.* New York: AMACOM, 1988.
4. Harrisberger, Lee. *Engineeringmanship—The Doing of Engineering Design,* 2nd ed. Monterey, Cal.: Brooks/Cole, 1982. Chapters 4 and 5.
5. Jones, J. Christopher. *Design Methods.* New York: John Wiley & Sons, 1981.
6. Keefe, William. *Open Minds.* New York: AMACOM, 1975.
7. Miller, G. Tyler Jr. *Living in the Environment,* 5th ed. Belmont, Cal.: Wadsworth Publishing Co., 1988.
8. Nadler, Gerald. *Work Simplification.* New York: McGraw-Hill, 1957.
9. *Value Analysis Value Engineering.* New York: Value Analysis Incorporated, 1968.
10. Also those references listed at the ends of Chapters 2, 3, and 4.

5.8 Exercises

1. List at least three ideas as possible solutions to each of the following "problem" statements:
   a) I don't have a date for the dance.
   b) My car won't start.
   c) I want to cook out and it is raining.
   d) I am nervous about taking the test.
   e) The computer is down and my program is due in two hours.
   f) How can I get to class on time?
2. A float in a carburetor for a gasoline engine is used to control the fuel flow valve. When the fluid level is high enough the float rises and closes the valve. One design of such a float is made from two brass stampings soldered together. The solder joint was not good in one sample and the float filled with gasoline, caused it to sink, and kept the valve open.
   a) What symptom might a lawn mower exhibit as a result of this failure?
   b) What other failure might result in similar symptoms?
   c) When the repairer took the carburetor apart, what clue do you suppose led to the discovery of the actual problem?
   d) List methods to remove the gasoline from the float. Give one advantage and one disadvantage of each method.
   e) Evaluate the ideas from (d) and order them, best to worst.
   f) Suggest ways to solve the problem of this non-floating float.

3. List subsystems of each of the following products.
    a) Bicycle
    b) Lock used at a dam site for boats
    c) AM radio
    d) Automatic toll booth on a toll road
    e) A residential home
    f) An adjustable vise grip
    g) A drafting machine

4. Record ideas that could be used for connecting a single power unit to a lawn mower, a snow blower, and a rotary cultivator.

5. Make a list of all the ways you could use:
    a) A single brick
    b) Ten bricks

6. In a group, brainstorm ideas for Exercise 5.4 or 5.5, or some other problem.

7. Make a list of any function that a connected telephone could be used for other than transmitting voice signals.

8. List as many ideas as you can on ways to "cover up" the front of a kitchen cupboard.

9. List ideas to be used to get students to begin assignments on time rather than the night before the due date.

10. List methods that could be used for determining the directions north, south, east, and west.

11. Suggest uses for football stadiums, many of which are used only 20–30 times a year.

12. Suggest ways to control traffic flow at road intersections.

13. List ideas for ways to use empty gallon plastic milk containers.

14. List ways to collect tolls on a toll road other than the stop-pay-start system which often causes traffic flow bottlenecks.

15. List ways of identifying pages of paper so they can be returned to their original order if dropped and scattered.

16. List some uses of old editions of books after new editions come out.

17. Suggest some uses for worn out car and truck tires.

18. Compare the differences in security measures that you would take if you were in charge of, or were designing, each of the following.
    a) A bank
    b) Ship cargo storage
    c) Computer power supply
    d) Educational facilities
    e) Health care facilities
    f) High rise office building
    g) An industrial plant
    h) A library
    i) Surface transportation

j) Air transportation
k) A personal residence
l) Single owner business building
m) A public building such as a post office

19. List ideas for solving each of the following problems.
    a) A sleepy driver wake up alarm.
    b) A method of locating a dropped contact lens.
    c) A way of displaying miles per gallon in a moving vehicle.
    d) A method of locking all the exterior doors in a house from one location.
    e) A method of determining if all those on board an airplane have their seat belt on and fastened.

20. One of the problems with older homes is lack of insulation. Discuss alternate methods of increasing the R factor of the walls and ceilings of an older home.

21. Gamma ray radiation can be attenuated by adding mass to the path that the radiation must pass to reach an object. Discuss ways to increase the mass of the walls and the ceiling of a residential home to make the inside safer from gamma rays.

22. Discuss alternate methods of collecting and returning golf balls on a driving range.

23. Discuss various methods of improving the lighting on drafting boards in a room where illumination is lower than minimum.

24. Discuss methods of determining which way is vertical and which way is horizontal.

25. Figure 5.6 shows a circular object with a nonhomogeneous mass at a distance R from the center. If the axis of the object is placed on knife edges the off center out-of-balance mass causes a torque, and the object rotates until the mass is at a low point. Devise a method of determining the out-of-balance mass and its location from the center so actions can be taken to balance the object.

FIGURE 5.6

Out of balance object for Exercise 5.25

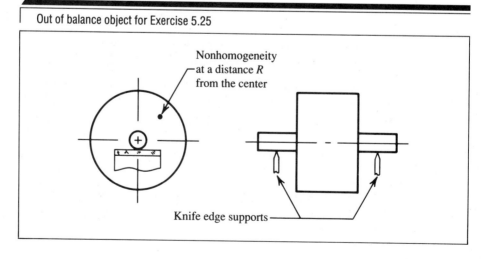

26. A dynamic damper, often called a dashpot, is a device that dissipates energy of motion to some other form so a moving system will come to rest without an impact load and without causing oscillation. A dashpot rating can be quantified by a constant C, with units of lbf·sec/ft, as the proportionality constant in the equation $F = Cv$. F is the force that resists motion and v is the velocity of motion. A damper is a common device to use when vibration is present. List some ideas for the design of such a damper.

27. Figure 5.7 is a layout of a traffic intersection. In the morning hours the majority of traffic comes from A going downtown to work, and from E following E'' to the high school. Some of the traffic from A follows A', but the significant amount follows A''. The traffic from B merges with that from A'' and is controlled by a yield sign. A little traffic comes from C and is controlled by a stop sign. Traffic that comes to D is also controlled by a stop sign. Any traffic coming from E can follow E' or E''. The traffic following E' is <u>not</u> controlled by any signs. The roads from E and A are wide enough for four lanes, but only as turning lanes. The lanes are not wide

FIGURE 5.7

Traffic intersection for Exercise 5.27

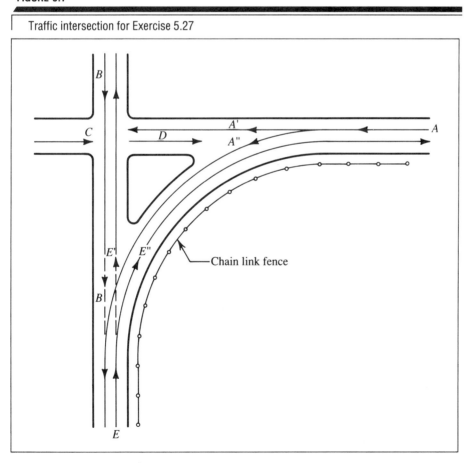

enough for four lanes of moving traffic. The speed limit is 25 miles per hour. In the afternoon most of the traffic comes from E with about one-third following E'. From E is uphill. Near-miss collisions occur between traffic from A'' and E'. A sidewalk and a six-foot high chain link fence follow the inside of the curve. Suggest alternatives to the traffic pattern to minimize the risk of accidents.

28. An accelerometer is a device that allows the measuring of acceleration. Various mechanisms are suited for such a device, but most of them make use of the basic relationship $F = ma$. The problem becomes one of determining the force applied to a known mass and calculating acceleration or calibrating the device so it reads acceleration directly. Sketch and discuss various devices that could be used for an accelerometer.

29. An extension ladder generally has a warning label on it that suggests it is unsafe to extend beyond a certain limit. This limit is to maintain a minimum overlap of the two sections. Most extension ladders do not have a device on them to actually prevent or deter overextension. Suggest ideas to assist in limiting overextension.

30. A yardstick is balanced on a fulcrum at the 12-inch mark by placing an object at the 1-inch mark. After some time has passed, and without being touched from an external source, the yardstick tips due to imbalance. What object could have have been placed at the 1-inch mark that would cause such an event to occur?

31. Perform a feasibility study on one of the following:
    a) Permit free downtown parking.
    b) Require each freshman college student to purchase a microcomputer.
    c) Change the law to allow polygamy.
    d) Number the parking places on campus and sell their use to the highest bidder.
    e) Allow one nuclear power plant per state.
    f) Market a push rotary lawn mower.
    g) Purchase and install sensors which will detect the presence or absence of a person in a room and turn the lights out if no one is present. Use your classroom as a sample room.

32. List all the verb-noun functions of something complicated, such as a bicycle. Note that some parts can have more than one function. A bike frame, for example, "supports axle" and "supports seat post."

The following exercises are intended for practice in creative and inventive thinking.

33. Design a device or method to allow an egg to be dropped from a third-floor window and reach the ground below without breaking.

34. Determine a way to light a match from a distance of ten feet.

35. Determine a way to fill a pop bottle located on the ground, from a third-story window.

36. Design a device to turn the pages of a book to be used by people who don't have hands.

37. Design a device to allow the opening of a flip-top pop can with only one hand.

38. Determine a method to join two pieces of string from a distance of ten feet.

39. Determine a way to drive a nail into a board from a distance of ten feet.

6 Refinement and Analysis

"Refinement is the first departure from unrestricted creativity and imagination. Practicality and function must now be given primary consideration."
◆ JAMES H. EARLE

Refinement and analysis are the next activities to undertake in the design problem solving procedure, and the first activities that many students consider to be real engineering. The stereotyped engineer is often pictured as an analyzer of data, and a user of formulas, handbooks, tables, calculators, and computers. Engineers do spend time performing these activities; however, engineers also spend time performing related activities—communication being perhaps the most important, but often the most neglected.

Refinement and analysis activities are performed with the goal of reducing the list of possible solutions to the ones that have a chance of being the most acceptable, and at the same time improving the workable ideas to be better than originally conceived. Decisions need to be made so that additional time and money can be spent developing the most promising idea. The problem solving activities to be discussed in this chapter are no different than activities previously suggested; they are intended to help you focus on the essentials. Collecting additional information through refinement and analysis is important so alternative ideas can be fairly judged and compared. Judgments will be clearer, and made with more confidence, when the best information available is being considered.

6.1 Levels Of Analysis

The ideas being considered for final problem solution must be analyzed and compared on many performance levels and will require input from many individuals. Keep the design team involved at all stages of project development. The emphasis placed on each performance level depends on the problem being solved.

6.1.1 Function Analysis

The first performance level to analyze is basic function. The basic function of each design idea must be analyzed and verified. The idea must meet the specifications and minimum requirements as stated under problem identification. If an idea doesn't meet minimum requirements but still has favorable attributes, consider modifying the idea, combining it with another idea to make it acceptable, or changing a basic requirement. A compromise between an otherwise good idea and a current specification may be required. Compromise means that the idea and the requirements shift toward common ground. An original performance goal may be too ambitious, such as a minimum weight requirement that cannot be met even by using plastics and aluminum. If an idea cannot be improved to meet a minimum standard, and the standard cannot be changed, then the idea must be dropped from further consideration.

Basic function also includes whether or not the product is strong enough. Will the design adequately resist the loads that might be applied? All loads and forces, such as static, dynamic, cyclic, impact, and vibratory, must be considered. Forces can be applied by the user as the product is being used, by the environment (wind, tides, water erosion), or by the manufacturing processes used to make the product. The analysis of forces can be done manually by using accepted design strength criteria, by computer simulation and strength analysis using finite element analysis, or by studying either full size or scale models. See Case Study #4 for an example of preliminary product strength analysis.

Basic function also includes the necessity to analyze reliability. A design engineer wants the final product to last an acceptable length of time providing it is used properly by the consumer. At some point during the product analysis, when the list of potential ideas is reduced to the most viable, reliability must be considered. An idea that is least expensive and has the shortest lead time may be the least reliable. Nothing will turn potential customers away faster than an unreliable product. Failure rate determination and reliability programs conducted during product development are essential. Physical model testing will produce information to help analyze reliability of products taken as a whole as well as component parts. Associated criteria for products and components derived from testing data are mean time between failures, those components most susceptible to failure, the effect of the failure of one component on the failure of other components, and the time it takes to return a product to operation after a failure. Reliability analysis uses statistical concepts such as mean, standard deviation, and coefficient of correlation. Specific situations are handled by using a normal, binomial, Poisson, or Weibull distribution.

6.1.2 Safety Analysis

Next to basic function, the most important level of analysis is safety. Safety must always be analyzed because performance without safety is risky at best. Safety issues include designing a product to protect the user and bystanders from such things as electric shock, cuts from sharp edges or moving parts, explosion and fire, burns from hot surfaces, injury from loose flying parts, entanglement in moving machinery, and exposure to radiation, noxious gases, air contamination, caustic fluids, and acids.

If a hazard such as a sharp edge can be eliminated, it should be. A sharp edge on sheet metal can be folded over. Sharp edges and burrs on parts that have been machined should be removed using processes such as grinding, tumbling in an abrasive slurry, or thermal deburring. Consumers can be protected from the edges of sheet metal and glass by using metal, rubber, plastic, or wood edging. If glass shattering is a possibility, then perhaps shatterproof glass or Plexiglas™ should be used instead of regular glass. Blades and other parts that need to be sharp to function, or machinery that has pinch points due to the movement of one part into and out of another, should be shielded to prevent accessibility unless power is shut off, not just covered with an easily removable shield. Examples include the power cutoff interlock on lawn mowers and snowblowers, and two-hand controls on hydraulic and mechanical presses. Electrical shock potential in electric power environments is reduced by using third wire grounding or plastic cases. Seat belts were designed, developed, and added to automobiles and airplanes to reduce the damage caused by an accident. Electric sparks are avoided by using ceramic or polymer components, as well as placing electrical contacts in airtight enclosures.

Some dangers to the users of products cannot be eliminated or physically shielded, as in the case of volatile fuels, dangerous chemicals (insect sprays and herbicides), and acids. These situations require warning labels to be placed in appropriate locations and specialized use instructions included. Chemical mixing and use instructions must be written clearly, including precautions and proper procedures as well as antidotal measures in case of emergency. When operation or use of a product can cause harm to bystanders as well as the user, as in the case of motor vehicles, rotary lawn mowers, and automatic door closers, special design problems exist. Product features such as signal lights, brake lights, seat belt buzzers, backup bells, wide turn signs on trucks, presence sensors, and automatic stops were designed to help protect people and devices from harm.

Care must be exercised when designing warning labels. As products become more complex, and the consumer develops a right-to-know attitude, there is a tendency to write warning messages that are not understandable, too long, too complicated, or too intricate. Additional problems arise when products are distributed and used by people with varying degrees of literacy, and who use different languages. Many companies now print and include instructions in different languages with their products, all in an attempt to avoid improper and/or dangerous use by the consumer. In very special cases licensing is required before a product can be used because of the specialized knowledge and skill needed to perform an operation safely, such as piloting an airplane or supertanker.

6.1.3 Human Factors

Safe use must not prevent a product from being comfortable for the consumer to use. Consider the human aspects of your design. Some areas of concern for user comfort are: convenience of use, noise levels, illumination (too much as well as not enough), light

TABLE 6.1

Aspect	Comfort Level Range	Discomfort or Dangerous Range
Humidity	30%–70%	0–10% & 90–100%
Temperature	65–75°F	<30 & > 100°F
Pressure	10–20 PSIa	< 8 PSIa & > 22 PSIa
Acceleration	0–0.1 G	> 1 G
Electric Current @ 60 Hertz	0–1 mA	> 10 mA
Radiation	0–0.2 REM/year	> 15 REM/year
Illumination	20–100 Footcandles	10,000 Footcandles
Noise	0–85 db	> 94 db
Vibration	0–1 cycle/sec	3–10 cycles/sec
Shock Waves	0–2.5 PSIg	> 7 PSIg
CO Level	0–10 PPM	> 10 PPM
CO_2 Level	0–1700 PPM	40,000 PPM
Oxygen Level	15–20%	> 60%
Moving Air	13–20 ft^3/min	0–5 & > 50 ft^3/min

Comfort and discomfort levels of selected human environmental aspects (11,12).

glare and reflection, vibration, temperature (too hot or too cold), humidity, air flow, oxygen and other gas concentration levels, and acceleration. Certain levels of these conditions are not only uncomfortable to us as we work, but can be dangerous to our health, as shown in Table 6.1 and references (11) and (12).

Comfort and environmental aspects of product design are also important on a community and country-wide level. These include the quality of air, land, and water as well as resource allocation and replacement. For an illustration of such a concern see Example 5.3.

Safe and comfortable use also relates to the size and shape of people. Reference (12) contains anthropometric charts that list data relating to the average size and shape of humans. A properly designed product can increase user comfort, accuracy, and speed of performance while decreasing fatigue, boredom, isolation, and frustration. Proper design includes good location of controls, readable dials, good instructions, ease of making adjustments, ease of maintenance, and warnings for emergency situations using red/green lights, blinking lights, horns, bells, buzzers or automatic shut-down mechanisms.

6.1.4 Service Requirements

A product that functions as required and is safe and comfortable for the user may have problems when service is required. A well designed product is easy to service. Often this aspect of product design is ignored until it is too late to make changes and still meet production schedules, so a hard-to-service product reaches the marketplace. Consider an electric motor on an air conditioner that requires annual lubrication, but is located inside the unit and requires the removal of a dozen screws to get at, and then requires an oil can with an unusually long access spout. How often will a consumer lubricate the motor? Consider the car designed with the spark plugs located in such a way that engine accessories must be removed before the spark plugs can be accessed, and even then only with a special wrench. This design adds to consumer cost. Sometimes the decision made about which parts will be normal replacement parts is made hastily and without regard for serviceability. Should a light bulb within a lamp assembly be replaceable? One particular manufacturer did not have the replacement bulb listed as a repair part, but listed only the lamp assembly.

6.1.5 Market Concerns

Engineers must also be concerned with the market and economic aspects of design. There is no point having a great product if it can't be marketed. Will the design meet the acceptance level of consumers? Does it fill their perceived needs? Is the cost acceptable to them? Market study prior to offering a product to the marketplace, and after introduction for follow-up analysis is appropriate so customer acceptance or rejection information can be fed back to the design environment. This is particularly important when a new product changes a custom or traditional pattern, such as push-button phones, or short-life, undersized spare tires. Designers are required to consider other market concerns such as those involving styling, for automobiles, houses, and buildings; potential health problems and community acceptance of the location of a land fill or hazardous waste disposal; a change in a banking procedure (instant access cards); a change from a hydraulic system to a pneumatic one, or vice versa; or a change in a mass transportation system.

6.1.5 Storage And Shipping

While designing a product, consideration must also be given to how it will be shipped and stored, and how to determine an acceptable shelf life. Though food most obviously has a finite shelf life, any number of products can deteriorate over time: colors fade in the sun, mixtures separate if left standing, metals corrode, wood warps, chemicals decompose, fuels lose volatility, grease may melt out of a critical joint, and rubber will deteriorate in the presence of ozone.

6.2 | Related Activities

The activities performed at this stage of the project should include anything that will help you make good judgments and decisions about the ideas under consideration.

6.2.1 Evaluate Ideas

One of the first activities to perform is to discard unworkable ideas. "Far-out" ideas, those for which no method exists to make them realistic or feasible, should be discarded. Revolutionary ideas ordinarily need state of the art changes to make them workable. An idea that might be discarded is the idea of a solar powered snowmobile. This does not imply that a snowmobile will never be solar powered, only that with the current state of technology, it is not commercially feasible. State of the art change involves research and development, which are jobs for the research engineer.

Other ideas might be discarded because of possible patent infringement, small market potential, or the amount of development and test time required to meet the current time and cost schedule. Poor ideas should be rejected before time and money are spent on development. Ideas and revision proposals can be rejected or postponed for many reasons, such as:

1. An idea is too similar to the competition. If a design is too similar to a competitor's, then your design needs to be lower priced, more reliable, available with shorter delivery time, or have a better warranty to be competitive.
2. An idea requires too many parts. If there are more parts in one design than another, the cost is also probably higher than necessary. More parts mean more drawings, more material, more labor (making and assembling), more complicated servicing and repair, and more repair parts in inventory, all of which add cost to the product.
3. An idea has parts that are difficult to manufacture. Parts can be so complex that standard methods of manufacture will not work.
4. An idea has a high estimated cost, product material and labor cost, high tooling cost, high facilities (buildings and machines) cost. An idea that is expensive must have features that make it better than the competition. It must be easier to use (changing a truck to four-wheel drive without getting out of the cab), last longer (ball or roller bearings instead of a sleeve bushing), perform a task that the competition can't (automatic re-dialing telephone), or have some other advantage.
5. Product development time is too long and schedule cannot be met. Product implementation time will increases if new state of the art is studied, or if a new product is being developed.
6. The new product is too similar to an existing product and there isn't any reason for the customer to trade up to the new model. A new product must have some advantage, such as cheaper operation, more work output in a shorter period of time, fewer skill requirements to operate, safer operation, or higher reliability.
7. An idea is too revolutionary, requiring a large increment in the state of the art or in the customer's adaptability: a three wheel car, a car with lever steering in place of wheel steering, or solar heating.

As ideas are reviewed, make notes about details to check or data to verify. These notes might include questions about clearance, amount of movement, fastening methods, or use of standard parts.

There is always the danger of discarding an idea that can be developed to be acceptable, so care must be exercised. Experience can be helpful. As you become more knowledgeable about ideas that work and those that don't work, you will become more confident about preliminary decisions. However, experience sometimes puts habit in front of logic and creativity. Don't get in the rut of thinking that the same solution will always be the best under all conditions.

6.2.2 Scale Drawings

Make scale drawings of the better ideas. Scale drawings help determine if an idea will really work as imagined, and help to determine specific product features. The specific details will be used later to compare the various design ideas. A comparison will be made to determine which of the most promising ideas, if any, are worth pursuing to final implementation and production. Scale drawings are generally started as complete product layout drawings. Then as each critical area is identified, sub-component layouts and details are drawn. Select a scale suitable for the amount of detail that needs to be shown. Scale drawings will help to determine specific details about your design. Sometimes the added detail will result in new ideas for solution.

Drawings will also aid in determining critical lengths, areas, volumes, and weights. In Case Study #4, the length of the arm and the slot dimensions can be determined from the scale drawings by measuring, or by calculations once the geometry is determined. Volumes, areas, and weights can also be calculated from drawing measurements. Computer-aided drafting systems will assist in the details of design layout and in many area and volume calculations.

FIGURE 6.1

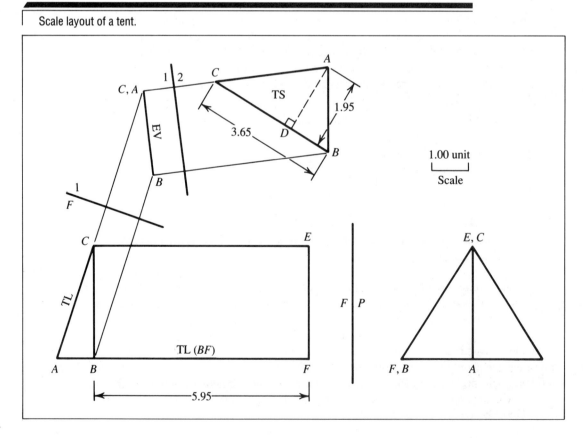

Scale layout of a tent.

EXAMPLE 6.1:
Layout analysis.

Figure 6.1 shows several views of a small tent; specific auxiliary views show the areas of material required. The area of surface $ABC = (BC)(AD)/2 = (3.65$ units$)(1.95$ units$)/2 = 3.56$ unit^2. Not only can the area of ABC be calculated from the second auxiliary view, but a full size enlargement would give a pattern that could be used for manufacturing. Likewise, the area of rectangle $BCEF = (BF)(BC) = (5.95$ units$)(3.65$ units$) = 21.72$ unit^2.

The next case study is an illustration of how drawings and mathematical analysis are often combined.

CASE STUDY #4:
Drawing table adjustment.

Consider a drawing table tilt mechanism design shown as a preliminary layout in Figure 6.2. The scale drawing shows the arm movement required, represented by line AB, and overall relationships so details can be accurately considered. One of

FIGURE 6.2

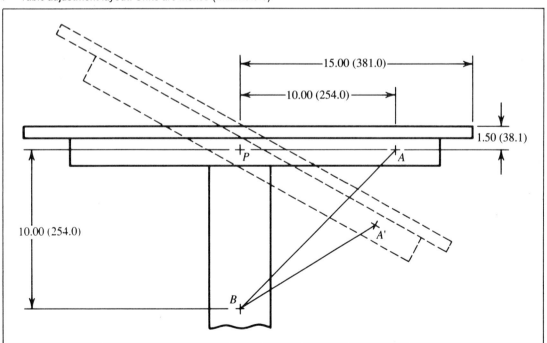

Table adjustment layout. Units are inches (millimeters).

FIGURE 6.3

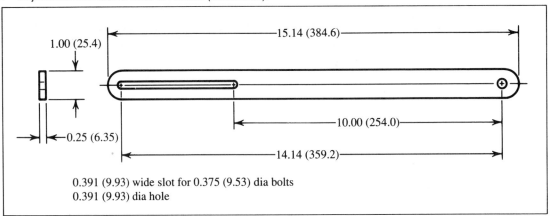

Adjustment arm details. Units are inches (millimeters).

0.391 (9.93) wide slot for 0.375 (9.53) dia bolts
0.391 (9.93) dia hole

the ideas for infinite angle adjustment from horizontal to 30° from horizontal is a slotted arm as shown in Figure 6.3. The preliminary arm dimensions of 0.25 x 1.00 inch (6.4 x 25.4 mm) are selected because this is a material size readily available from steel suppliers. The layout also permits analysis of other arm notch designs such as shown in Figure 6.4.

Drawings also assist in the analysis of forces. Let the force on the edge of the drawing table from a person leaning on it equal 100 lbf (444.8 N) as shown in Figure 6.5. (The 100 lbf was determined by placing a bathroom scale on the edge of the table and leaning on it.) Vector analysis can be used to determine the force in the arm, and how much force must be supported at the clamp. If F equals the force in the arm, determine the sum of the torques about point P as:

$$100(15 \cos \theta + 1.50 \sin \theta) - F[10 \cos ((90 - \theta)/2)] = 0, \quad \text{and}$$

$$F = \frac{1500 \cos \theta + 150 \sin \theta}{10 \cos ((90 - \theta)/2)}, \quad \text{where } 0° \leq \theta \leq 30°$$

$F_{0°} = 210$ lbf (934 N), and $F_{30°} = 160$ lbf (712 N)

If a bracket is placed on both sides of the table, then the force in each arm is one-half the amount calculated, 105 lbf (467 N) and 80 lbf (356 N).

The size of some components can be determined by measuring the scale drawing. Other sizes can be calculated by using data from the layout and applying appropriate mathematical models. Consider the drawing table arm. Once the compressive force is determined, the cross section dimensions can be analyzed based on a material's compressive strength, and its modulus of elasticity using column analysis. This would be most critical at $\theta = 0°$ because the force is the largest when the length of the moment arm is the greatest, and column failure is most likely when the column length is greatest.

FIGURE 6.4

Alternate arm design.

If a mild steel with $S_{yc} = 36{,}000$ psi (248 MPa) and $E = 30 \cdot 10^6$ psi (207 GPa) is selected, then the Euler column model, $P_c = \pi^2 EI/A^2$, can be applied as follows.

$$105 \text{ lbf} = \frac{(\pi^2)(300 \cdot 10^6)(I)}{[(0.65)10\sqrt{2}]^2} \text{ and } I = 0.00003 \text{ in}^4 \text{ (12.5 mm}^4\text{)}$$

The minimum moment of inertia in the arm shown in Figure 6.3 will occur in the area where the slot material is removed; $I = 2(0.30)(0.25)^3/12 = 0.00078$ in^4 (325 mm^4). This is a conservative value because the slot is not continuous throughout the

FIGURE 6.5

Table adjustment vector analysis. Units are inches (millimeters).

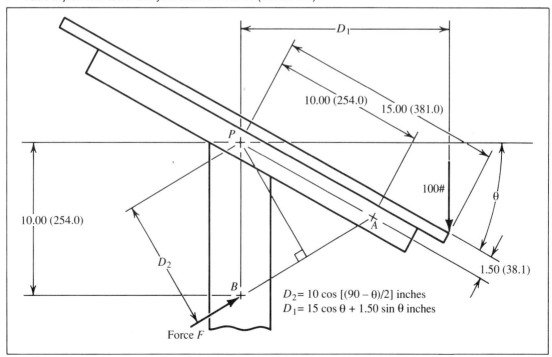

whole arm. The preliminary dimensions of 0.25 × 1.00 inches (6.35 × 25.4 mm) are adequate from a column analysis standpoint, even with the slot material removed, because 0.00078 > 0.00003.

A scale layout may also reveal that parts and assemblies do not work as originally conceived. If the problem is bad enough the idea may have to be dropped. However, the problem discovered may be solvable by using a new idea or by modifying an existing idea. An idea previously considered and rejected might be usable in a modified application. Review ideas previously discarded to see if one will be appropriate now. You did write them all down, didn't you? Solving new smaller problems will go on throughout the project. New ideas will be needed at all stages of the project, right up to the day of first production, and beyond.

Equally good ideas sometimes occur, and by studying drawings it may be possible to offer a product with optional features. Products with optional features include radios that run on batteries or plug into a wall socket, automobiles with a choice of engines, a power drill that can be used as a sander or a grinder, and a boat designed to run with sails or an engine. Scale drawings may also point out features that are particularly favorable when compared to the competition, such as larger fuel storage, greater maneuverability, fewer parts to service, more standard parts, or easier servicing.

6.2.3 Material and Manufacturing

Component requirements that are obtained from the scale drawings lead eventually to the selection of material. Materials are selected based on:

1. Their mechanical properties such as strength, toughness, ductility, or hardness.
2. Their physical properties such as specific weight, melting temperature, heat conductivity, thermal conductivity, or thermal expansion coefficient.
3. Their chemical properties such as corrosion resistance, state of matter at a given pressure and temperature, or the atomic lattice structure.
4. Their cost and availability.

Materials and their properties are important to make developmental design decisions. From a practical standpoint, materials must also be selected based on what is available. The drafting table support arm is an example. Regardless of how small a cross section is possible from a strength standpoint, an efficient choice of material will be a size that is readily available and can be handled by the existing manufacturing capabilities.

Component sizes and required details, coupled with an available material, will lead to the selection of manufacturing processes. Decisions made during the design of a product have a great effect on the method of manufacture. Always consider available processes and their relative costs when designing component parts. Processes fall into the following main categories.

1. Casting and Molding: Confining liquid or plastic material in a mold during solidification.
2. Shaping forms: Moving metal by forming, hammering, pressing, and forging.
3. Cutting from larger pieces: Shearing, sawing, nibbling, flame cutting, plasma arc cutting, and laser cutting.

4. Machining: Material removed by the cutting action of a tool bit such as milling, drilling, boring, and grinding.
5. Joining: Welds, glue, staples, rivets, screws, and nuts and bolts.
6. Finishing: Plating, painting, anodizing, and metal spraying or sputtering.

 CASE STUDY #4 CONCLUDED:

The drawing table arm in Figure 6.3 is 15.14 inches (384.6 mm) long. If the double radius cutoff shear blade wastes 0.50 inch (12.7 mm) between cuts, then a standard 192.00 inch (4876.8 mm) bar will yield 192.00/15.64 = 12 pieces as illustrated in Figure 6.6.

Alternative methods can be used to manufacture the slotted arm shown in Figure 6.3.

Method #1:

Operation 10	Shear to length with double radius cutoff blade from a bar.
Operation 20	Punch the slot using a special punch.
Operation 30	Punch the hole using a standard punch.
Operation 40	Flatten as required.

Method #2:

Operation 10	Shear length with double radius cutoff blade from bar.
Operation 20	Drill (2) holes, one for slot beginning.
Operation 30	End mill slot.
Operation 40	Pencil grind machining burrs.

FIGURE 6.6

Bar cutoff operation. Units are inches (millimeters).

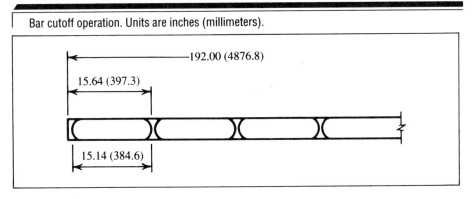

Product cost is determined from the material requirements; the time required for each manufacturing step; the hourly wage rates; the overhead rates of labor costs, product development, and material storage; an expected rate of return from the capital investments; and an adequate profit. Scale drawings are also the basis for this activity. The material costs are generally obtained through the efforts of the purchasing department, while the labor estimates are generally obtained through the efforts of a manufacturing or industrial engineer working with the accounting department. The manufacturing methods proposed for the support arm need cost analysis data so the least cost method can be selected. A judgment is also needed to determine which method gives the most consistent quality control of the parts. Low cost without quality is false economy. Drawings are needed to make parts, whether they are for full-size or scale models.

6.2.5 Models

Some designs may be so complex that there are no mathematical models to analyze stresses and deformations. These products will be checked by building full-size or scale models. Scale drawing details can verify the concepts that in the idea stage of development were idealized with sketches, but models must be built before final appearance design can be established. The looks of a product can also have an effect on its marketability. Automobiles, home appliances, furniture, and houses are examples of products for which appearance contributes a great deal to marketing success.

Models are also useful to verify clearances between members, lengths of hoses and wire, and room for safe and efficient assembly and repair procedures. If an awkward assembly or repair procedure cannot be simplified, the need for a special tool may result, leading to another design project for the engineer and more cost for the company and the consumer.

There are alternatives to consider when building models for product evaluation. One alternative is to build and test components and variations on them. When the best of the components have been selected, they are assembled together for the final product. The advantage of this method is that individual effort can be concentrated on a small portion of the design. The disadvantage is that the interrelationship of various components may be overlooked and the final assembly will not perform as desired. The best of each of the components may not be the best for the total assembly.

A second alternative is to build a complete product based on preliminary component test, and then test and improve the product as a whole unit. A disadvantage of this method is that a particular component may not be improved to its full potential because its performance is adequate and major emphasis is placed on the less than adequate components. Advantages are that the whole product will be working. Changes will be made for total product improvement, not just a single component improvement, and there will be less total development time.

A third method is to build a complete product without any component test and then make changes to "make the product work" and improve its performance. The disadvantage of this method is that a product may be built that will not work, or will take a long time to get working, and money and time will be wasted. An advantage is that work on the final product starts immediately and if it works, time and money will be saved. Each project has to be considered separately because the procedure that worked last time may not work this time. Stay flexible and adaptive, and be ready to compromise.

Regardless of the activities undertaken to model and analyze ideas, the computer is likely to play an important role. Computer Aided Drafting (CAD) is commonly used for product development. Existing parts can be easily changed or used as the basis for a new but similar part. A computer search for an existing part to meet a given condition is also possible if Group Technology software is used to categorize existing parts. Many CAD systems will also calculate weights and volumes of parts as they are designed. If the CAD system is connected to material and production reference information, computer cost analysis can be performed. Connecting a CAD system to a computer aided manufacturing system allows process routing, including tooling, machine settings, and direct computer control procedures.

The finite element method of structural analysis is also practical using commercially available software and computer modeling techniques. "The finite element method is a computer-aided mathematical technique for obtaining approximate numerical solutions to the abstract equations of calculus that predict the response of physical systems subjected to external influences." (1)

Many activities related to component design can be enhanced by using a computer, from report writing on a word processor to statistical analysis of experimental data. It is beyond the scope of this text to go into details on computer software available to assist the engineering design function. The reader should, however, remember one thing: just because the answer comes from a computer, and has eight-digit accuracy, the result is not necessarily correct or applicable. The engineer must still judge the validity of the results. The following case study is an example of implementing a design solution before completing adequate analysis.

CASE STUDY #5:
Room cooling problem.

A laboratory on the first floor of a two-story building is used as a physics laboratory. The room is approximately 30 feet wide x 46 feet long x 10 feet high (9.15 x 14.02 x 3.05 m). Two walls have exposure to the outside, are constructed of 8-inch (0.203 m) block faced with 4-inch (0.104 m) brick, and have eleven, 3 foot-6 inch (1.067 m) x 6 foot-0 inch (1.829 m) single pane windows in them. The room is directly over the boiler room, which supplies heat, steam, and other utilities to the building. The lab is uncomfortably hot during the summer semester and during many weeks of the fall and spring semesters. Students have difficulty concentrating and doing good work. Opening windows for ventilation does little good, as outside temperature and humidity in the summer are also high, and breezes insufficient for cooling.

The first solution many people would implement is to install air conditioning, a common cure for an uncomfortable environment. And indeed this was done. Two 13,000 Btu/hour window air conditioners were installed. The result was no reduction in temperature, a slight reduction in humidity, and stale air.

When the problem was presented to an engineering class, an analysis of the situation revealed that the two window air conditioners could not compensate for the heat transfer through the floor, let alone the walls and ceiling. The boiler room was a nearly constant heat source at 85°F (29.4°C), and to keep the lab at 70°F

(21.1°C) would require the removal of approximately 52,000 Btu/hour. *Note:* The concrete ceiling of the boiler room has an R value of approximately 0.4 ft$^2\cdot$°F·hr/Btu (0.4 m$^2\cdot$°C·hr/Btu.).

Students had two primary solutions to the problem.

1. Open the windows in the boiler room to let hot air out. Every degree below 85°F (29.4°C) in the boiler room means a reduction of approximately 3500 Btu/hr transfer to the lab above. This solution is practical because the windows open into a metal grated window well.

2. Increase the R value of the boiler room ceiling by installing insulation. Only then do the window air conditioners (the habit solution to the problem) have a chance to cool the lab.

Solution #1 was a no-cost solution and was easily implemented. Solution #2 has not been implemented because of lack of funds.

Case Study #2 is continued here to further illustrate the idea analysis procedure.

CASE STUDY #2 CONTINUED:

All parts moving problems are not the same because of variations such as the configuration of the parts, the number of parts to be moved, the protection required, and the facilities available. Each of the ideas for moving parts has advantages and disadvantages that must be considered based on the limitations and restrictions known. Following is a brief discussion of the advantages and disadvantages of the ideas listed in Section 5.3, followed by a judgment about whether the idea should be considered for the final solution based on the limitations listed in Section 4.1.

Idea	Advantages	Disadvantages
1. Belt conveyor	Continuous operation, load and unload from many locations, can handle large volume.	Takes up floor space, parts may fall off, parts are exposed to environment, parts may bump into each other, power drive is required.
2. Roller conveyor	Continuous operation, load and unload from many locations, can handle large volume, no power drive required.	Takes up floor space, parts may fall off, parts are exposed to environment, parts may bump into each other, parts may fall between rollers.
3. Overhead conveyor	Continuous operation, load and unload from many locations, will handle moderate volume.	Parts may fall off, parts are exposed to environment, parts may bump into each other, power drive required, dangerous to pedestrians.

4. Gravity chute	No power required, continuous operation, high volume capacity, operator not required.	Parts may fall off, parts are exposed to environment, parts may bump into each other, need difference in elevation great enough to overcome friction, chute material may damage parts.
5. Liquid float	Continuous operation, operator not required, moderate volume.	Parts may bump into each other, need difference in elevation, need to dispose or pump return of fluid, possible fluid corrosion on parts.
6. Pneumatic tube	Fast, operator not required, parts protected from environment.	Parts may bump into each other, air compressor required, need load and unload station.
7. Hand carry	Use on demand, each operator can move own parts, low floor space requirement.	Personnel fatigue, may drop parts, weight limit controls parts per load.
8. Tote pans	Use on demand, each operator can move own parts, low floor space requirement.	Personnel fatigue, may drop parts, low weight limit, need supply of tote pans.
9. Wheelbarrow	Use on demand, each operator can move own parts, low floor space requirement.	Personnel fatigue, may tip over and damage parts, low weight limit, need supply of wheelbarrows.
10. Sled	Use on demand, each operator can move own parts, low floor space requirement, pull force in place of lift force.	High friction, need supply of sleds.
11. Two wheel dolly	Use on demand, each operator can move own parts, low floor space requirement.	Weight limit, lift and push load.
12. Push dolly	Use on demand, each operator can move own parts, more floor space required.	Push or pull force, need supply of dollies.
13. Forklift	High weight capacity, faster than by hand, use for other applications.	Need forklift and operator, need maneuvering space, more expensive.
14. Robot	Continuous operation, no operator needed, reliable, high load capacity.	Most expensive, need electronic control of some kind.
15. Catapult	Fast.	Need catapult device, part contact damage, need clear path, flying parts dangerous.

Comments and judgments:

1. Conveyor belt: Parts can be protected by using expandable plastic mesh protective sleeves. If standard conveyor units are used, the cost should be within budget. Unless conveyor has side spurs to each machine, a person is needed to load and unload parts on and off the conveyor. Reject because a conveyor will create more parts moving capacity than required, and the conveyor uses up floor space. Reconsider if volume of parts to move increases.

2. Roller conveyor: Parts can be protected by using expandable plastic mesh protective sleeves, or by placing parts on small skids that will ride the rollers. If standard roller units are used the cost should be within budget. Unless conveyor has side spurs to each machine, a person is needed to load and unload parts. Reject because a conveyor will create more parts moving capacity than required, and the conveyor uses up floor space. Reconsider if volume of parts to move increases.

3. Overhead conveyor: Easier to pass spurs to each machine. Reject because highly finished parts are hard to hang from a moving conveyor, and because of the danger to the parts and pedestrians.

4. Gravity chute: Lower cost than ideas 1, 2, or 3. Reject based on elevation drop required to make idea work, as departments are on the same floor.

5. Liquid float: Reject based on possible fluid leak, possible part corrosion, and elevation difference needed to make fluid flow, or the cost of a pump and plumbing to return fluid.

6. Pneumatic tube: Easy to route tube to each machine. Unanswered question: How much air pressure or vacuum is required to move 25-pound part? Reject because of time limit of ten days and unanswered question. Could research idea for future applications.

7. Hand carry: Reject because it is less efficient than present method.

8. Tote pans: Reject because it is less efficient than present method.

9. Wheelbarrow: Reject because it is less efficient than present method.

10. Sled: Reject because it is less efficient than present method.

11. Two wheel dolly: Reject because it is less efficient than present method. More danger of loads tipping over.

12. Push dolly: Reject because it is less efficient than present method.

13. Forklift: This is an expansion of the present system that will increase speed and increase load limit. Will the increase in speed also increase part damage? Will a new job classification be required for a forklift operator? Are there other jobs the forklift can do when not moving parts in this department? Is there a forklift already available that can work at this location part of the time? How much time is available for this use? Where can a forklift be stored?

14. Robot: Reliable, route can be easily reprogrammed. Reject based on high cost and higher capacity than required. Consider at a later time for a plant-wide parts moving system.

15. Catapult: Reject because of danger to parts and people.

It is not unusual for the majority of ideas to be rejected, nor is it unusual for the idea that surfaces as being "best" to be an extension of a present design. Unless a present design is totally inadequate, or a state of the art change comes along, most design changes are modifications of existing designs.

In this case the real problem turned out to be one of excessive part damage and poor inventory control, caused primarily by parts coming in contact with each other on the flat pallet, and the few number of parts moved per pallet. After additional discussion and problem understanding, it was decided to concentrate efforts on protecting the finished parts and increasing the number of parts per load.

Parts protection can be improved by using plastic or cardboard protective sleeves, but several sizes are required, and they would have to be returned and reused to keep cost down. Increasing the number of parts moved per load can be achieved by putting sides on the pallets. But perhaps there is a better solution. Figures 6.7 and 6.8 illustrate two of the more promising design proposals that will protect the parts, allow for more parts per load, and allow for movement to the next department by hand, hand pallet mover, or fork lift.

Full size test models of both designs were built and used for a short time to see how they worked. The upright model was easier to load and unload and was selected as the idea to implement even though the layered method allowed more parts to be moved per load. The layer separators were awkward to handle and they

FIGURE 6.7

Preliminary solution idea for Case Study #2.

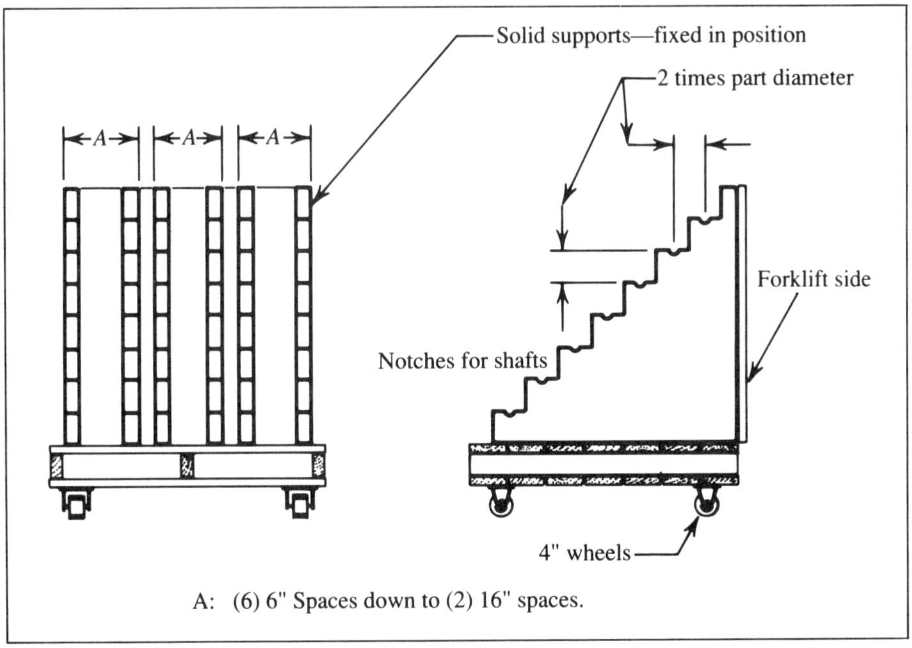

A: (6) 6" Spaces down to (2) 16" spaces.

FIGURE 6.8

Preliminary solution idea for Case Study #2.

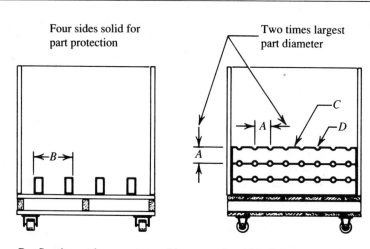

B: Spacing set by operator stacking parts, 6 to 16 inches.
C: Wood or plastic dividers
   1. Separate set for each major diameter range, or
   2. Make A slightly larger than largest diameter and do not use every notch for large diameter parts.
D: Radius equal to largest part radius, or notch only one side of separators equal to largest part diameter. Notch can be circular or rectangular.

didn't stay with the pallets for reuse. The upright model was modified to a vertical rack rather than slanted, and a center column with four sides was used to increase the number of parts per load (Figure 6.9).

The wheels were selected based on an estimated maximum weight as follows.

2	2.50 x 2.50 x 40 inch hardwood cross members (50 lb/ft^3)	15 lbs
54	2.50 x 2.50 x 24 inch hardwood cross members (50 lb/ft^3)	234 lbs
28	1.50 dia x 16 inch steel parts (490 lb/ft^3)	225 lbs
1	40 x 40 inch hardwood pallet	90 lbs
	Total	564 lbs

The wheel selected was Globe #308504, a 4 inch diameter by 1.50 inch wide swivel rubber wheel, 5 inch overall height, with a 240 lb per wheel capacity, $9.20 each.

The final cost of $300 per carrier allowed an adequate number to be built. The carriers were not all built and in operation in ten days, but the project was under way, and enough people had been involved in the trial so that no major problems occurred during final implementation.

FIGURE 6.9

Final solution, Case Study #2.

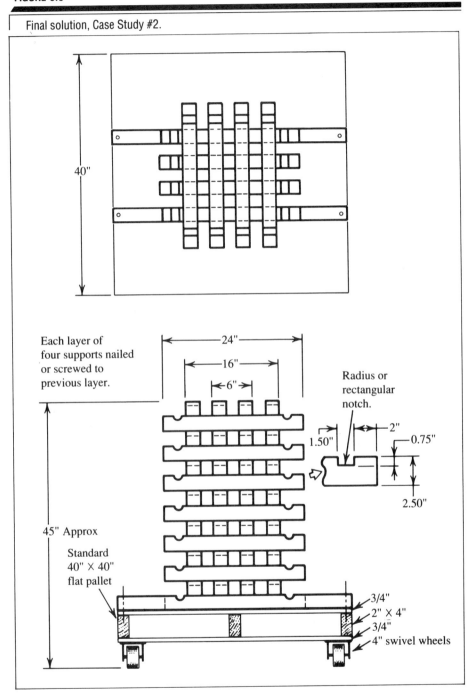

6.3 Summary

As refinements of the project ideas are made, and details of manufacture, assembly, storage, and shipment are clarified, requirements for machine tools, material handling, painting, assembly, shipping, and storage must also be determined. Planning for the construction or procurement of plant facilities is as important to a successful project as a good product design. These requirements take design time, money, and a schedule of their own. The money and time must be incorporated into the overall project plan and schedule.

While refining ideas, it is common to think of new ideas, to realize that original specifications are too restrictive and must be relaxed, or that the original specifications can be improved. Modifying goals is a common occurrence. An early change in a specification that cannot be met will save time and money. An improvement in a specification will help evolve the product into a better than expected design. New ideas occur because of the increased concentration that accompanies refinement. New ideas might be a combination of the ideas being refined, or result from an improvement that becomes apparent while creating accurate drawings or while building a model. Additional thoughts may also suggest a change in the basic problem statement and specifications. Maybe your layout will show that a digging depth specification can be increased because of the linkage design you have proposed.

Do not suppress ideas that may change the scope of specifications of your project, but be cautious. A change in scope can require additional time, money, or staff support. These must be considered before a change is requested or approved. On the other hand, you may propose a change that will use less time, money, and staff. Those changes are the best kind. If you suggest a project change, be sure you have the answers to the questions that will certainly come up regarding project resources and the effect on the project time schedule.

6.4 References

1. Burnett, David S. *Finite Element Analysis*. Reading, Mass.: Addison-Wesley, 1987.
2. Chow, William Wai-Chung. *Cost Reduction in Product Design*. New York: Van Nostrand, 1978.
3. DiLavore, Philip. *Energy: Insights from Physics*. New York: John Wiley & Sons, 1984.
4. Dreyfuss, Henry. *The Measure of Man*. New York: Whitney Library of Design, 1967.
5. Fisk, Marian J. and Anderson, H. William. *Introduction to Solar Technology*. Reading, Mass.: Addison Wesley, 1982.
6. French, M. J. *Invention and Evolution*. New York: Cambridge University Press, 1988.

7. Lucas, Henry C. *The Analysis, Design, and Implementation of Information Systems*. New York: McGraw-Hill, 1981.
8. Schenck, Hilbert. *Theories of Engineering Experimentation*. New York: McGraw-Hill, 1979.
9. Stefanides, E. J. "Piezo Film Reduces Cost of Single-axis Accelerometer." *Design News*. Sept. 4, 1989, 212-213.
10. Tuve, George L. and L.C. Domholdt. *Engineering Experimentation*. New York: McGraw-Hill, 1966.
11. Wilcox, Alan D. *Engineering Design: Project Guidelines*. Englewood Cliffs, N.J.: Prentice Hall, 1987. Emphasis on electronics.
12. Woodson, Wesley E. & D.W. Conover. *Human Engineering Guide for Equipment Designers,* 2nd ed. Berkeley: University of California Press, 1965.
13. Woodson, Wesley E. *Human Factors Design Handbook*. New York: McGraw-Hill, 1981.

6.5 Exercises

1. In his book *Invention and Evolution,* M. J. French says,

   A large part of design consists of choosing from a very large number of possible ways of performing the various functions required of the thing designed, a single way for each function. The choice is complicated by the interactions of one function with another and by the many different ways in which the functions may be combined.

   Discuss his comment and how it affects the design process.

2. Think of other ideas for a drawing table support arm other than those shown in Figures 6.3 and 6.4.

3. Think of an idea for a support arm that would more likely be in tension than compression. This would eliminate column buckling as a possible mode of failure.

4. Name a product or circumstance that rates favorably, and one that rates unfavorably, for each of the following items.

   a) Sound
   b) High illumination
   c) Vibration
   d) Low illumination
   e) Radiation
   f) Low temperature
   g) Pressure
   h) High temperature
   i) Shock waves
   j) Moving air

5. List products that have "failed to function" as advertised or have disappeared from the marketplace, and give a reason why.

6. Assume that 20-foot long cars maintain one car length space between themselves for each 10 miles per hour of their traveling speed.
   a) What speed allows the maximum number of cars per minute to pass a given point?
   b) What is the maximum number of cars per minute that can pass a given point?
   c) What speed will allow 18 cars per minute to pass a given point?

7. For a particular period of time, aircraft accidents were recorded and tabulated based on the time of day they occurred. The representative data follows. Does the data support the conclusion that it is safer to fly at dawn? Explain.

Time of Day	% of total accidents
Daylight hours	86.0
Dark at night	7.0
Light at night	1.6
Dusk–twilight	4.0
Unknown	0.4

8. a) Analyze the apparatus in Figure 6.10 and explain how it can be used for an accelerometer when placed in the plane of motion. Be specific.
   b) Do both vertical tubes have to have the same area? What happens if the area of the vertical tube is one-half the area of the horizontal tube?
   c) Discuss advantages and disadvantages of having different cross-sectional areas for the vertical and horizontal tubes.

FIGURE 6.10

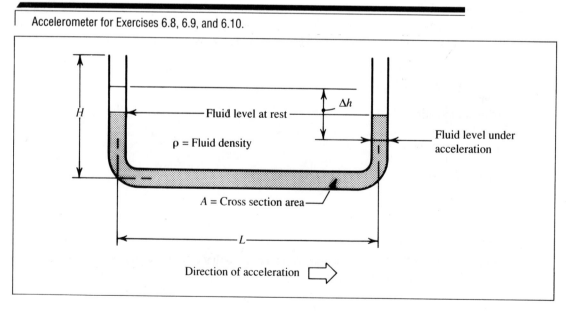

Accelerometer for Exercises 6.8, 6.9, and 6.10.

FIGURE 6.11

Accelerometer for Exercise 6.11.

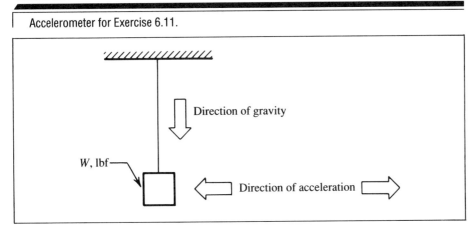

9. How does the fluid selected for the apparatus of Figure 6.10 affect its operation? Some possible fluids are: water, density of 1.0 gm/cm³; glycerin, density 1.26 gm/cm³; ethyl alcohol, density 0.806 gm/cm³; and mercury, density 13.6 gm/cm³.

10. Analyze and refine the design in Figure 6.10 for an acceleration range of 1–5 G's. Complete scale drawings so the device can be built.

11. Analyze and refine the idea for an accelerometer sketched in Figure 6.11. Complete scale drawings so the device can be built.

12. Analyze and refine the idea for an accelerometer sketched in Figure 6.12. Complete scale drawings so the device can be built.

13. The ideas for accelerometers shown in Figures 6.10 and 6.12 are directional in concept. If they are not lined up with the direction of acceleration, the reading will be too low. Make additional refinement and analysis so the principles can be used regardless of the direction of acceleration.

FIGURE 6.12

Accelerometer for Exercise 6.12.

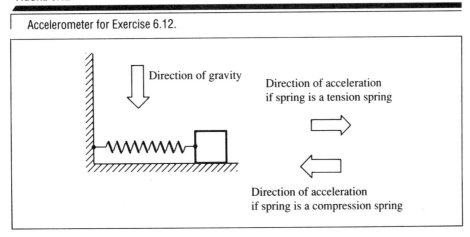

FIGURE 6.13

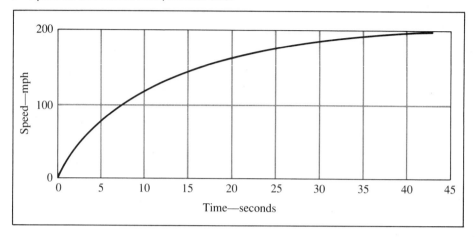

Sports car acceleration curve, Exercise 6.15.

14. A particular jet plane can accelerate from 0 to 575 miles per hour in 65 seconds. The speed-time graph is nearly a straight line. What force in G's does the pilot feel during the 65 seconds?

15. A particular sports car can accelerate from 0 to 200 miles per hour in 40 seconds. Due to various factors it cannot accelerate at a constant rate (Figure 6.13). Determine the maximum G's the driver feels.

16. From the data and answers from Exercises 6.14 and 6.15, determine the length of a race that would end in a tie for the jet and the race car.

17. A particular sports car can accelerate from 0 to 60 miles per hour in 6.08 seconds, and brake from 60 to 0 miles per hour in 126 feet. Which causes the driver to feel the higher number of G's, accelerating or braking? Quantify your answer.

18. What force would a seat belt system have to be designed for so the driver of the sports car in Exercise 6.17 would be safe?

19. Repeat Exercises 6.17 and 6.18 for a 5.29 second acceleration and 158 foot braking distance.

20. Refer to Figure 6.14. Water wells are located at A, B, C, D, E, and F. The water from the wells has to be pumped to the treatment plant before distribution. The distances between each well, and between each well and the treatment plant, are shown in the figure. Pipe for a direct line costs P $/meter, pipe for a double size pipe to handle the water from two wells costs $1.5P$ $/meter, and pipe for a triple size pipe to handle the water from three wells costs $1.8P$ $/meter. Find an appropriate network of pipe for this system for lowest cost.

21. An engineer is working on a project to design a method of moving sand from bulk ground storage to an overhead truck loading and weighing system. During the

FIGURE 6.14

Pump-pipe network for Exercise 6.20. Distances are meters.

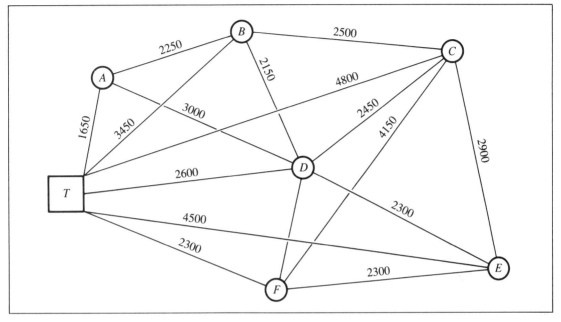

course of development the following items have been listed as important to consider. The object is to keep cost down to meet the system needs.

  a) Analyze how each of the following items is related to cost. Consider how the items listed are related to each other. Add items to the list as required.

  E, Energy cost, $/kwh
  T, Time equipment is used, hr/yr
  C, Capacity of one bucket, lb/bkt
  F, Material flow rate, lb/sec
  I, Annual interest rate, %/yr
  BC, Bucket cost, $/lb/bkt

  BS, Bucket spacing, ft/bkt
  S, Structure cost, $
  P, Power required, hp
  N, Number of buckets
  CS, Conveyor speed, ft/sec

  b) List questions that should be answered before proceeding with a design proposal for this project.

22. There are many products that have been designed to avoid catastrophe in the case where something goes wrong. For each of the following items specify a potential problem for the customer, specify what commercial designs are available to prevent the danger, and then think of other solutions to the problem. Example: Electric hedge trimmers may be used outside when it is wet. This creates the potential for electric shock. Grounded (three-wire) wire and grounded plugs are

available on commercial models. The problem can also be avoided by using double insulated cases, battery operated units, an electric circuit protected with a ground fault interrupter circuit breaker, or by growing bushes that do not have to be trimmed.
  a) Rotary lawn mower blades
  b) Snow blower augers
  c) Automatic garage door opener/closer
  d) Barrier arm at a toll gate
  e) Manually operated paper cutters

23. The following products have inherent dangers while being used. What is the danger, and discuss some ideas to overcome the danger. Example: An empty light bulb socket invites the insertion of a finger and an electric shock is likely. A smaller diameter base would make this less likely. Another possible solution is to have a safety seal that rotates to fill the opening of the hole when the bulb is removed.
  a) Power steering on an automobile
  b) Power brakes on an automobile
  c) Electric toaster
  d) Electric stove
  e) Incandescent light bulb
  f) Chain saw

24. An engineer is designing a ladle hook for supporting a ladle of molten steel in the steel making process. Figure 6.15 shows a sketch of the problem and of the proposed design solution. The engineer specified the following items as part of the solution. Analyze each item by itself and in conjunction with all of the items listed. Comment on the validity of each item. Be quantitative when possible.
  a) The method of construction proposed is five layers of 1.40 inch thick steel.
  b) Each of the five pieces is to be flame cut from plates.
  c) Twenty-six, 1.00 inch diameter rivets, made from steel with S_{ys} = 15,000 psi, will be used to hold the plates together. The safety factor for the rivets is given as 2.
  d) In the critical areas around the 8-inch support pin and the 15-inch ladle shaft, the plates will be joined by butt welding.
  e) The engineer claims that if the 6-inch central web width is changed to 12 inches and the thickness changed accordingly, the hook will hang so the center of the ladle shaft hole is directly under the upper shaft hole.
  f) The safety factor for the hook design is given as 3 when using material with S_{yt} = 37,500 psi.

25. Draw a fully dimensioned scale drawing of the ladle hook as proposed.

26. Determine an alternate design for the ladle hook and make complete scale drawings of the idea.

27. If electric heat is 100% efficient, that is, all the purchased energy is converted into heat, and a natural gas furnace is only 65% efficient, why is a gas furnace still the "best" buy?

FIGURE 6.15

Data for ladle design, Exercise 6.24.

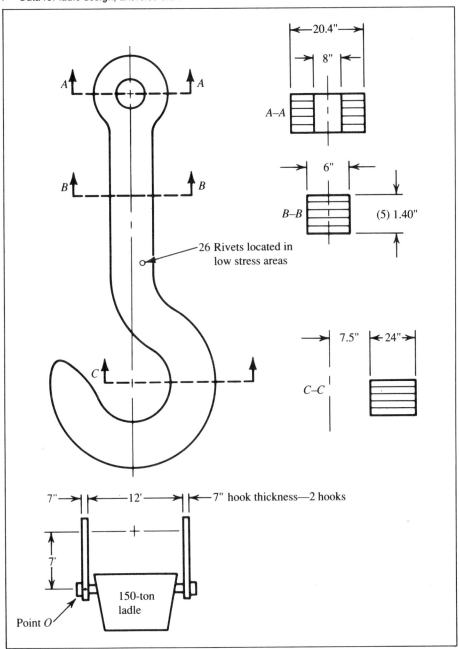

Refinement and Analysis

FIGURE 6.16

Data for Exercise 6.30.

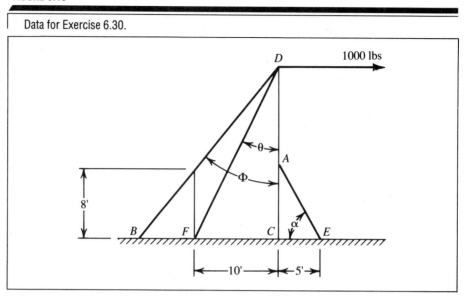

FIGURE 6.17

Schematic for Exercise 6.31.

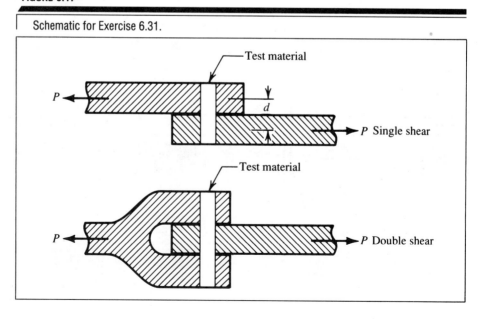

28. List advantages and disadvantages of the following methods of heating a house.
    a) Natural gas
    b) Wood
    c) Electricity
    d) Fuel oil
    e) Coal
29. The frequency of "small" oscillations, in Hertz (cycles per second), of a simple pendulum of length L is $f_n = \sqrt{g/L}$, and the time of cyclic motion $t = 2\pi\sqrt{L/g}$.
    a) If a pendulum controlled clock is running slow, should the length of the pendulum be lengthened or shortened? Explain.

FIGURE 6.18

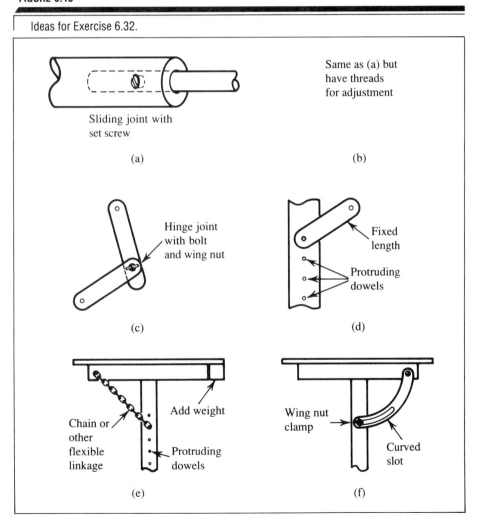

Ideas for Exercise 6.32.

b) Explain why the effective length of a pendulum can be made shorter by adjusting the mass along the length of a fixed length pendulum bar.

c) Explain why pendulum clocks with long pendulum arms are easier to adjust to an accurate time compared to short pendulums.

30. Refer to Figure 6.16. A 20-foot high post is embedded in the ground at point C and has a 1000-pound pull on the top as shown. The post is beginning to bend over and it is desired to add a support. Space limitations prevent any near side brace further than 5 feet from C, and far side braces are impeded by an 8-foot high fence 10 feet from the base of the post. Several suggestions have been made.

    a) Add a brace on the near side to some point A on the post.
    b) Add a brace on the far side within the 10-foot limit.
    c) Add a brace from point B or beyond that passes over the fence.

    Analyze these ideas and others that you may have for adding support to the post. Consider cost of material, total structure stability, and installation methods.

31. The concept of a machine to test the shear strength of materials is not particularly complicated. However, the simplest of single shear machines might induce bending stress in addition to shear stress. Analyze and compare the differences between a single shear area testing machine and a double or triple shear area testing machine. Refer to Figure 6.17.

32. Figure 6.18 shows some possible designs for an adjustable drawing table support arm. Refine each of the ideas to workable solutions, including drawings of all component parts.

33. Suggest other possible solutions for the problem discussed in Case Study #5.

7 Decision and Implementation

"Intangible factors such as social values, political influences, and psychological effects are extremely difficult to quantify and measure. Yet, as all practicing engineers soon learn, these factors often control the acceptance or rejection of a design."
✦ DALE D. MEREDITH

Decisions are made during all phases of project activities. Decisions already made include the details of specifications, market potential, and sales forecasts, as well as whether to reject or develop various solution ideas. The decision necessary at this stage of the problem solving procedure is to determine whether the project should be canceled, modified, or implemented, and if implemented, which idea is most acceptable.

Canceling a project means that there is no solution available that has a chance of being successful under the present restrictions. A number of possibilities can lead to the canceling of a project. Perhaps the cost cannot be reduced to an acceptable level and sales projections at a high price indicate that too few products will be sold to make a profit. Perhaps the market has changed since the project was started. Maybe a competitor's product solves the problem better, or at less cost. Maybe a technological breakthrough has occurred and the state of the art has advanced beyond the present design: a more efficient slide rule doesn't have much market potential. Other reasons for canceling a project include:

1. No money left, the budget has been depleted. Perhaps research and development costs were higher than anticipated, and no acceptable solution was developed.

2. No time left, the due date arrived without a solution. The timing of product completion is often a critical concern, and perhaps project bid data arrived late and no proposal bid was possible.

3. No trained personnel to complete the project. If critical state of the art development depends on a few key people, and they leave, become ill, or die, a project may have to be canceled or postponed.
4. The project is badly timed, there is low customer interest, there is a saturated market or even a disappearing market. When the price of oil on the world market fell in the mid 1980s, and the number of oil wells drilled was reduced, the market for new oil-well drilling equipment disappeared.
5. The state of the art has outdistanced the expertise of the company. When transistors, integrated circuits, and printed circuit boards were developed, companies had to cancel or modify projects that involved vacuum tubes.
6. The consumer isn't ready for the product. The project is ahead of its time. Chester Carlson invented the process of xerography in 1938 but struggled to find commercial acceptability, rejected by companies such as RCA, General Electric, and IBM. It wasn't until 1949 that a small company in Rochester, New York marketed the first commercial machine. That company became the Xerox Corporation in 1961.
7. Public safety may be at risk. Perhaps the side effects of implementing an idea are worse than the effects of not implementing it. Refer to the discussion of the Aswan dam in Section 2.3.

There is nothing wrong with canceling a project, no matter whose favorite it is. Many projects should have been dealt this fate and weren't. One person's position of power and lust for success should not precipitate ludicrous spending of time and money on poor projects. Philanthropists can do as they like, but competitive businesses can't afford to. Fortunately, as the Xerox story indicates, persistence and determination can often overcome many problems.

If there are still questions of meeting market competition, maybe the best design can be modified to make it viable. Perhaps specifications need to be changed and the problem solving activities repeated. A large capital investment may be involved to make a power unit available for rototilling, grass cutting, and snow blowing, and inadequate sales increase is projected to justify the expense. A design specification adjustment may be required. An idea previously discarded may now be the most acceptable solution to the problem, or perhaps research has come up with new technology to apply to the problem. The most acceptable solution will always include compromises between conflicting alternatives.

Many product development projects are evolutionary in nature, that is, the weaker ideas are dropped during analysis and refinement, and the time and money go toward developing the most promising idea. If that is the case, then at this point the decision needed is whether or not the product presently developed meets enough of the design criteria to continue. It is hoped that the idea that has been refined and analyzed as workable will meet all of the problem identification criteria, will be clearly better than the other ideas, and be profitable. If it passes the test, the direction of implementation is clear.

If there are still several ideas that qualify under all the conditions set forth, and one answer is not clearly the best, then a decision must be made on the direction of the project. This will also be the situation when a feasibility study has been done, because one of the goals of a feasibility study is to propose several methods of solution to the problem.

7.1 Idea Selection

At any stage of an engineering project, decisions are made about which idea(s), if any, are to be further developed. The list of feasible solutions to a problem needs to be reduced to the one idea that seems to have the best chance of succeeding, or reduced to the best three or four ideas, any one of which may turn out to be most acceptable. Reducing the list to just one idea allows all available resources to be concentrated on developing that concept. If the list is reduced to three ideas and they are all developed, then either more money or more personnel must be assigned to the project or the project will take longer to complete. The next example compares resource allocation if three ideas are in the running for further development work.

EXAMPLE 7.1:
Idea development schedule.

In Figure 7.1, the horizontal axis represents periods of time, and the size of each rectangle represents the proportion of money and personnel being used during each time period. Figure 7.1(a) represents the method of idea selection where one idea is kept and developed. The project is shown to be completed in five periods of time and five units of money and personnel. During the first period of time one-third of the money and personnel time is spent on each idea to determine which should be continued.

Figure 7.1(b) illustrates what happens if each idea is studied longer to collect more information to support the final selection. If the total amount of money and personnel for each time period is the same as in Figure 7.1(a), then the project will not be completed in the same amount of time, it will take eight periods of time.

Figure 7.1(c) illustrates the approach where the project gets done in the same number of time periods as in Figure 7.1(a), but requires more money and personnel, three times more during time period two and twice as much during time period three. Unless other projects are available for the extra people to work on at the conclusion of this project, employee utilization will be poor, and a stable work force hard to maintain. It may be possible to utilize the extra personnel on ideas #2 and #3 after idea #1 is discarded. Difficulties exist if the time frame for idea #2 includes some tests that are time dependent, such as fatigue or creep tests. The total time may not be reducible even with additional personnel.

The main advantage of the methods shown in Figures 7.1(b) and (c) are that alternate ideas are being developed, and can be used if the idea that was first thought to be the best fails to succeed. If all the time and money goes into one idea and it doesn't work, backing up to develop an alternate idea will cause a product introduction delay.

FIGURE 7.1

Alternate time-personnel schedules.

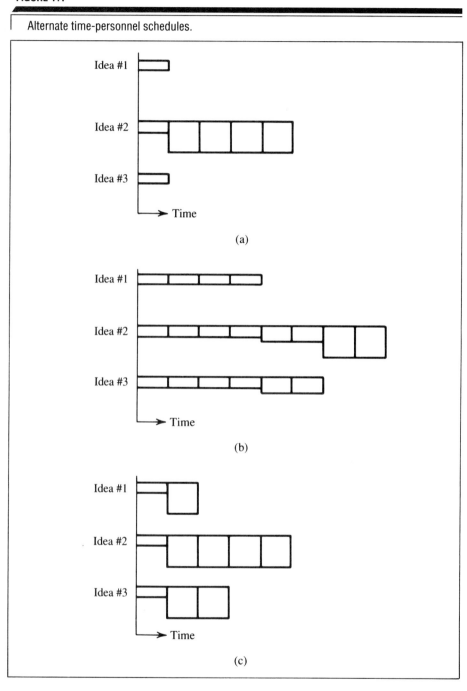

If several preliminary ideas have been developed and are still feasible, the time will come when the preferred idea must be selected for implementation. One method for making the selection is by using a matrix, weighted-value technique as follows.

The first step is to make a **list of all the criteria** that you think could be used to compare the design solutions. Though this list can include anything that comes to mind, to be effective the list must include those things required for a fair comparison of the various proposals and must include the original specification goals. Some criteria to consider are:

- weight specification
- size specifications
- operation specifications
- cost to manufacture
- appearance
- number of parts
- ease of assembly
- ease of repair
- return on investment
- cost of use by the consumer
- time needed for delivery
- environmental specifications
- ease of use by the consumer
- cost of tooling
- possibility of harm to the user
- possibility of market penetration
- shelf life
- estimated life before failure

When the criteria list is complete, **assign a priority value** to each item on the list. Assigning a priority identifies the items on the list that are most important, and those that are the least important. Use a scale of 1 to 10 or 1 to 100 and assign a numeric value to each criterion. Most people prefer to associate a high value as good, so assign a 10 to the most important criteria and a 1 to the least important. If appearance is valued 10 and cost 8, then appearance is considered 25% more important than cost. More than one item on the list can have the same priority value if they are judged to have equal importance, or decimal values can be used if close refinement is necessary. To do an unbiased job of assigning criteria values, do not consider any of the solution ideas when assigning the values. Criteria values can also be assigned by several people on an independent basis and the results averaged.

After the value of each criterion has been determined, **normalize the values** by dividing each of the individual values by the total of the values. Each item will then have a normalized value between 0 and 1, and the total of the normalized values will equal 1. Normalizing is not mandatory but it gives range values more suitable for intuitive inspection because the values are easily compared as percentages.

Now **make a chart** with the criteria items to be compared as row entries, in order of their normalized priority values, and the idea solutions to be compared as column headings. For each criterion on the list **compare and rank the ideas.** The ranking values can cover any range—0 to 10 or 0 to 5 are common. Use high values for good rankings. Zero is included because an idea may not meet a particular criterion at all. Ranking can also be done by several people and the results averaged.

When each idea has been ranked for each of the criteria, multiply the ranked value by the normalized criterion value factor. Add the products and get a total score for each idea. The idea with the largest total is the "best" solution, that is, the most acceptable with the criteria and ranking selected. Ordinarily several ideas will have totals that are appreciably higher than others. Any solution idea with a very low total indicates that it does not measure up to the required criteria and can be discarded.

This system of evaluation is a tool to help weed out the poorest ideas and to focus remaining effort on the better ones that remain. Choices still must be made and optimization continued. The idea with the highest score can be selected and implemented, or the chart can be studied for the highest ranking in each category and the idea with the majority of high rankings selected. Time and energy can also be focused on improving the low ranked criteria of the "best" idea. This is simply a way to be sure the most acceptable idea is used. Another approach is to take the highest ranked idea from each criterion and generate a new idea that incorporates all the better features.

Obtaining numeric values of the total scores is not the purpose of this method because assigning numeric values to design criteria cannot be done in a consistent way. It is the relative values that are important. The procedure is recommended for a project that has several solutions under consideration, all of which seem workable. The procedure does take time, and it is difficult to agree on values while doing the evaluation. The main purpose of the procedure is to focus attention on the features of the design that are really important and to help decide which idea best meets the design criteria. Don't be surprised if new concerns are proposed during the procedure, because concentrated thinking is still taking place, and creativity hasn't been shut off. It may also happen that other comparison criteria will come up during this idea analysis and a new comparison will need to be done. Recall that one of the tasks of an engineer is to iterate so improvement can continue. It is essential that the design team agrees that the idea solution implemented is the "best." Another method of selection optimization is discussed in Exercise 7.14. The following case study illustrates the criteria evaluation and ranking procedure just discussed.

CASE STUDY #6:
Baseball field revision.

An engineer was assigned a design problem involving a baseball field. There was an existing field, but it had a left field foul line of only 240 feet (73.2 m). An investigation was made and alternatives proposed that would increase the foul line to 300 feet (91.4 m). Table 7.1 lists the design criteria selected to compare the ideas. The table includes the criteria values assigned. The criteria values listed in Table 7.1 indicate that cost has been selected as the most important factor, and shower and locker facilities as least important. The normalized values are calculated by dividing each value by 54, the total of the assigned values.

Table 7.2 shows each criterion ranked for each of five possible solutions. The present field was included as a basis because nothing has to be done. The best solution may be to make no change. The present field and the off-campus field both rank highest in cost because there is no reconstruction cost associated with these two ideas. Idea #1 is most costly to implement. All the on-campus ideas are equally close to locker facilities, whereas the off-campus idea is not close to locker facilities.

If the chart results are taken at face value, no change will be made because the present field ranks highest with a 7.74. Idea #2 is the poorest and probably doesn't

TABLE 7.1

Criteria and assigned values for baseball field project.

Criteria	Priority Values	Normalized Values
1. Cost of implementation, moving dirt, fences, retaining walls, etc.	10	0.19
2. Room around the diamond for benches, warm up, and spectators	6	0.11
3. Sun in the eyes of the players	5	0.09
4. Houses and cars in the location of home runs and foul balls	5	0.09
5. Meet minimum requirement of 300 foot foul lines	8	0.15
6. Shadows on the field from buildings	3	0.06
7. Near student dorms, convenience of stopping by to watch the game	6	0.11
8. Close to shower and locker facilities	2	0.04
9. Ground suitable for good grass and infield skin maintenance	4	0.07
10. Interference with other sports fields	5	0.09
Totals	54	1.00

Underlined words are keywords used in Table 7.2.

TABLE 7.2

Ranking comparison for five ideas for the ballfield revision.

Criteria	Normalized Priority Values, N	Idea #1 On Campus R_1	NR_1	Idea #2 On Campus R_2	NR_2	Idea #3 On Campus R_3	NR_3	Idea #4 Off Campus R_4	NR_4	Idea #5 Present Field R_5	NR_5
1. Cost	0.19	2	0.38	3	0.57	4	0.76	10	1.90	10	1.90
5. Minimum foul lines	0.15	10	1.50	10	1.50	10	1.50	10	1.50	0	0.00
7. Student convenience	0.11	10	1.10	10	1.10	10	1.10	1	0.11	10	1.10
2. Room	0.11	7	0.77	4	0.44	7	0.77	8	0.88	8	0.88
3. Sun	0.09	9	0.81	3	0.27	8	0.72	5	0.45	9	0.81
4. Foul balls	0.09	9	0.81	1	0.09	8	0.72	8	0.72	8	0.72
10. Interference	0.09	3	0.27	3	0.27	4	0.36	10	0.90	10	0.90
9. Ground	0.07	8	0.56	6	0.42	5	0.35	9	0.63	9	0.63
6. Shadows	0.06	8	0.48	6	0.36	9	0.54	7	0.42	8	0.48
8. Locker facilities	0.04	8	0.32	8	0.32	8	0.32	1	0.04	8	0.32
Total scores			7.00		5.34		7.14		7.55		7.74

The values in the R_1, R_2, R_3, R_4, and R_5 columns are the comparative rankings. For example, ideas #4 and #5 have minimum cost, idea #1 has highest cost, and ideas #2 and #3 are nearly as expensive.

warrant further consideration. If idea #1 could be done with less money than idea #3, and earn a 5 ranking in cost, idea #1 would be second best.

The final solution proposed for the project was a compromise. Did you expect anything different? It was decided to reduce the cost of idea #3 by eliminating the excavation for a full 300-foot (91.4 m) foul line and settle for 285 feet (86.9 m). This compromise increased the ranked value cost to an 8, and reduced the rating for minimum foul lines to an 8. This resulted in a new total value of 7.60. One could argue that a 285-foot (86.9 m) foul line rates a 0 because it doesn't meet minimum requirements, but recall that a compromise has been judged the most acceptable alternative under the imposed restrictions, not the utopian best. To date no action has been taken on the project because funds have not been made available. The final recommendation for the ball field revision project is given in Figures 7.2 and 7.3.

Follow-up to Case Study #6. In the fall of 1988 the existing baseball field was converted into a soccer field, and baseball games are now played at an off-campus field managed by the city. This is an example (not an uncommon one) of a problem that was resolved in a far different way than the original data, study, and analysis suggested.

FIGURE 7.2

Report on Case Study #6.

```
To: Coach Appleton
From: Jim Student
Re: Changing the baseball field configuration
Date: May 10, 1986
    I recommend a change in the baseball field layout to the one
shown on the attached drawing. This proposal is an answer to your
request to find a reasonable cost method of increasing the length
of the left field foul line. This proposal increases the length
from the present 240 feet to 285 feet. The cost of the proposal
is estimated at $2500. This cost covers relocating the fences and
backstop, and re-sodding the present infield, which becomes part
of the outfield of the proposed change. No excavation is
required.
    Other ideas were considered. One idea would have given a full
300-feet foul line, but excavation would have added an estimated
$7500 to the project.
    There is also a field in town that could be used for a rental
fee of $100 per game, but there are no shower facilities there.
Players would have to come back to campus to shower and change
clothes. An off-campus field would also discourage students from
''stopping by'' to watch a game.
     Please call me after you have studied the attached drawing.
I want to answer any questions you may have as well as discuss
possible implementation. If I have not heard from you by May 24,
1986, I will contact you.

cc: Members of the Athletic Committee
```

FIGURE 7.3

Final ball field proposal.

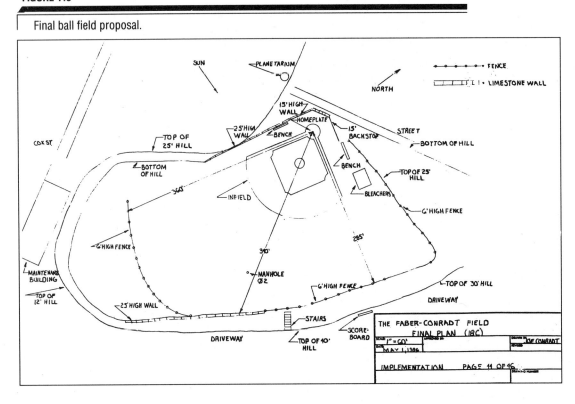

7.2 Implementation

Implementation includes the activities that are required to convert the "best" idea into a functioning product in the hands of consumers in an economical manner. Once a project has the green light, in some sense the hard work is yet to come. Good communication, that necessary ingredient to keep everyone's effort concentrated toward the common goal, is just as important now as before. More people will be involved once the decision to implement is made.

The final version of the project requires **detail and assembly drawings**, but additional development may be required, so changes continue to be made to drawings and related bills of materials. Preliminary drawings will already have been created, either for a full size prototype to be built, or for an accurate and complete product cost to be calculated. A complete cost analysis requires material and manufacturing labor costs, both of which come from detailed drawings and assembly specifications. For implementation, the drawings must be reviewed to be sure they include all the latest changes and incorporate appropriate company standards. Ordering 10,000 of an old version of a part is not only costly, it is demoralizing.

Product testing will continue during implementation, and last minute changes may be proposed. Some changes are mandatory, as when a part begins to fail under test. Other changes may be just the nice-to-do type and will be incorporated only if they can be done without disrupting the schedule, such as a change in appearance. Some proposed changes are saved for the next version or for a different model. A proposed change that cannot be implemented within the present time schedule may be offered later as an option. For example, a trailer pulling package option for a passenger car (heavy duty shock absorbers, special mirrors, higher horsepower engine, heavy duty transmission), or a software update for a computer system.

During implementation the **schedule for product release** generally becomes the controlling factor. An exception is a problem that may cause the consumer harm or place others in a dangerous position. Don't compromise a quality product just to get it out the door and on the way to the customer. The cost of warranty claims and recalls can offset short gains in sales. Lawsuits might occur that could force your company into financial disaster or out of business. Good ethics requires you to keep management notified about any known or potential problems. For the record and for future litigation possibilities, put all such findings in writing and send them to the appropriate management levels. Read reports on the April, 1986 space shuttle disaster and subsequent hearings, which appeared in newspapers and magazines, to learn how critical and important communication and timely action are.

Many **production activities** must be completed using the details from final drawings. Proper tooling, including dies, jigs, fixtures, gaging, and machine tools must be provided. The cost and complexity of these items should have been estimated for budget purposes, and now they must be ordered, designed, built, and tested. The testing of processes and tooling is no less important than the testing of product components. Production tooling must be capable of making parts to drawing specifications in a consistent manner. Time for the testing of processes and tooling is particularly important if a new process is being used. An engineer may have to spend many days at the factory of the vendor supplying the new process. It is easier to check out a machine in the factory of the supplier than in the user's plant. For complex operations on a new machine or process, the first run of production parts will likely be made at the vendor's factory.

Plant facilities such as assembly lines, paint booths, and storage areas must be prepared and made ready for full scale production. Will the new parts fit through doorways and under overhead lights? Many problems can arise as production starts. Methods of assembly that were no problem on a custom model building scale now become major obstacles. Forklifts used during development to assist with large component assembly must be replaced with cranes and hoists. The special touch of a technician's assembly procedure must now be specified at a particular torque or pressure setting, assembly procedures confirmed, and proper assembly tools and equipment provided.

A project involving system changes requires **personnel training**. Perhaps new forms and methods of processing them need to be introduced. Set up workshops in advance of full scale use to train users properly. Because some learning curves have very low slopes, adequate training time is essential. Plan for parallel system operation if all the portions of a new system have not been completely tested, because bugs will occur. This is particularly important when converting from a manual system to a computer system, or changing computer systems. A computer will not make a poor system run smoothly, it will only generate errors and chaos faster. New systems and products also require **procedures or operating manuals**. An operator's manual must be accurate and specific, and if a license is not required for operation of the product, the manual must include a warning that operation in any mode not specified in a list of safe modes can result in injury to both people and property, and possibly death.

Before a product is released to the consumer final **product parts listing and service manuals** need to be prepared. Parts lists and repair parts listings are important for support of dealers who make repairs and sell service parts, as well as for customers who need to know normal service operations. Poor customer relations can result from poor service information. Imagine the frustration on the part of a consumer who, after requesting a new light bulb for a light generator system, was told that the entire light bulb-light shroud-lens assembly had to be purchased because that was what was listed in the service manual.

Prior to product release, **market advertising** must be readied with product specifications and features highlighted. This includes information on available options, accessories, and cost. If a product unveiling show will be held to publicize the new product to the dealers, all the preparations must be planned and scheduled. Sometimes first production units are loaned to preferred customers for their use, so shakedown testing can be done that more accurately reflects customer applications.

When the first product is shipped, or the new system is put on its own, personal emotions run high. Many questions run through the minds of project staff. Will the product work as tests have predicted? Will the production units be as good as the test units? Was the marketplace reached in time? Were enough tests run so the reliability and confidence levels are high? Is the selling price too high? Too low? The questions will be answered by the consumer.

 CASE STUDY #3 CONCLUDED:

The special pallet racks for moving ground shafts worked so well that their use was expanded. Empty racks start at the lathe department, where the parts are turned and center-drilled. The operator loads the parts on the racks and then the parts go to heat treatment. From heat treatment they go to the finish grind and then to assembly. As the racks are emptied, they are sent back to the lathe department and the cycle is repeated. Inventory control is improved and the surface of the parts does not get nicked or scratched any more. More racks were needed with this approach, but the increase was justified because of the improvement in inventory control and the reduction of scrap. The parts are now protected at each stage of production, parts are not lost in transit, and a more accurate count is maintained.

7.3 References

1. Amos, John M. and Bernard R. Sarchet. *Management for Engineers.* Englewood Cliffs, N.J.: Prentice Hall, 1981.
2. Andreasen, M. Myrup et. al. *Design for Assembly*. New York: Springer-Verlag, 1985.

3. Bronikowski, Raymond J. *Managing the Engineering Design Function*. New York: Van Nostrand Reinhold, 1986. A summary of the steps of the design procedure from a manager's point of view; chapter on patents.
4. Golembeski, Dean J. "Struggling to Become an Inventor." *Invention And Technology*, Vol. 4, No. 3, Winter 1989, 8-15.
5. Hajek, Victor. *Management of Engineering Projects*. New York: McGraw-Hill, 1984.
6. Lentz, Kendrick W. *Design of Automatic Machinery.* New York: Van Nostrand Reinhold, 1985.
7. Lipson, Charles and N. J. Seth. *Statistical Design and Analysis of Engineering Experiments*. New York: McGraw-Hill, 1972.
8. Meredith, Dale D. et al. *Design and Planning of Engineering Systems*. Englewood Cliffs, N.J.: Prentice Hall, 1973.
9. Morris, George. *Engineering—A Design Making Process*. Boston: Houghton Mifflin, 1977.
10. Pahl, Gerhard and Wolfgang Beitz, edited by Ken Wallace. *Engineering Design*. London: The Design Council, 1984. Many mechanical design hints.
11. Papanek, Victor. *Design For Human Scale*. New York: Van Nostrand Reinhold, 1983.
12. Schmidt, Frank W. and Robert E. Henderson. *Introduction to Thermal Sciences*. New York: John Wiley & Sons. 1984.
13. Spotts, M.F. *Design Engineering Projects*. Englewood Cliffs, N.J.: Prentice Hall, 1968.
14. Tichy, H.J. *Effective Writing for Engineers-Managers-Scientists*. New York: John Wiley & Sons, 1967.
15. *Consumer Reports* magazine.

7.4 Exercises

1. List comparison criteria you would consider if you were going to purchase each of the following products and wanted to compare brands.
    a) Rowboat
    b) Bookshelf
    c) Card table
    d) Automatic clothes washer
    e) Microcomputer
    f) Bicycle
    g) Camping tent
    h) House
2. List comparison criteria you would consider if you were going to make a decision about each of the following items.
    a) College to attend
    b) Vacation to take
    c) Engineering job to take

3. Select the product from Exercise 7.1 with which you are most familiar, place values on the criteria you have listed, and then normalize the values.

4. Select at least two products that are currently on the market for the product selected in Exercise 7.3 and compare them. Complete the calculation and select the most acceptable. If the selection comes out contrary to a selection you or someone you know has made, discuss reasons why.

5. Discuss specific feedback data that indicate if a mass-produced product, such as a moped, has been successful.

6. Discuss specific feedback data that indicate if a new system, such as inventory control, has been successful.

7. Suppose a new machine tool, a key process in a new product, was delayed in shipment well past the proposed production date. Discuss ways to work around the problem and meet the original date.

8. The following list includes some products that have critical market introduction dates. List the products in order, from those that have the most critical market introduction dates to those with the least critical introduction dates. List reasons for your ranking.
    a) Christmas trees
    b) Lawn mowers
    c) Snow blowers
    d) Water skis
    e) New car
    f) A new watch
    g) New residential house light controls
    h) A new typewriter
    i) A new college textbook

9. Discuss the problems that occur if the products listed in Exercise 7.8 arrive too early or too late.

10. Consider a new graduate engineer who is faced with the following decisions all of which are viable choices:
    A) Going into the consulting business as a sole proprietor,
    B) Going into the consulting business as a partner with two experienced engineers,
    C) Going to work for a local engineering firm, or
    D) Going to work for a national engineering firm.

    Assume that the probability for success depends upon the economy, which could:
    1) Enter a boom cycle,
    2) Continue at a steady pace, or
    3) Enter a recession.

    The following chart gives the engineer's estimated relative earnings expected for each of the choices. For example, going into business with two partners in a steady economy will yield half the earnings of going into business alone in a boom economy.

	1	2	3
A	100	40	0
B	80	50	10
C	85	60	15
D	90	75	15

a) Which choice should the engineer make if the economy has an equal chance at a boom, a steady pace, or a recession? Why?

b) Which choice should the engineer make if the economy has a 50 percent chance at a steady pace, 25 percent chance at a boom, and 25 percent at a recession? Why?

c) What choice should the engineer make if recession is certain? Why?

11. Table 7.3 represents a comparison made on five different cars by automotive experts.

    a) Determine a criterion evaluation value for each category and determine which car best suits your need if you were going to buy a new car.

    b) List other items you would add to the list if you were doing the analysis for yourself.

TABLE 7.3

Comparison data for five cars, Exercise 7.11.

Car #	1	2	3	4	5
a. Styling	20	19	14	19	14
b. Quality control	18	21	18	20	15
c. Instrumentation	16	19	17	20	16
d. Ride comfort	14	12	19	19	17
e. Visibility	18	20	21	20	16
f. Load hauling	18	12	21	21	16
g. Handling	13	19	14	17	15
h. Fuel economy	5	20	10	15	25
i. Acceleration ability	13	6	17	21	14
j. Dollar value	13	18	14	21	19

12. Table 7.4 lists five materials that are being considered for the outside covering of a building. Three criteria are to be used for evaluation:

    1) High density, which is good for shielding from γ rays

    2) High thermal resistance, which is good for reducing heat loss

    3) Low cost, which is less expensive to install

    a) Assign importance values to the three criteria.

    b) Rank each of the five materials for each criteria.

    c) Select a material.

    d) List other criteria you might want to consider if you were in charge of this project.

TABLE 7.4

Comparison data for five materials, Exercise 7.12.

Material Number	Density lb/ft³	Thermal Resistance R value/inch of thickness	Cost $/ft²
1	169	0.0007	17
2	120	0.20	5
3	140	0.10	3
4	170	0.02	30
5	62	0.25	0.50

13. List and compare the advantages and disadvantages of:
    a) a three-legged chair
    b) a four-legged chair
    c) a five-legged chair

14. An alternate method of comparison to determine the "best" of many ideas is to compare all possible combinations of ideas two at a time. No detailed breakdown of selection criteria is done, each idea is treated as a whole. For example, if six proposals are viable, and they are coded A, B, C, D, E, and F, then 15 comparisons can be made. If you keep track of the number of times that each proposal is preferred, then the ranking of "best" can be done based on that number. In the example shown in Figure 7.4, proposal C comes out on top followed by E, F, A, D, and B. In the case of a tie for "best," then further analysis of those tied needs to be done. Use this method of comparison on Case Study #6 or any of the other comparison problems. The procedure can also be performed three ideas at a time to confirm the two at a time decision. Comparing six things three at a time results in $_3C_6 = 6!/3!3! = 20$ combinations, while comparing six things two at a time results in $_2C_6 = 6!/2!4! = 15$ combinations.

FIGURE 7.4

Example of idea evaluation by pairs.

The idea at the top of the second column is compared with each idea in the first column, and the most acceptable one selected. Then the idea at the top of column three is compared with each idea in column one and the best idea recorded. This is continued until all combinations have been compared.

	A vs.					
B	A	B vs.				
C	C	C	C vs.			
D	A	D	C	D vs.		
E	E	B	E	D	E vs.	
F	F	F	C	F	E	

Idea	A	B	C	D	E	F
Number of "wins"	2	1	4	2	3	3

15. Compare buttons, zippers, and Velcro™ as methods of keeping a jacket closed. Consider such features as:
    a) Original cost
    b) Ease of use by two-handed people
    c) Ease of use by one-handed people
    d) Ease of repair
    e) Appearance
    f) Reliability
    g) Noise

16. Compare four modes of transportation: automobile, train, bus, and airplane. Consider such concerns as:
    a) Cost
    b) Ease of access
    c) Comfort in transit
    d) Safety
    e) Taking children
    f) Sleeping on route
    g) Eating on route
    h) Taking luggage

    Make the comparison for:
    a) A trip of 10 miles
    b) A trip of 100–500 miles
    c) A trip of 1000–2000 miles

17. Compare the advantages and disadvantages of a wooden sharpenable pencil and a mechanical pencil.

18. An alternate approach to lists of comparison criteria is to create a checklist with very specific items for a very narrow use, such as a checklist for inspecting a house to purchase, a car to buy, or a college to attend. Select a specific task, such as buying a car, and list the specific items to check for such a project.

Design Problems

The following project problems are proposed to allow you to practice the design procedure. There is no limit to the type of problem that can be considered, and you should be capable of leading any problem solving investigation. You may not be an expert in the area of the specific problem, but you know how it has to be approached. The problems are separated into two groups. Group I contains problems that do not have many details given, which is typical of open-ended design problems. Designs should be complete, with detail drawings, complete specifications, and either a scale model or a full-size working model.

Group II problems represent another approach to design, as they begin with a solution idea. It is not unusual for an engineer to be given a preliminary sketch and told, "Make it work." Someone has an idea, a brainstorm if you will, and wants to see if the idea can be made functional. This beginning idea could be a sketch or a picture of a competitor's product. In fairness to the person giving you the idea, and by necessity if the person is your superior, an analysis will be made. This is a case where the problem solving starts with the idea, and loops to problem identification, analysis, and then back and forth as required.

While working on any of the problems, remember there is no single best solution. That can be frustrating if you have not had practice in solving open-ended problems, but that is the real world of engineering. Real problems have no single answer. Good engineers know this and stay flexible and imaginative in their thinking. If your work on any of these problems leads you to a fortune-making idea, all the better.

Group I—Incomplete Problem Identification

1. Design a saddle support winch system for removing boats from the water for dry dock work.
2. Design a coin sorter.
3. Design a system and components that will make coffee on a continuous basis.
4. Design an automatic garage opener when only one inch of clearance exists between the open door and the ceiling of the garage.
5. Street and parking lot repair companies often use a gasoline powered abrasive cutoff machine to cut out sections of pavement for repair. Some are pushed by hand. Design a self-propelled machine that will cut through the concrete or asphalt and have a variable feed rate so the power used for cutting can be held constant.
6. Design a 1000 gallon water tank.

7. An area 200 feet by 150 feet is available for a parking lot. The lot is on a corner and has an average grade of 10 percent parallel to the 200 foot dimension. The local code limits the grade in parking lots to 6 percent. Design a parking lot for this property.

8. Design a system to allow the pilot of an airplane to know if everyone is buckled up before takeoff.

9. Design a log splitter.

10. Design a portable traffic signal light.

11. Discuss a device and a system to monitor and control the direction of movement of objects (people, vehicles, etc.) for an application such as in-only and out-only traffic flow.

12. Develop a conceptual design for an automatic machine that will accept boards that are 0.750 ± 0.030 inches thick, 3.00 ± 0.50 inches wide, and 16.00 ± 1.00 inches long, cut the boards to 14.00 ± 0.03 long, and drill two dowel holes in each end spaced 1.00 ± 0.01 inches apart and on center ± 0.01 inches.

13. Blast shelters generally have valves in their air intake ducts to prevent shock waves from entering. Springs or hydraulic cushions are often used in these mechanisms. Design such a valve for a 24-inch diameter opening. Check appropriate literature to determine the pressures involved.

14. A preliminary design for a manually operated, bicycle pedaled air ventilator system has the following specifications: 55 rpm pedal speed, 785 mm diameter fan, 480 rpm fan speed, 75 watt output, 50 m^3/min air flow for a 300 meter duct, or 100 m^3/min for a 25 meter duct. A two operator design specifies 100 m^3/min air flow for a 125 meter duct. Determine if the specifications seem reasonable, and suggest a possible design. Include an estimate of the human effort required to operate the device.

15. An abandoned railroad right-of-way has been taken over by a group and they intend to convert the property into a bicycle and hiking trail. They are worried that motorcycles and other powered vehicles will go on the trail and tear up the surface. Determine a solution to their problem to insure that powered vehicles will not have access to the trail.

16. Design a system of filling bags with 50 ± 2 pounds of granular material coming from a bulk storage system. Determine how many bags per hour can be filled with the device you design.

17. Design a child's chair that can double as a two-step step stool.

18. Design a device for controlling the on-off cycle of a series of lights. The on and off time should be variable.

19. Design a bicycle rack for home use.

20. Design a vise that will automatically center a part clamped under a drill press spindle.

21. Design an experiment to verify the value of the modulus of elasticity of a material using one of the beam deflection models. Reference your strength of materials textbook for various equations.

22. Design a bimetal temperature sensor that will control an electrical circuit for a temperature difference of 10°F.
23. Design a fixture that aligns pipes for welding. Select appropriate size ranges from commercially available sizes.
24. Design an educational toy. Part of the identification activity will be to determine the educational objective desired.
25. Design an apparatus to test the torque needed to unscrew a bolted assembly.
26. Design a windmill to convert wind energy to another form of energy.
27. Design an apparatus and a system to measure the efficiency of various windmill blade designs.
28. An experiment is to be conducted on the life of incandescent light bulbs as a function of ambient temperature. Design an apparatus that can be used to control the ambient temperature between 100°F to 500°F for individual bulbs ranging from 25 to 100 watts.
29. Design an experiment to test the effect of vibration on the life of incandescent light bulbs. Vibration amplitude and cycles per second should be variable.
30. Design a machine to weigh objects up to 500 pounds. A "high" number of weighings per minute is desired.
31. Design a test apparatus to test the compressive strength of wood dowels up to 1.00 inches in diameter.
32. Design a test apparatus to test the shear strength of wood dowels up to 1.00 inches in diameter.
33. Design an experiment to test the time to set for a given bonding strength for a wood glue. Include in the experiment such variables as: face grain joints, end grain joints, combination grain joints, butt joints, lap joints, type of wood, clamping pressure, time, humidity, and applied tensile and shear loads.
34. Design an experiment to study the pullout force of a wood screw. Include in the experiment such variables as: type of wood, diameter of screw, length of screw, size of pre-drilled hole, and direction of wood grain.
35. Design an apparatus that converts rotary motion into linear motion. The length of the linear motion should be variable, but be able to be fixed at any desired length. The stroke should be faster in one direction than the other. One-direction travel time should be at no more than 40 percent of the total stroke cycle time: 4.00 inches ≤ stroke length ≤ 10.00 inches.
36. An experiment is to be conducted to determine water-caused soil erosion in various soils growing various plants. The samples of plants will be grown in boxes 2 feet x 2 feet x 8 inches containing 6 inches of soil. A wire mesh box liner will allow the soil-planting sample to be removed from the box for testing purposes. Design an apparatus to hold the wire mesh sample, allow a measured amount of water to be applied, vary the angle of the sample so runoff angles can be changed, and accumulate the runoff for study. Type of soil and variety of plants are also variables.

37. Design a single assembly apparatus to determine the percent grade of land over a 10 foot length. The range of grades to measure is 0 to 100 percent with an accuracy of ±1 percent.

38. Design an experiment to study the variation of cable pull force of a rope or wire cable over a pulley. Allow for changes in pulley diameter, bearing changes, the angle of contact between the cable and the pulley, and the weight being lifted. How can this experiment be expanded to study coefficients of friction?

39. Design an experiment to study the variation of tension force of a V belt running at various speeds over a sheave. Allow for changes in sheave diameter, bearing changes, the angle of contact between the belt and the sheave, and the tension in the belt. How can this experiment be expanded to study coefficients of friction?

40. Design a device that can be used by a carpenter to support and hold a door in position while hinge and latch locations are being determined and cut.

41. Design an apparatus for sorting individual ball bearings used in bearing assemblies, in size intervals of 0.0001 inch in the range 0.2500–0.2505 inches ball diameter.

42. Wood and some other materials have directional strength properties. If a load is applied at some angle θ from a primary strength direction, an adjustment for safe loading must be made. Design and conduct an experiment to determine the maximum allowable angular load F_n as a function of the angle θ, F_c, and F_{cp}; F_c is the maximum allowable strength of the material in the primary direction and F_{cp} is the maximum allowable strength perpendicular to the primary direction. It has been suggested that the form of the answer is $F_n = (F_c)(F_{cp})/ (F_c \sin^m θ + F_{cp} \cos^m θ)$ where m is dependent upon the type of wood.

43. In October, 1988 the Consumer Product Safety Commission banned the game of lawn darts from the marketplace, saying, "What limited recreational value lawn darts may have is far outweighed by the number of serious injuries and unnecessary deaths." (*Des Moines Register*, October 29, 1988). Design a similar game that would be safer for the public.

44. Design a mechanism that will automatically feed bars to a bar lathe operation. Consider bars from 1.00 to 6.00 inches in diameter and 8 to 16 feet long.

45. Design a "vandal proof" trash collection receptacle for use in parks and other public places.

46. Design a device that will shut off the flow of a fluid after a given amount has flowed past the device.

47. Design a device for sampling rainwater that will activate when rain begins to fall and will collect a sample of rain at regular intervals. The interval can be timed to the amount of rain. This device would be helpful for monitoring the acid content in rain.

48. The packing of a number of equal diameter cylinders together inside a larger cylinder is often required in engineering problems. The question that needs answering is "What is the smallest large cylinder for n small cylinders as $n = 2, 3, 4, \ldots$?" When $n = 7$ or 13 the configuration is easy to calculate. However, when $n = 5$ or 8 and for many other values the most efficient layout is not obvious. Design a device that will allow the smallest large cylinder to be determined for a given

number of smaller equal diameter cylinders. The problem is to have a device with a variable large diameter within which a given number of small cylinders can be arranged.

49. Checklists are good to have so that important items are not forgotten. Any activity can be assisted by using a checklist. Create a checklist for items to be considered when undertaking a design project. The more general it is, the wider the scope it will have.

Note: Exercises 6.10, 6.11, 6.12 are also for design exercises.

Group II—Preliminary Problem Solution Given

1. Figure D1 is an idea for a bar clamp. Complete the details of such a design.
2. Figure D2 is an idea for a monkey wrench. Complete the details of such a design.
3. Figure D3 is an idea for a parallel jaw clamp. Complete the details of such a design.

FIGURE D1

Bar clamp concept.

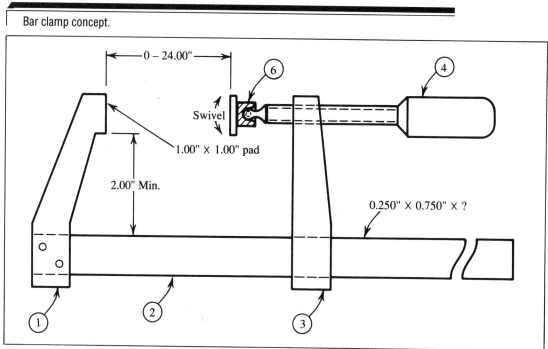

FIGURE D2

Monkey wrench concept.

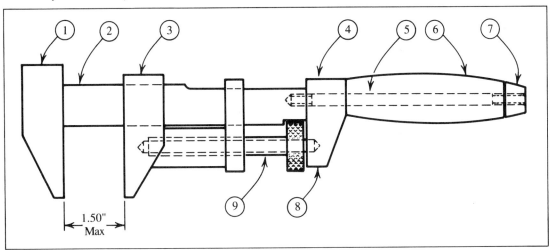

FIGURE D3

Parallel jaw clamp concept.

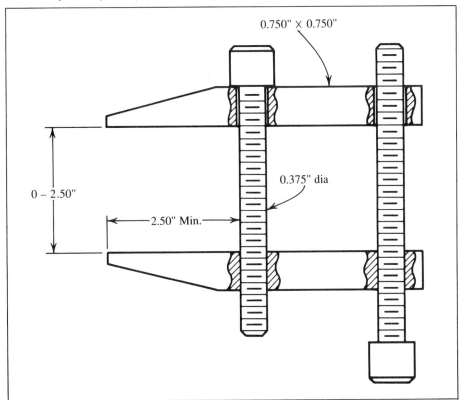

FIGURE D4

Rotating control layout.

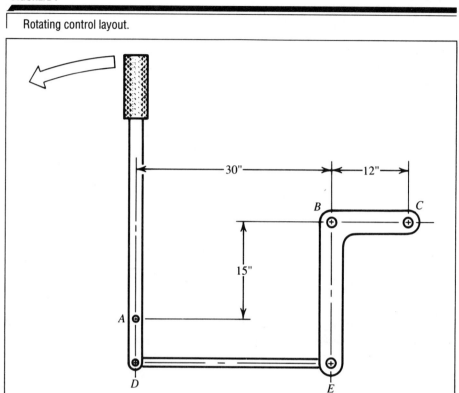

FIGURE D5

Sliding handle concept.

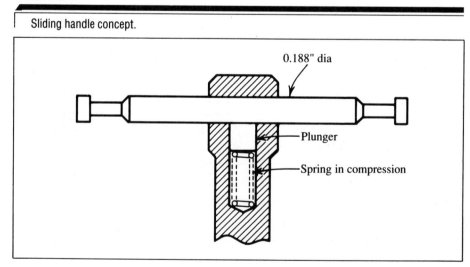

FIGURE D6

Hand punch concept.

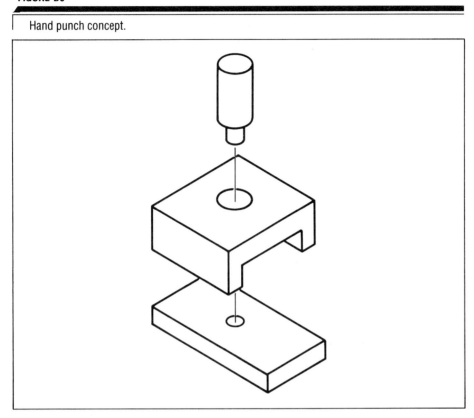

4. Figure D4 shows the layout of a mechanism whose purpose is to move a pin located at point C vertically 3 inches while the handle moves through an angle of 70°. The location of points A, B, and C are fixed, but D and E are not. The resistance to movement at point C is 50 pounds. Design a mechanism that will efficiently satisfy the requirement.

5. Figure D5 is an idea for a sliding handle for tools so the handle can be offset in tight locations. The handle should slide freely back and forth, but not come off when used. Complete the details of such a design.

6. Figure D6 is an idea for a hand punch. The device should hold a belt and allow buckle holes to be punched in the center of a leather belt and spaced appropriately. Complete the details of such a design.

FIGURE D7

Ink roller concept.

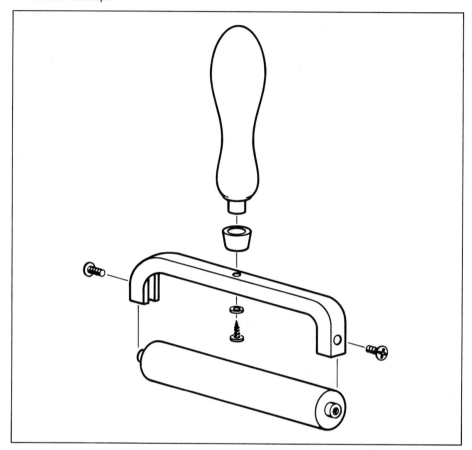

7. Figure D7 is an idea for an ink roller for a small printing operation. Complete the details of such a design.//
8. Figure D8 is an idea for a hand fruit masher. Complete the details of such a design.
9. Figure D9 is an idea for a child's wagon. Complete the details of such a design.
10. Figure D10 is an idea for a movable and adjustable lawn lounge. Complete the details of such a design.

FIGURE D8

Fruit masher concept.

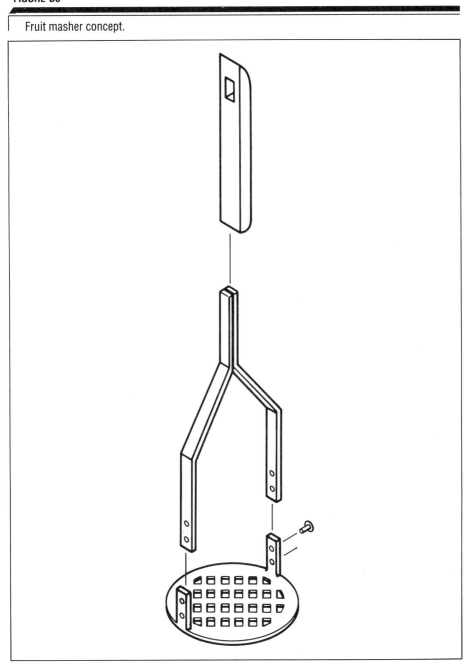

Design Problems

FIGURE D9

Wagon concept.

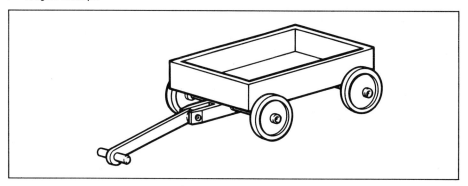

FIGURE D10

Lawn lounge concept.

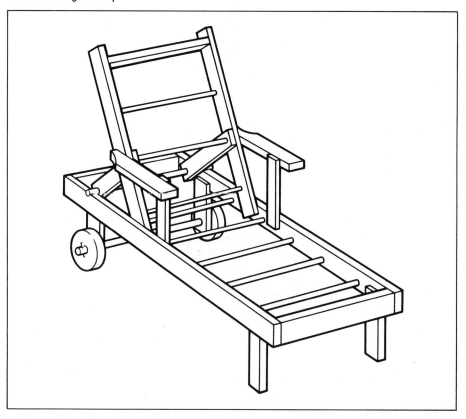

FIGURE D11

Gear puller concept.

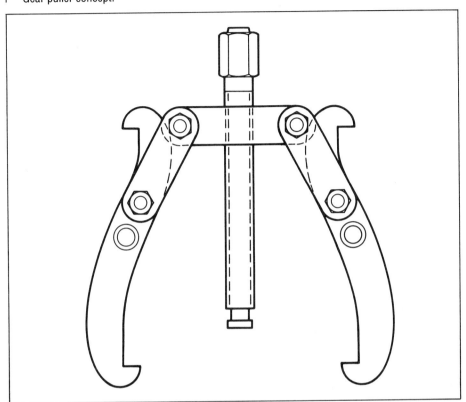

11. Figure D11 is an idea for a gear puller. The arms have two holes to allow for a larger range of gears. Complete the details of such a design. Consider a two-arm and a three-arm puller.
12. There are differences between gear pullers and ball and roller bearing pullers. Can the design in Exercise 11 be altered to be used for both? If yes, modify the design. If no, design an appropriate puller.
13. Figure D12 is an idea for modular retaining wall construction. Consider other ideas and complete the details of this design.

FIGURE D12

Retaining wall concept.

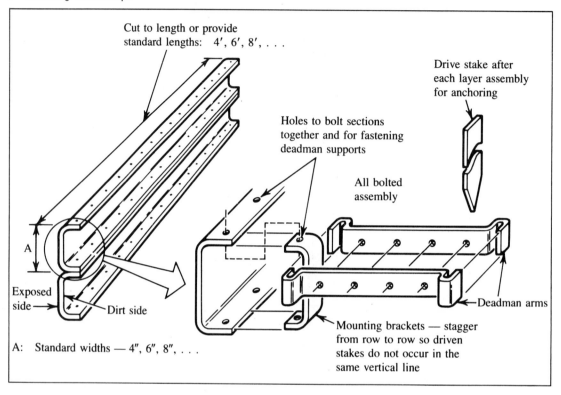

14. Complete the details of other ideas for a modular retaining wall design and compare it to the one in Problem 13.

15. Figure D13, dimensioned in inches, represents a conceptual design for a pin locator assembly. The purpose of the assembly is to locate the $\frac{1}{2}$ inch radius at five different locations $\frac{3}{4}$ inch apart. The slot is used to adjust an idler arm for five speed settings of a machine. Complete the design.

Group II—Preliminary Problem Solution Given

FIGURE D13

Machine adjuster concept.

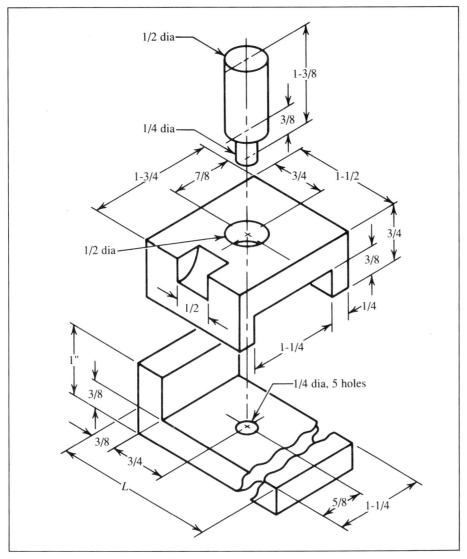

SECTION TWO

DETAILS OF REFINEMENT AND ANALYSIS

Refinement and analysis activities, which were summarized in Chapter 6, are the link between creative ideas and the practicalities of making an idea work as intended. The objective of refinement and analysis is to improve the ideas to create a functional product, and to make the product do its intended job in an efficient way. Poor ideas have to be discarded and good ideas must be improved, refined to work efficiently, and detailed to allow economical manufacture.

Development of the final design through refinement and analysis must be done to ensure proper function and to ensure the stated problem is solved. For some projects development activities may take years, particularly if they involve a change in the state of the art. Developments in medicine, hybrid corn, or a space station are examples of long-term projects. Many iterations of ideas, tests, redesigns, retests, and compromises are required. Test criteria and procedures must be established. Data taken must be studied and the results incorporated into redesign. There are so many details to be taken care of, and so much knowledge to be applied, that it is impossible to discuss a specific sequence of events for all projects. This is not to say a plan is not needed; indeed, without a plan the most noble of projects will flounder. Unknown and unplanned things do occur, and engineers have to deal with them. The more details considered, and the better use made of existing data, the better will be the chance for success and an optimum design.

Activities of the engineering design problem solving procedure have been discussed in Chapters 4 through 7; the specific steps for any project, however, will depend on the circumstances that exist at the time the project is being worked on. Always keep in mind such constraints as budget, market condition, time, and the state of the art. Chapters 8 and 9 go into more detail on many of the specific tasks to perform related to the concepts presented thus far.

As technology changes and theories and methods are altered, you too must change and alter your approach. To be a good engineer, you must be ready to change with a growing technological society. State of the art changes must be incorporated in new products as well as in product revisions. Lifelong learning is a necessity for engineers.

The following chapters discuss specific criteria to be considered during the solution of a design problem. The criteria discussed are themselves not design problems, but they include the details needed to judge different ideas and make good choices during the design process. Do not confuse the single answer exercises included in Chapters 8 and 9 with design problems. The exercises are intended to acquaint you with the concepts, and are necessary for practice and understanding. It is up to you to apply the concepts during the refinement and analysis steps of the design activity. The order of topics is not to suggest a specific sequence to be used for any given project. It may not be the sequence you will use on any project, but the points discussed are common to a variety of engineering projects. Even if some of the ideas are never used, just knowing that something might be considered increases your value to a project and to anyone you work for.

8 Engineering Models

"Modeling is the means we use to ignore what we cannot understand and to consider what we do understand."
✦ ROBERT F. STEIDEL AND JERALD M. HENDERSON

Models are an important and useful tool for the engineer and are used to analyze products, systems, and technical phenomena so the design process results in a safe, reliable product. Models include any visualization analysis tool that assists the clarification of solution ideas. The objective is to gain an understanding of the problem and of potential solutions so the good ideas can be improved, the poor ones rejected, and questions answered.

8.1 Types of Models

Models may look like a real product but not necessarily operate like one, such as a full size model of an automobile to show styling. Models may operate like a real product but not be the size of one, such as an airplane model in a wind tunnel. Models may look completely different from the real product but give precise data and information about the product, such as an electric analog simulation of a mechanical device, or the volume formula for determining how much oil will fit in a storage tank. The model, $V = \pi r^2 h$, doesn't look at all like a cylindrical storage tank, but it accurately predicts the volume based on a given radius and height; units, of course, must be consistent. A model may also be completely different from the product under test but behave the same. A mechanical system and an electrical system can be analyzed using the same differential equations; only the meanings of the variables change (see Table 8.1).

TABLE 8.1

System model analogies.

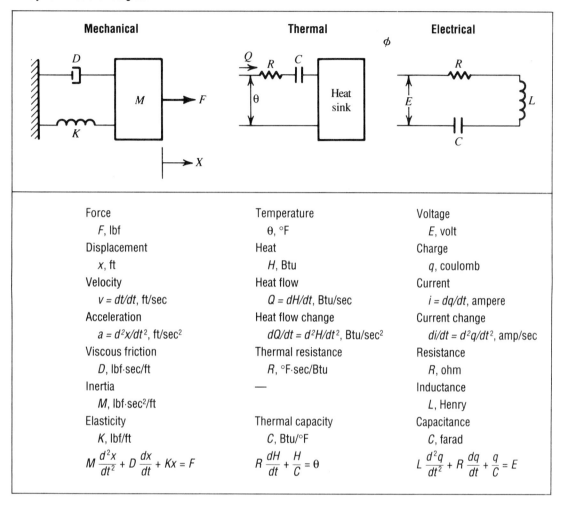

Mechanical	Thermal	Electrical
Force	Temperature	Voltage
F, lbf	θ, °F	E, volt
Displacement	Heat	Charge
x, ft	H, Btu	q, coulomb
Velocity	Heat flow	Current
$v = dt/dt$, ft/sec	$Q = dH/dt$, Btu/sec	$i = dq/dt$, ampere
Acceleration	Heat flow change	Current change
$a = d^2x/dt^2$, ft/sec^2	$dQ/dt = d^2H/dt^2$, Btu/sec^2	$di/dt = d^2q/dt^2$, amp/sec
Viscous friction	Thermal resistance	Resistance
D, lbf·sec/ft	R, °F·sec/Btu	R, ohm
Inertia	—	Inductance
M, lbf·sec^2/ft		L, Henry
Elasticity	Thermal capacity	Capacitance
K, lbf/ft	C, Btu/°F	C, farad
$M\dfrac{d^2x}{dt^2} + D\dfrac{dx}{dt} + Kx = F$	$R\dfrac{dH}{dt} + \dfrac{H}{C} = \theta$	$L\dfrac{d^2q}{dt^2} + R\dfrac{dq}{dt} + \dfrac{q}{C} = E$

An **iconic model** looks like the thing it is representing. It can be the same size, smaller, or larger. The model may show all of the features of the object being modeled or just some. The features chosen to show are the ones to be checked or emphasized for a presentation or test study. Figure 8.1 is an example of an iconic model.

An **analog model** behaves like the thing it represents in that it obeys the same laws of action. In some cases it may also look like the thing it is representing, like a model ship being tested for reaction to water turbulence. At other times it may not look like the thing at all, such as an electrical system that follows the same equations as a mechanical system as illustrated in Table 8.1. The equations shown in Table 8.1 are themselves models.

FIGURE 8.1

Three-dimensional non-functional model.

A **symbolic model** represents in an abstract way the thing it stands for. Symbolic models include mathematical formulas and graphic drawings, situations where one symbol arbitrarily stands for something else. Symbolic models are intended to condense principles to a form that can be manipulated and studied on paper or with a computer. The level of abstraction can be very high compared to the iconic and analog models. This is because the symbols used can be associated with many different physical problems. Following are four examples of symbolic models.

EXAMPLE 8.1:
Multi-use symbolic model.

Consider the symbolic model $A = bh/2$ and what the model can be used for.

One use is to calculate the area of a triangle, where b is the base and h is the altitude upon that base, where the units of each quantity must be the same. Another possibility is to calculate the net price of b number of items, selling at unit price h, which are on the half-price sale table. A third possibility exists if A is known instead of b and h. Then $bh = 2A$ represents the equation of a degenerate hyperbola (two straight lines).

Symbolic models are frequently used by engineers to assist in performing their job of design. A symbolic model can be applied whenever it matches a physical problem. Engineers are familiar with, and use, many symbolic models. Some symbolic models are exact models, while others are only approximations of the real thing.

EXAMPLE 8.2:
Exact symbolic model.

A common exact symbolic model is $A = \pi r^2$, used for finding the area of a circle when the radius is known. No one doubts the validity of the model, and it is used by engineers and others without hesitation, but π itself is inexact to any number of decimal places because it is an irrational number.

EXAMPLE 8.3:
Exact symbolic model.

The exact symbolic model for calculating the base of natural logarithms is an infinite series:

$$e = 1 + 1/1! + 1/2! + 1/3! + \cdots + 1/(n-1)! + \cdots$$

Any degree of precision can be obtained by evaluating successive terms of the series, but it is only an approximation (inexact) when a finite number of terms are used.

EXAMPLE 8.4:
Exact symbolic model.

The symbolic model $S_e = 0.50(S_{ut})$ is an exact statement used to estimate the fatigue strength of a steel, but it does not necessarily give exact results. It is an model to use in place of more accurate data that would take time and test to determine. Recall all the variations that can occur and affect the process of determining fatigue limits. Is the model less usable? No. Can the model be used any time for any situation? No. *You* will determine if the model is usable in a given situation. The ability to determine when a model is usable is one of the abilities of a good engineer.

The following case study is an example of how existing mathematical models can be used to help analyze and quantify raw data. This case is an example of an exact model used in an inexact situation.

CASE STUDY #7:
Tree Age Analysis.

To illustrate the use of exact models in inexact applications, consider the problem of determining the age of trees. To begin the analysis, an assumption about the growth pattern of trees is necessary. Trees grow faster in wet weather than dry, and in general have a fast growth period in the spring followed by a slow growth period the rest of the year. This alternating growth pattern causes a variation in wood density that makes rings visible when a tree is cut. Counting growth rings is the accepted model used to estimate the age of a tree. This method is just an estimate, however, because the effects of unusual climate conditions and/or other environmental conditions can cause unusual ring patterns. Variations in ring patterns include incomplete rings and multiple rings during a one year period.

Counting growth rings is a straightforward model to follow and is easy to apply, but a problem exists. How is the age of a tree that is still standing to be determined without cutting the tree down? Some enterprising person, perhaps an engineer, solved the problem by designing the Swedish Increment Borer. The Borer is a hand instrument that drills a hole about 0.1875 inches (4.76 mm) in diameter in a tree and removes a core sample. This core sample is then studied, and the rings counted. Carefully done the procedure causes no harm to the tree, and the core can be replaced and waxed over to prevent any infection This is a nice technique and uses the same modeling scheme, counting the rings. It has some difficulties, however. The boring tool must be aimed at the center of the tree to pick up all the growth rings, and it is not an easy task to drill by hand into a wet tree. There is also some potential for dropping the core and losing the ring count. Multiple samples from the same tree are required to verify the count and to average out irregularities in ring growth. On a cut tree multiple counts can be done with little additional effort. With a borer, additional time and effort are required to obtain multiple counts.

A second method to estimate the age of a tree also uses the growth ring principle and assumes that the rings are circles. In fact, some tree ring patterns are nearly perfect circles. If the circumference of the largest growth ring circle and the spacing between rings are known, an estimate of a tree's age can be made using the following analysis. Let D = the diameter of the largest ring, and C = the circumference of the largest ring, then $C = \pi D$. If d = the distance between rings, the age of the tree can be modeled as:

$$\text{Age} = D/2d \text{ or Age} = C/2\pi d \tag{8.1}$$

Either model can be used, depending on whether D or C is known. Using Age = $C/2\pi d$ will give more consistent results because when the circumference is measured, diameter variations are averaged out. If Age = $D/2d$ is used, a single measure of diameter may be an extreme value and give less accurate results. These models also assume a constant ring spacing and this is probably never true, ring spacing being dependent on annual growing conditions. An average value for d can be obtained from cut trees or boring samples. A large number of samples for values of d should be used to average out growth variations. The model, Age = $C/2\pi d$, is an exact one, but the physical situation being modeled is an inexact one. The engineer must always be prepared to make decisions about how a model will be used.

To use the model Age = $C/2\pi d$, two pieces of data are required, the circumference of the tree and the ring spacing. The circumference of a tree can easily be measured, but be sure to make a circumference reduction for bark thickness recalling that each half inch of bark thickness will change the circumference by π inches. The ring spacing, d, can be estimated from other trees in the area of the same type that have been cut, or with core samples. It is not surprising to find trees from different regions that have ring distances twice the spacing of trees of the same type from other regions. The ring spacing is a signature of the type of tree and how it was grown, a reflection of the species and the local growing conditions.

To use these tree-age models requires data collection. Many engineering design models fall into the same situation. Table 8.2 lists sample data collected from trees that have already been cut. The data is graphed in Figure 8.2 and a ring spacing grid overlaid on the graph. The straight line overlay assumes a 0.250 inch bark thickness. This figure illustrates the wide variation in ring spacing.

Using Age = $C/2\pi d$ will only be as valid as the accuracy of d, so a decision must be made on its value. This is another example of the necessity of making a decision and accepting the results. Always record the assumptions and decisions made. Careful data collection is also important, such as restricting measurements to one species of tree.

How old is a tree with a circumference of 40 inches? Using $d = 0.06$ inches, and Equation 8.1, the estimate of age is 106. If $d = 0.10$ inches, the age would be only 64 years.

Suppose a piece of wood has been processed through a mill and all that is visible is a portion of the ring pattern, and the center of the tree is not on the piece of wood. How can the age of the tree that the piece of wood came from be determined? An extension of the previous model is appropriate, where equally spaced circular rings was assumed.

The rings can be counted to get a minimum age, but without the center of the tree shown to obtain a total count; further analysis is needed. Recall from geometry that three points not on a straight line determine a circle. The approach used here is to locate three points on one ring and extend this to a full circle to obtain the circumference. One mathematical model that can be used for a circle is:

$$X^2 + AX + Y^2 + BY + C = 0. \qquad (8.2)$$

Whose radius is: $r = \sqrt{-C + (A/2)^2 + (B/2)^2}.$ \qquad (8.3)

This calculated r and an estimated value for d can then be used in the model Age = r/d. Substitute three pairs of (x, y) coordinates into Equation 8.2 and solve simultaneously for A, B, and C. Observe that the data points are only estimates of true values, but they are substituted into an exact symbolic model.

Using Equation 8.2 an estimate can be made of the age of the tree from which the board in Figure 8.3 was cut. Superimpose an X, Y coordinate system on a ring that is the longest and most regular, and identify three points on it. The farther apart the points are, the more accurate the calculations will be. Placing the origin of the coordinate system on the ring simplifies calculations because one ordered pair will be $(0, 0)$ and then $C = 0$. For this example the three points selected are $(x_1, y_1) = (0, 0)$, $(x_2, y_2) = (2.90, 0.46)$, and $(x_3, y_3) = (4.30, 0.06)$.

TABLE 8.2

Tree ring data from cut samples.

Circumference of tree including bark, inches	Number of growth rings counted
$18\frac{1}{2}$	17
$6\frac{1}{2}$	14
$7\frac{7}{8}$	11
$9\frac{7}{8}$	14
$6\frac{3}{4}$	7
$7\frac{1}{2}$	10
$16\frac{1}{8}$	15
$6\frac{7}{8}$	4
$10\frac{3}{8}$	9
$7\frac{1}{4}$	9
$7\frac{1}{2}$	10
$7\frac{1}{2}$	10
$7\frac{1}{4}$	7
$8\frac{3}{4}$	8
$8\frac{1}{4}$	6
$8\frac{1}{8}$	12
$7\frac{5}{8}$	5
$6\frac{1}{2}$	12
$25\frac{1}{4}$	30
$29\frac{7}{8}$	6
$22\frac{1}{8}$	41
$48\frac{1}{2}$	18
$49\frac{1}{2}$	26
6	8
$7\frac{3}{8}$	11
$9\frac{3}{4}$	14

Substitute the values into Equation 8.2 and solve for A, B, and C.

$$0^2 + 0(A) + 0^2 + 0(B) + C = 0 \tag{1}$$

$$2.90^2 + 2.90A + 0.46^2 + 0.46B + C = 0 \tag{2}$$

$$4.30^2 + 4.30A + 0.06^2 + 0.06B + C = 0 \tag{3}$$

From (1), $C = 0$, and (2) and (3) become:

$$8.62 + 2.90A + 0.46B = 0 \tag{2}$$

$$18.49 + 4.30A + 0.06B = 0 \tag{3}$$

which are satisfied when $A = -4.43$ and $B = 9.13$; so $r = \sqrt{0 + (-4.43/2)^2 + (9.13/2)^2}$ = 5.07. An estimate for d can be determined from the figure, $6d = 0.68$ and $d = 0.11$, and the tree's age is approximately equal to $5.07/0.11 = 46$ years.

FIGURE 8.2

Sample tree growth data.

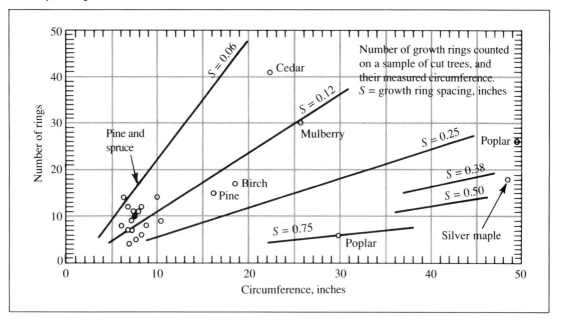

A mathematical model can be applied to a physical situation any time the assumptions and accuracy are within acceptable limits. One of the biggest errors an engineer can make is to believe an exact answer is obtained just because an exact model is used.

8.2 | Boundary Conditions

When a system is being modeled, the replacement system often has variations created by the assumptions made. Matching boundary conditions is one method to use for model-sample compatibility.

FIGURE 8.3

Sample tree section.

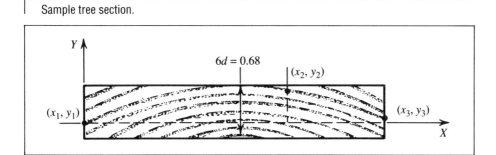

EXAMPLE 8.5:
Matching boundary conditions.

It is desired to model a uniformly loaded cantilevered beam, Figure 8.4(a), with a cantilevered beam with a single load at the free end, Figure 8.4(b). Possible model assumptions include:

1. Equal free end (maximum) deflection
2. Equal midpoint deflection
3. Equal bending moment at the support end, or
4. Equal free end slope angle.

This example will assume equal free end deflection. The free end deflection of a uniformly loaded beam $\delta_{max} = wl_u^4/8E_uI_u$, and for a singly loaded beam $\delta_{max} = Fl_s^3/3E_sI_s$, therefore $wl_u^4/8E_uI_u = Fl_s^3/3E_sI_s$, and $F = (3\,E_sI_swl_u^4)/(8E_uI_ul_s^3)$. If the material, length, and section moduli of the two beams are equal, then $F = 3wl/8$. If modeling of scale and material are also to be considered, then the ratios E_s/E_u, I_s/I_u, l_u^4/l_s^3 all become important in determining F.

8.3 Full-Scale Models

Some engineering projects do not lend themselves to the use of scale models. Scaling effects are often more difficult and costly to undertake than are the full-sized models. This is often the case when a component or assembly is used in a variety of ways, some of which are unknown.

Large machinery such as a bulldozer or motor grader are hard to test on a scale basis because it is difficult to scale the applied loads and component strengths. Testing a full-sized model is more reliable, and more information about the machine can be determined, including such items as performance specifications, noise levels, fits and clearances, operator comfort, alternate uses, and data for future production, such as manufacturing processes and assembly procedures. Certain specifications such as weight, digging depth, available horsepower, and traveling speeds can also be more

FIGURE 8.4

Schematic for Example 8.5.

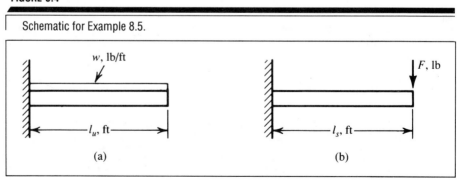

reliably determined from a full-size model. NASA sends full-sized unmanned spacecrafts into space to test rocket components and performance. Automobile engines are tested at full size, partly to be certified for exhaust emission standards and fuel economy requirements. Automobile companies crash test full-size models to determine actual standards. Scale models don't answer the questions with the necessary reliability. *Note*: Recent advances in computer simulation have reduced the need for some real product crash testing.

8.4 | Scale Models

Many products are tested at reduced size because of cost, time to build, size of test apparatus, and the data to be obtained from the model. Some products modeled to scale are dams and buildings. Part of the reason is that a full-sized model would be expensive and impractical to site. The models built in these cases are not generally used to test for product strength (see Figure 8.5). They are built to study the interrelationships of systems such as heating, electric, and plumbing, and to permit visualization by those who do not have the time or ability to study detail drawings. A scale model of a dam is easily understood, and this is particularly important when presenting a project for approval and financing. Three-dimension visualization is much easier to do with a model, even for people who can read drawings.

When a scale model is built for test, different concepts and procedures are required than if a model is being used just for conceptual analysis. Dimensional analysis plays a role in scale model testing. Dimensional analysis not only helps determine the test requirements such as size, loads, or fluid media, but it allows consolidation of experiments by controlling dimensional groups.

If a model is going to be scaled and used for visual analysis only, then any material can be used. Only the dimensions of the objects are of concern. If a product will be scaled in half, all linear dimensions are reduced by one-half. In some cases very small details are not built at all, but are painted or stenciled on the model. This is common for small details such as wire, nuts and bolts, nails, or floor tile layouts on models of plant layouts or houses.

If, however, a scale model is being tested for such things as pressure or heat transfer, care must be exercised. If a model is built to one-half size, then areas will be one-quarter size. The results obtained from mathematical models using area, such as Force = Pressure × AREA, Strength = Stress × AREA, or Heat Transfer = Rate × AREA, must be properly analyzed, as the next example illustrates.

EXAMPLE 8.6:
Scale model adjustments.

A model of a boat sail is built to one-quarter actual size. If the test force measured on the model is 50 lbf (222 N), what force would you expect on the full-size sail?

A model built to 1/4 size will have areas $(1/4)^2 = 1/16$ size, so the expected force on the full-sized sail is 16(50) = 800 lbf (3559 N).

FIGURE 8.5

Three-dimensional model showing the details of a continuous caster line.

If the volume of a scale model is of concern for testing, then mathematical models using such variables as weight, material cost, and buoyancy force need to be compensated for by volume ratios. Volume ratio analysis requires the cube of a linear dimension, as the next example illustrates.

EXAMPLE 8.7:
Scale model adjustment.

A one-tenth scale model of a spherical water tank is being built. If the full-size tank is intended to hold 10,000 gallons, how much will the model hold?

The volume of a sphere is $V = (4/3)\pi r^3$, where r is the radius. Therefore, the volume of the model is $(1/10)^3$ of the full-sized tank, or 1/1000 of 10,000 gallons = 10 gallons. This reduces the load of the water to 1/1000 of the actual load for the model.

8.5 Dimensional Analysis

More complicated modeling problems can benefit from dimensional analysis. Dimensional analysis is a method used to formulate a mathematical model that can then be used as a basis for theoretical analysis, or for experimental analysis. Dimensional analysis is used to create and analyze functional relationships based on the requirement that the Newtonian dimensions of length, L, time, T, and mass, M, must be the same if terms are to be added or subtracted; and the powers of the L, T, and M factors must be the same on both sides of an equation.

Checking dimensions on existing models is not difficult, as the following example illustrates.

EXAMPLE 8.8:
Dimension and unit verification.

The formula $A = h(b_1 + b_2)/2$ is used to find the area of a trapezoid, where b_1 and b_2 are the lengths of the parallel bases, and h is the perpendicular distance between the bases. The dimension of b_1 and b_2 must be the same because they are being added; and the dimension of A must be the same as the product of the dimension of h and $(b_1 + b_2)$. In this case the dimension of both sides of the equation is L^2. The constant 1/2 does not have a dimension.

Having equal dimensions does not guarantee that the units are correct, however. Consider the case where b_1 and b_2 are measured in inches and h is measured in feet. A conversion factor is required depending on the desired units for A. If the value of A is desired in square inches, then the formula must be adjusted to $A = 12h(b_1 + b_2)/2$; if the value of A is desired in square feet, then the formula must be adjusted to $A = (h/12)(b_1 + b_2)/2$. In both cases the constant 12 is dimensionless, but it does have units of in./ft.

Constants can exist in any one of three conditions:

1. They can be dimensionless and have units, such as material strain, which is dimensionless but has units of in./in.

2. They can be dimensionless and unitless such as e, the base of the natural logarithms, and

3. They can have dimensions and have units, such as the gravitational constant, which has dimensions of LT^{-2}, and units of ft/sec² or m/sec².

Dimensional analysis is also helpful when trying to establish new mathematical models to predict or explain physical phenomena. The basic Buckingham Theory, used to analyze dimensional relationships between variables, states that there exists (Π_1, Π_2,..., Π_n) dimensionless groups such that $F(\Pi_1, \Pi_2,..., \Pi_n) = 0$. One valuable result of the theory is that if a variable is forced to appear in only one of the Π_i groups, then it can be separated and solved for as a dependent variable.

Product functions are an attempt to use dimensional analysis to create a mathematical model which can be verified or rejected by experiment. The theorized model becomes a working model used to try to explain variable interrelationships, but there are no guarantees. The basic procedure used to obtain the Π groups is explained in the next example using a familiar physical model, that of a structural beam with a single load.

EXAMPLE 8.9:
Dimensional analysis of a beam

Step #1: Identify the variables that can have an effect on the problem. This is not always an easy thing to do, because sometimes the very thing that is making experimental results fluctuate will be the variable omitted from the analysis. On the other hand, the addition of a variable that is not related to the physical situation will only confuse the calculations and possible outcomes. Table 8.3 lists the relevant variables for the beam problem.

Step #2: Specify the dimensions of each of the variables; these have been included in Table 8.3. Refer to Appendix A for a list of commonly used variables and their units. Variables that are not listed in the appendix can generally be determined from combinations of entries.

Step #3: Assume a product and exponent function that relates all the variables, including a proportionality constant: $1 = kP^aX^bI^cE^d\delta^e$ (8.4)

This form requires that 1 be dimensionless. The value of k will eventually be determined from experiment, mathematical theory, and/or unit conversion if the variables are measured in mismatched units, such as feet and inches. The values of the exponents must now be selected so that each of the basic dimensions, $L, T,$ and M cancels and leaves a dimensionless equation.

Step #4: Establish a matrix showing the exponents and their respective units as shown in Table 8.4.

Step #5: Write linear equations with the exponents as variables and the matrix entries as coefficients.

For M: $a + 0(b) + 0(c) + d + 0(e) = 0$ (8.5)

For L: $a + b + 4c - d + e = 0$ (8.6)

For T: $-2a + 0(b) + 0(c) - 2d + 0(e) = 0$ (8.7)

Step #6: Solve the equations simultaneously. If there are more unknowns than equations, it is necessary to solve the equations parametrically. From Equation 8.5,

$a = -d$; Equation 8.7 is redundant and gives the same relationship. Substituting $a = -d$ into Equation 8.6 gives $b = -4c + 2d - e$. Substitute the values for a and b into Equation 8.4 and $1 = k(P^{-d})(X^{-4c+2d-e})(I^c)(E^d)(\delta^e)$. (8.8)
Group the bases with each of the exponents, and $1 = k(I/X^4)^c (X^2E/P)^d (\delta/X)^e$. (8.9)
This results in three dimensionless groups: $\Pi_1 = I/X^4$, $\Pi_2 = X^2E/P$, $\Pi_3 = \delta/X$. Alternate groups can be created by multiplying or dividing these groups together. For example, Π_2 can be replaced with $\Pi_2 \times \Pi_3 = XE\delta/P$.

By assigning arbitrary values to the exponents in Equation 8.9 various relationships can be obtained, some helpful, some not. For example, if $e = d = 0$ and $c = 1$, then $I = X^4/k$, not a very meaningful relationship because no functional constant k can be found that relates cross sectional moment of inertia with beam length. In general the exponents should not be set equal to zero because critical variables may disappear from the analysis. If $e = d = a = 1$, then $\delta = (1/k)(PX^3/EI)$ and this should look familiar. The value of k can be determined from experiment by varying the value of PX^3/EI and measuring δ; PX^3/EI is treated as a single variable. The values of P, X, E, and I can vary individually as long as the value of PX^3/EI is held constant. If several physical models are made from the same material, then E is constant and the variable becomes PX^3/I. For a simply supported center loaded beam the value of k turns out to be 48 when X and δ are in inches, I is in^4, and E is lb/in^2. For an end loaded cantilever beam, $k = 3$.

For consistent experimental results, relating scale model results to the full size object, the value of each Π group should be the same for the full-size real product (p) and the scale model (s), that is: $I_p/X_p^4 = I_s/X_s^4$, $X_p^2E_p/P_p = X_s^2E_s/P_s$, and $\delta_p/X_p = \delta_s/X_s$.

TABLE 8.3

Variables and dimensions for a beam analysis.

Variables	Dimensions
P—Load	ML/T^2
X—Beam length	L
I—Cross section moment of inertia	L^4
E—Material modulus of elasticity	M/LT^2
δ—Deformation	L

TABLE 8.4

Beam variables and exponent coefficients.

Variable	P	X	I	E	δ
Exponent	a	b	c	d	e
M (Mass)	1	0	0	1	0
L (Length)	1	1	4	−1	1
T (Time)	−2	0	0	−2	0

> **EXAMPLE 8.10:**
>
> ## Use of a dimensional group from Example 8.9.
>
> If a full size beam has a length of 10 meters and a section moment of 8 mm^4, a model beam 2 meters long should have a section moment = $(2^4)(8)/10^4$ = 0.0128 mm^4. The value of 0.0128 does not determine a specific shape.
>
> A circular cross section, $I = \pi D^4/64$, will have $D = [(64)(0.0128)/\pi]^{1/4}$ = 0.71 mm.
>
> A square cross section, $I = S^4/12$, will have $S = [(12)(0.0128)]^{1/4}$ = 0.63 mm.

The following case study further illustrates the dimensional analysis procedure and some of the problems encountered when it is used.

CASE STUDY #8:

Wind Force Modeling.

Consider the problem of trying to determine a mathematical model to quantify the force created on a flat plate exposed to the wind or other moving fluid.

Trial #1: Assume the related variables are A, the flat plate area, V, the fluid velocity, and F, the force on the plate, as listed in Table 8.5.

Assume a product function $1 = kF^a A^b V^c$, which implies $(MLT^{-2})^a (L^2)^b (LT^{-1})^c = M^0 L^0 T^0$. The exponent of each variable, (M, L, or T), must be the same on both sides of the equation, so:

for M: $a = 0$

for L: $a + 2b + c = 0$

for T: $-2a - c = 0$

Solve these three equations simultaneously and $a = b = c = 0$, so no function exists. This means the correct variables were not selected for the problem, if a product function is to be used.

Trial #2: Include another variable, ρ, the density of the air, and consider the new list of variables shown in Table 8.6.

Assume a product function $1 = kF^a A^b V^c \rho^d$, which implies that $(MLT^{-2})^a (L^2)^b (LT^{-1})^c (ML^{-3})^d = M^0 L^0 T^0$. Analyze exponents of the variables M, L, and T:

for M: $a + d = 0$

for L: $a + 2b + c - 3d = 0$

for T: $-2a - c = 0$

Solve these three equations simultaneously and $d = -a$, $c = -2a$, and $b = -a$, so

$1 = kF^a A^b V^c \rho^d$ becomes $1 = kF^a A^{-a} V^{-2a} \rho^{-a}$

If $a = 1$, then $F = A\rho V^2/k = C_1 A \rho V^2$, where C_1 is a constant. If this model is tested experimentally, using the units of measure listed in Table 8.6, and the flat

TABLE 8.5

Variables and dimensions for Case Study #8, trial 1.

Variable	Dimensions
F—Force, Newtons	MLT^{-2}
A—Area, m²	L^2
V—Wind velocity, m/sec	LT^{-1}

TABLE 8.6

Variables and dimensions for Case Study #8, trial 2.

Variable	Dimensions
F—Force, Newtons	MLT^{-2}
A—Area, m²	L^2
V—Wind velocity, m/sec	LT^{-1}
ρ—Air density, kg/m³	ML^{-3}

plate is perpendicular to the wind, then C_1 turns out to be 1/2.

Trial #3: Experimentation shows that if the plate is not oriented perpendicular to the wind, C_1 is not 1/2, but some other value, and if the angle with the wind is not held constant, then C_1 is not constant, and another variable should be considered.

The angle θ that the plate makes with the wind has no dimensions, and if included in the procedure of trial #2, will drop out of the analysis. Experimental data indicates that the angle does affect the force, and so it must be included. At this point the more generalized theory must be employed and θ is included as an additional Π group. Because the form of the additional group is not known, it must be included as a function, $h(\theta)$. The model for the force, extended from trial #2, now becomes $F = C_2 A \rho V^2 h(\theta)$. The constant will likely be a different value so subscripts are used to differentiate them from model to model. The function $h(\theta)$ must be estimated from experimental data, from mathematical theory, or intuition; is $h(\theta) = \cos \theta$ a good choice? Consider the area of the plate projected perpendicular to the wind as shown in Figure 8.6.

Trial #4: Consider the possibility that fluid viscosity might influence the results. Table 8.7 shows the listing of variables and dimensions necessary to include viscosity in the analysis.

Assume a product function $1 = kF^a A^b V^c \rho^d \mu^e$, which implies that $(MLT^{-2})^a (L^2)^b (LT^{-1})^c (ML^{-3})^d (ML^{-1}T^{-1})^e = M^0 L^0 T^0$. Analyze the exponents of M, L, and T, and

$$\text{for } M: a + d + e = 0 \quad (1)$$
$$\text{for } L: a + 2b + c - 3d - e = 0 \quad (2)$$
$$\text{for } T: -2a - c - e = 0 \quad (3)$$

FIGURE 8.6

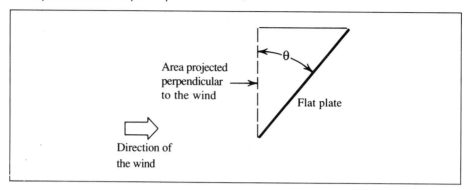

Projected area of a flat plate exposed to the wind.

TABLE 8.7

Variables and dimensions for Case Study #8, trial 5.

Variable	Dimensions
F—Force, Newtons	MLT^{-2}
A—Area, m²	L^2
V—Wind velocity, m/sec	LT^{-1}
ρ—Air density, kg/m³	ML^{-3}
μ—Air viscosity, N·sec/m²	$ML^{-1}T^{-1}$

From (1), $a = -d - e$, substitute in (3), $2d + 2e - c - e = 0$, and $c = 2d + e$. Substitute a and c in (2), and $-d - e + 2b + 2d + e - 3d - e = 0$ and $b = (e + 2d)/2 = e/2 + d$.

$1 = kF^a A^b V^c \rho^d \mu^e$ becomes $1 = kF^{-d-e} A^{e/2+d} V^{2d+e} \rho^d \mu^e = k(\rho A V^2/F)^d (V\mu(\sqrt{A})/F)^e$, so

$\Pi_1 = \rho A V^2/F$ and $\Pi_2 = V\mu\sqrt{A}/F$.

The problem now is that F occurs in both groups and cannot be separated unless d or e are set equal to zero. Setting an exponent equal to zero is risky because an important variable can be lost. The problem of separating F can be resolved by creating a new Π group to replace an existing group using a combination of existing groups. Combinations of groups are permissible because parameters other than d and e could have been selected for the final form, and other Π groups derived. Replace Π_1 with $\Pi_1' = \Pi_1/\Pi_2 = \rho V \sqrt{A}/\mu$, and $1 = k(\rho V \sqrt{A}/\mu)^d (V\mu(\sqrt{A})/F)^e$. Now let $d = e = 1$ and solve for F: $F = C_3 \rho V^2 A$, and viscosity drops out. If viscosity is still deemed an important factor, then it must be included as a separate product function $g(\mu)$ as was done with angle variation, and $F = C_3 \rho V^2 A \, g(\mu) h(\theta)$.

Trial #5: Suppose experimental observation still shows a variation of the force. What other variables might be included? How about temperature? If temperature is included as a variable, an additional product function, $k(t)$, must be included because temperature, whether measured in °C or °F, has no units of M, L, or T, and would not be part of other Π_i groups. The model becomes $F = C_4 A \rho V^2 h(\theta) k(t)$.

If only one of the variables is associated with the temperature changes, such as density, then that variable could be made the temperature function. If that is done, the model becomes $F = C_5 A \rho(t) V^2 h(\theta)$.

Another variation that has not been considered is geometry, because the assumption was that the exposed area is flat. If the projected area perpendicular to the direction of the wind is held constant, but its shape changed, the force will vary. The problem is how to incorporate a geometry change into the model. Should A become A(geometry)? The solution, as students of fluid mechanics are aware, is to use a factor C_D, called the drag coefficient, which is a function of plate geometry, fluid viscosity and velocity, and surface smoothness. The final model becomes $F = [C_D A_p \rho(t) V^2]/2$, where A_p is the area projected perpendicular to the wind.

Some dimensionless groups common to particular fields of engineering are used regularly without needing to be derived for each problem. Table 8.8 lists some of the groups used frequently in fluid mechanics. Most of the groups were derived to permit the running of valid scale model experiments.

TABLE 8.8

Common dimensionless groups used in fluid mechanics.

Group Name	Force Ratio	Quantity Ratio
Reynolds number	Inertia/viscosity	$\rho VD/\mu$
Pressure coefficient	Pressure/inertia	$\Delta p/(\rho V^2)/2$
Froude number	Inertia/gravity	$V\sqrt{gL}$
Weber number	Inertia/surface tension	$V\sqrt{\rho L/\sigma}$
Drag coefficient	Drag/inertia	$D/(\rho V^2 D^2)/2$
Lift coefficient	Lift/inertia	$L/(\rho V^2 D^2)/2$
Mach number	Inertia/inertia	$V\sqrt{K\rho}$
Thrust coefficient	Thrust/inertia	$F_t/(\rho V^2 D^2)/2$
g	Gravity conversion factor	
μ	Absolute fluid viscosity	
σ	Fluid surface tension	
V	Fluid velocity	
D or L	Critical dimension in direction of fluid flow	
ρ	Fluid density	
Δp	System pressure variation	
K	Fluid compressibility	

The Reynolds number, for example, is used commonly in fluid flow analysis. It is symbolized by the dimensionless quantity $\rho VD/\mu$, where ρ is the fluid density, V is the fluid velocity, D is the diameter of the pipe in which the fluid is flowing or the critical dimension of an object exposed to the flowing fluid, and μ is the absolute viscosity of the fluid. If the value of the calculated Reynolds number for a pipe flow model is less than 2700–4000, the flow is laminar; if the value is above that range, the flow is turbulent. The transition region is typical of physical systems that have surface roughness variation. Two systems with the same Reynolds number are said to be dynamically similar.

The Mach number is another common reference group ratio, usually used in relating the speed of an airplane or other missile to the speed of sound. Mach 2 is a speed twice the speed of sound. *Note*: The speed of sound is not constant, but usually taken as 1120 feet/sec (321.4 m/sec) at sea level for average atmospheric conditions.

8.6 Experimental Data Analysis

When experiments are performed engineers anticipate that some consistent relationship will exist between input variable change and output variable change. If this anticipated relationship follows certain criteria, an empirical equation relating the variables can be established. The most common relationships include the following models, where X is the independent variable and Y is the dependent variable.

The model of a straight line on **linear coordinate graph paper** is:

$Y = mX + b$, where (8.10)

   m = the slope of the line, and

   b = the Y-axis intercept, the value of Y when $X = 0$.

The models for straight lines on **semilog coordinate graph paper** are:

$Y = B(10)^{m_{10}X}$, $Y = B(e)^{m_e X}$, or $Y = BM^X$, where (8.11)

   m_{10} is the slope of the line using common logarithms,

   m_e is the slope of the line using natural logarithms, $M = (10)^{m_{10}} = (e)^{m_e}$, and

   B = the Y-axis intercept, the value of Y when $X = 0$.

Often, to obtain a straight line on semilog paper, it is necessary to interchange the axis of the variables, and two choices are possible, either $\log Y$ versus X, or $\log X$ versus Y.

The model for a straight line on **log-log coordinate graph paper** is:

$Y = B(X)^M$, where (8.12)

   M = the slope of the line, and

   B = the Y-axis intercept, the value of Y when $X = 1$.

8.6.1 Least Squares Analysis

A mathematical method for determining the values of the parameters for slope and intercept is based on the method of least squares analysis. This method seeks a function such that the sum of the squares of the residuals is a minimum. The generalized method of least squares analysis is as follows.

Start with a function that is a linear combination of common functions that has a chance of closely fitting the experimental data. Any function can be used, but you have to decide which may be best. Some examples include:

$Y = c$

$Y = aX^3 + bX^2 + cX + d$

$Y = a \sin X$

$Y = a \cos X + b \exp(X^2)$

Write an expression for the sum of the squares of the residuals. A residual is the difference between the value of Y as calculated from the function and the experimental value of Y. If the function chosen in step 1 is $Y = mX + b$, a straight line on linear coordinate graph paper, then the sum of the residuals $R = \Sigma[(mx_i + b) - y_i]^2$, where (x_i, y_i) are the data for $i = 1, 2, \ldots, n$ number of data pairs.

Make the value of R as small as possible by taking first partial derivatives and setting them equal to zero. In this example the unknowns are m and b, so:

$$\partial R/\partial m = 2\Sigma(mx_i + b - y_i)x_i = 0 \quad \text{and} \quad \partial R/\partial b = 2\Sigma(mx_i + b - y_i) = 0.$$

Solve the differential equations simultaneously: $\partial R/\partial m$ becomes $m\Sigma x_i^2 + b\Sigma x_i = \Sigma x_i y_i$, and $\partial R/\partial b$ becomes $m\Sigma x_i + nb = \Sigma y_i$, so

$$m = \frac{n\Sigma(x_i y_i) - \Sigma x_i \Sigma y_i}{n\Sigma x_i^2 - (\Sigma x_i)^2}, \quad \text{and} \tag{8.13}$$

$$b = \frac{\Sigma y_i - m\Sigma x_i}{n} = \frac{\Sigma x_i^2 \Sigma y_i - \Sigma x_i \Sigma(x_i y_i)}{n\Sigma x_i^2 - (\Sigma x_i)^2} \tag{8.14}$$

The values of M and B for a straight line on semilog paper are found by modifying Equations 8.13 and 8.14 as follows.

$$m_{10} = \log M = \frac{n\Sigma(x_i \log y_i) - \Sigma x_i \Sigma \log y_i}{n\Sigma x_i^2 - (\Sigma x_i)^2} \tag{8.15}$$

$$\log B = \frac{\Sigma x_i^2 \Sigma \log y_i - \Sigma x_i \Sigma(x_i \log y_i)}{n\Sigma x_i^2 - (\Sigma x_i)^2} \tag{8.16}$$

The values of M and B for a straight line on log-log paper are found by modifying Equations 8.13 and 8.14 as follows.

$$M = \frac{n\Sigma(\log x_i \log y_i) - \Sigma \log x_i \Sigma \log y_i}{n\Sigma(\log x_i)^2 - (\Sigma \log x_i)^2} \tag{8.17}$$

$$B = \frac{\Sigma(\log x_i)^2 \Sigma \log y_i - \Sigma \log x_i \Sigma(\log x_i \log y_i)}{n\Sigma(\log x_i)^2 - (\Sigma \log x_i)^2} \tag{8.18}$$

8.6.2 Graphical Methods

An alternate method to the analytical method can be used if the data is plotted on appropriate graph paper and a "best" straight line is drawn through the points. "Best" is best in your judgment, so keep the line as close to as many points as possible.

On linear coordinate graph paper the value of b can be read directly from the graph where the line crosses the Y-axis. The value of m can be found in two ways:

1. Measure Δx and Δy with a scale, adjust for any differences in X-axis and Y-axis divisions, and divide: $m = \Delta y/\Delta x$; or
2. Select two points on the straight line drawn, (x_a, y_a) and (x_b, y_b), and use the slope equation $m = (y_b - y_a)/(x_b - x_a)$.

The graphical solution for a semilog "best" straight line is similar to the method for a linear coordinate graph. Draw a straight line through the data keeping the line as close to as many data points as possible. The intercept B is read from the graph where the line drawn crosses the Y-axis. The slope is determined by selecting two data points (x_a, y_a) and (x_b, y_b) on the line and calculating M. Do not measure Δx and Δy using a scale and calculate $\Delta y/\Delta x$ for slope, as this does not work for semilog paper.

$$\log M = \frac{\log y_b - \log y_a}{x_b - x_a}, \quad \text{or} \quad \ln M = \frac{\ln y_b - \ln y_a}{x_b - x_a} \tag{8.19}$$

The graphical solution for log-log paper locates B where the line drawn crosses the line $Y = 1$. The value of M can be found in two ways.

1. Select two data points on the line drawn and use the equation:

$$M = \frac{\log y_b - \log y_a}{\log x_b - \log x_a}, \text{ or}$$

2. Measure Δy and Δx with a scale and calculate $M = \Delta y/\Delta x$.

Measuring Δy and Δx is permissible as long as the length of one log cycle on both axes is the same. Commercial log-log paper is printed this way.

8.6.3 Additional Adjustments

On semilog graph paper, if the data plots close to a straight line but is still slightly curved, an adjustment can be attempted to obtain a better straight line; Equation 8.11 becomes:

$$Y - c = BM^X \tag{8.20}$$

The value of c is estimated as follows.

1. Select two points, (x_a, y_a) and (x_b, y_b), from the *curve* on semilog paper that are closest to the data. These points may be the first and last data pairs because points farther apart will improve the accuracy of the calculation.
2. Calculate $x_3 = (x_a x_b)/2$. (There's that $ab/2$ model again!)
3. Read y_3 from the *curve* at the value of x_3.
4. Calculate $c = \dfrac{y_a y_b - y_3^2}{y_a + y_b - 2y_3}$.

On log-log graph paper, if the data plot is close to a straight line but is still slightly curved, an adjustment can be attempted to obtain a better straight line; Equation 8.12 becomes:

$$Y - c = BX^M \tag{8.21}$$

The value of c is estimated as follows.

1. Select two points, (x_a, y_a) and (x_b, y_b) from the *curve* on log-log paper that comes closest to the data.
2. Calculate $x_3 = \sqrt{x_a x_b}$.
3. Read y_3 from the curve at x_3.
4. Calculate $c = \dfrac{y_a y_b - y_3^2}{y_a + y_b - 2y_3}$.

An alternate method for determining c for the log-log model is made by estimating the Y-axis intercept on the *linear coordinate plot* of the data.

For both the semilog and the log-log model, the value of c is negative if the original curve is concave down and positive if the original curve is concave up. Keep in mind that the value of c is an estimate. A better value for c, and a better "best" straight line, may be possible by trying other values of c in the neighborhood of the estimated value. An iterative computer program would be a good tool to use to improve the results. To verify that some other value for c is "best," calculate data residuals relative to the empirical equation and/or a coefficient of determination, and/or a percent error at each data point.

8.6.4 Polynomial Model

If none of the preceding models give a satisfactory fit to the experimental data, it may be worthwhile to try to fit data to the basic polynomial equation:

$$Y = a_n X^n + a_{n-1} X^{n-1} + \cdots + a_1 X + a_0.$$

The number of data pairs must be one or more larger than the degree of the equation model sought. If there are 10 pairs of data, polynomials up to degree 9 are possible. The higher degree models are rarely used. This is not because of the calculation required, for computers have simplified that step for us. The main reason is that if the degree must be high to get a good fit, the "best" fit is probably being forced, and the data are really not statistically related at all. Additionally, a polynomial model of high degree can deviate significantly from a smooth curve representative of the data between data values and if the data are extrapolated. See reference (3) for additional discussion on the errors of polynomial fits.

Using the least squares procedure from Section 8.6.1, the equations to use for n pairs of data to estimate a second degree polynomial, $Y = aX^2 + bX + c$, are:

$$a \Sigma x_i^2 + b \Sigma x_i + nc = \Sigma y_i \tag{8.22}$$
$$a \Sigma x_i^3 + b \Sigma x_i^2 + c \Sigma x_i = \Sigma x_i y_i$$
$$a \Sigma x_i^4 + b \Sigma x_i^3 + c \Sigma x_i^2 = \Sigma x_i^2 y_i, \text{ where } i = 1, 2, 3, \ldots, n$$

These equations are solved simultaneously for a, b, and c. This procedure minimizes the sum of the squares of the vertical residuals from each data point to the parabola determined.

A polynomial equation of degree $\leq n$ can be found that passes through n pairs of data using techniques such as the Newton interpolating polynomial, but there is no control of what happens to the equation between data points or outside the range of the data.

8.6.5 Coefficient of Correlation

As previously mentioned, following a precise model does not guarantee meaningful results. Models obtained using least squares analysis are no different. One indicator that helps judge the validity of the results of linear regression analysis is the coefficient of determination, r^2, or the coefficient of correlation, r. If the least squares analysis yields $y' = mx + b$, and $\mathbf{y}$ is the mean of the dependent data, then the coefficient of determination is defined as follows.

$$r^2 = \frac{\Sigma(y_i' - \mathbf{y})^2}{\Sigma(y_i - \mathbf{y})^2} = \frac{\Sigma(mx_i + b - \mathbf{y})^2}{\Sigma(y_i - \mathbf{y})^2}, \text{ where } i = 1, 2, 3, \ldots, n \tag{8.23}$$

The value of r^2 is the part of the data variation explained by the least square model; an r^2 value of 0.85 means that the least squares equation explains or accounts for 85 percent of the variation of the data. An r^2 value of 1 means that there is a perfect match between the variables in an increasing relationship (positive slope). An r^2 value of -1 represents a perfect match with a decreasing relationship (negative slope). The ability of r to determine the correlation depends on the number of data pairs used. As the number of data pairs increases, a lower value of r predicts the same correlation reliability. Appendix K is a tabulation of coefficients of correlation for two variables. If the calculated value of r is larger than the table entry, then correlation is probable to the significance level indicated. The coefficient of correlation r can be calculated for any empirical equation.

Alternate forms for calculating r^2 are available; three are given here for analysis of a straight line on linear coordinate paper.

$$r^2 = \frac{(n\Sigma x_i y_i - \Sigma x_i \Sigma y_i)^2}{[n\Sigma x_i^2 - (\Sigma x_i)^2][n\Sigma y_i^2 - (\Sigma y_i)^2]} = \frac{[\Sigma(x_i - \mathbf{x})(y_i - \mathbf{y})]^2}{\Sigma(x_i - \mathbf{x})^2 \Sigma(y_i - \mathbf{y})^2} \tag{8.24}$$

$$r^2 = \frac{(\Sigma x_i y_i - n\mathbf{x}\mathbf{y})^2}{[\Sigma x_i^2 - n\mathbf{x}^2][\Sigma y_i^2 - n\mathbf{y}^2]}$$

For semilog and log-log analysis make the same substitutions made in Equations 8.13 and 8.14 to obtain Equations 8.15 and 8.16, and Equations 8.17 and 8.18. The next example illustrates the empirical equation method.

EXAMPLE 8.11:

Linear coordinate analysis.

Consider the following data pairs taken from a test on a coil tension spring within the elastic limit of the material. The data = (load on spring, N; spring length, mm): (0, 1.45), (1, 2.10), (2, 2.45), (3, 3.03), (4, 3.38), and (5, 4.09).

The data plots reasonably straight on linear coordinate graph paper (see Figure 8.7). The form $Y = mX + b$ is appropriate, where X represents the load and Y represents the spring length. The "best" straight line has been drawn on the graph. There is good reason to draw the line through the point $(0, 1.45)$. This is the free length of the spring and the spring would have been manufactured to this length. A more complete test would check this length after the test to be sure that the test did not exceed the elastic limit of the material, and to see if the original spring length has been influenced by coating materials or temperature variation. Variations from linearity occur because the material is less than perfect in reacting to a load, and measuring instruments are less than perfect in recording the true length of the spring. The slope is calculated by using two points on the line drawn,

$$m = \frac{3.75 - 1.45}{4.45 - 0} = 0.52 \text{ mm/N, and } Y = 0.52X + 1.45.$$

A purely mathematical solution, using Equations 8.13 and 8.14, can also be determined. It is convenient to tabulate the data for ease and accuracy of calculations (see Table 8.9).

$$b = \frac{55(16.5) - 15(50.06)}{6(55) - (15)^2} = 1.49 \qquad m = \frac{6(50.06) - 15(16.5)}{6(55) - (15)^2} = 0.50$$

$Y = 0.50X + 1.49$, and the coefficient of determination is:

$$r^2 = \frac{[6(50.06) - 15(16.5)]^2}{[6(55) - 15^2][6(49.84) - 16.5^2]} = 0.993, \quad \text{and } r = 0.997$$

FIGURE 8.7

Graph for Example 8.11.

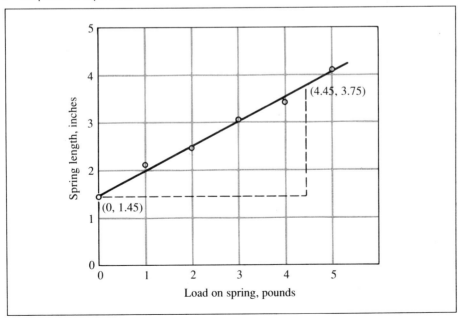

TABLE 8.9

Calculations for linear coordinate least squares fit, Example 8.11.

	x	y	x²	xy	y²
	0	1.45	0	0	2.10
	1	2.10	1	2.10	4.41
	2	2.45	4	4.90	6.00
	3	3.03	9	9.09	9.18
	4	3.38	16	13.52	11.42
	5	4.09	25	20.45	16.73
Totals	15	16.50	55	50.06	49.84

It is not surprising that the mathematical slope and intercept differ from the graphical values because the graph line was forced through (0, 1.45). Your judgment must decide which is "best."

For a problem with six data pairs, Appendix K shows a value of $r = 0.9172$ for a 99 percent confidence level. Remember to enter the table at $(n - 2)$. If the calculated value of r is greater than 0.9172 there is less than 1 chance in 100 that no real correlation between the variables exists. This example has good correlation because $0.997 > 0.9172$.

The following case study illustrates how analysis refinement and optimization can be performed using the empirical formula method.

CASE STUDY #9:
Empirical formula refinement.

Consider the following data representing the amount of energy consumed in certain years.

Year	1	11	16	21	26	27	28	29	30	31
Energy, Btu's × 10^{12}	24	34	40	45	54	57	59	62	66	69

The data plotted on a linear coordinate graph, Figure 8.8, does not show a good straight line and would ordinarily be passed over for something better. But let's see what a mathematical analysis shows using the least squares models. Table 8.10 contains the preliminary data calculations required.

$$m = \frac{10(12{,}495) - 220(510)}{10(5710) - 220^2} = 1.47 \qquad b = \frac{510 - m(220)}{10} = 18.76$$

$$Y = 1.47X + 18.76,$$

$$r^2 = \frac{[10(12495) - 220(510)]^2}{[10(5710) - 220^2][10(27964) - 510^2]} = 0.956 \text{ and } r = 0.978$$

FIGURE 8.8

Linear coordinate graph for Case Study #9.

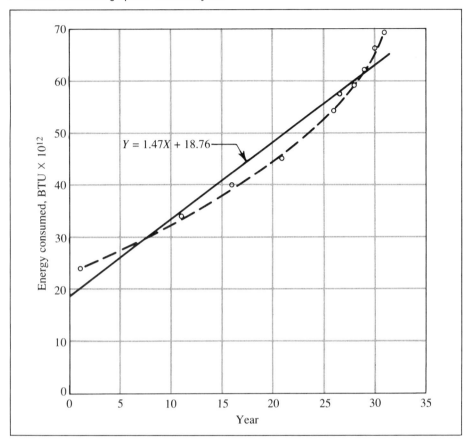

TABLE 8.10

Calculations for linear coordinate least squares fit, Case Study #9.

	x	y	x^2	xy	y^2
	1	24	1	24	576
	11	34	121	374	1156
	16	40	256	640	1600
	21	45	441	945	2025
	26	54	676	1404	2916
	27	57	729	1539	3249
	28	59	784	1652	3481
	29	62	841	1798	3844
	30	66	900	1980	4356
	31	69	961	2139	4761
Totals	220	510	5710	12495	27964

The coefficient of correlation might suggest that the equation $Y = 1.47X + 18.76$ is a good one. There is, however, a residual at each point as shown in Table 8.11. *Note*: Even though the sum of the residuals is small because of + and − values, the sum of the squares of the residuals is quite large. Observe in Figure 8.8 that the line that comes closest to all the data points is curved, and the end points lie on one side of the best straight line while the middle points lie on the other side. Can this "answer" be improved?

Consider the same data plotted on semilog paper as shown in Figure 8.9, and the related calculations in Table 8.11.

$$\log M = \frac{10(384.18) - 220(16.87)}{10(5710) - 220^2} = 0.0150, m_{10} = 1.035$$

$$\log B = \frac{5710(16.87) - 220(384.18)}{10(5710) - 220^2} = 1.3573, B = 22.76, \text{ and}$$

$$Y = 22.76(1.035)^X$$

The value of 22.76 agrees well with the graphical answer of 23.0.

$$r^2 = \frac{[10(384.18) - 220(16.87)]^2}{[10(5710) - 220^2][10(28.67) - 16.87^2]} = 0.929, \text{ and } r = 0.964$$

FIGURE 8.9

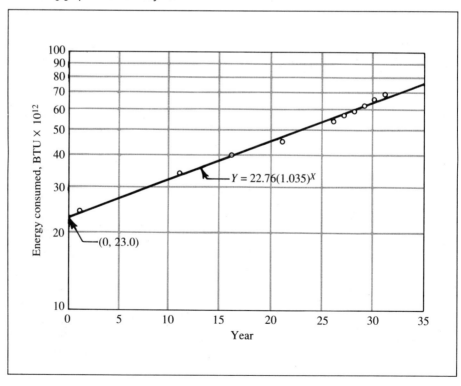

Semilog graph for Case Study #9.

TABLE 8.11

Residual calculations, linear coordinate least squares fit, Case Study #9.

x	y	$y_i = 1.47x + 18.76$	$\Delta y = y_i - y$	$(\Delta y)^2$
1	24	20.23	− 3.77	14.21
11	34	34.93	0.93	0.86
16	40	42.28	2.28	5.20
21	45	49.63	4.63	21.44
26	54	56.98	2.98	8.88
27	57	58.45	1.45	2.10
28	59	59.92	0.92	0.85
29	62	61.39	− 0.61	0.37
30	66	62.86	− 3.14	9.86
31	69	64.33	− 4.67	21.81
Totals			1.00	85.58

TABLE 8.12

Calculations for semilog least squares fit, Case Study #9.

	x	y	log y*	x log y*	(log y)²*
	1	24	1.38	1.38	1.90
	11	34	1.53	16.85	2.35
	16	40	1.60	25.63	2.57
	21	45	1.65	34.72	2.73
	26	54	1.73	45.04	3.00
	27	57	1.76	47.41	3.08
	28	59	1.77	49.58	3.14
	29	62	1.79	51.98	3.21
	30	66	1.82	54.59	3.31
	31	69	1.84	57.00	3.38
Totals		220	16.87	384.18	28.67

* Values rounded from 7 digit calculations.

Table 8.13 summarizes the residuals and their squares. Recall that the theory requires that the sum of the squares of the residuals is a minimum; 21.02 is as small as the sum of the residuals can be with the semilog model, which is approximately one-fourth the amount determined using the linear coordinate model. The data in Table 8.13 suggests that the plotted data is still curved, however, because the residuals go from − to + to − as X increases. Can the residuals be made smaller, or the − to + to − error pattern eliminated?

TABLE 8.13

Residual calculations for semilog least squares fit, Case Study #9.

x	y	$y_i = 22.76(1.035)^x$*	$\Delta y = y - y_i$	$(\Delta y)^2$*
1	24	23.56	−0.44	0.20
11	34	33.23	−0.77	0.59
16	40	39.47	−0.53	0.29
21	45	46.87	1.87	3.51
26	54	55.67	1.67	2.79
27	57	57.62	0.62	0.38
28	59	59.64	0.64	0.40
29	62	61.72	−0.28	0.08
30	66	63.88	−2.12	4.48
31	69	66.12	−2.88	8.30
Totals			0.38	21.02

* Values rounded from 7 digit calculations.

To try for a better fit, an estimate of a data offset value (discussed in Section 8.6.3) for the Y variable is needed. This requires estimating a value for a new point x_3. For x_3 to be within the range $1 \leq x_3 \leq 35$, a careful selection of x_a and x_b is needed. Let $x_a = 1$ and $x_b = 31$, then $x_3 = 1(31)/2 = 15.5$. From the dashed line in Figure 8.8, y_3 is approximately equal to 39, and

$$c = \frac{24(69) - 39^2}{24 + 69 - 2(39)} = 9.0$$

The curve is concave upward, so c is positive and subtracted from the Y values. Table 8.14 lists the adjusted data and preliminary calculations.

$$\log M = \frac{10(366.849) - 220(15.912)}{10(5710) - 220^2} = 0.01929, m_{10} = 1.045$$

$$\log B = \frac{5710(15.912) - 220(366.849)}{10(5710) - 220^2} = 1.16675, B = 14.68$$

$$Y - 9 = 14.68(1.045)^X$$

$$r^2 = \frac{[10(366.849) - 220(15.912)]^2}{[10(5710) - 220^2][10(25.647) - 15.912^2]} = 0.9878, \text{ and } r = 0.994$$

Additional accuracy was used on this calculation because r is approximately equal to one. The maximum value of r should be one, but round off error may make $r > 1.00$. When only two-digit accuracy is used in Table 8.14, $r = 1.054$. Additional refinement could be made by using additional decimal accuracy. Write and run a computer program with double precision accuracy to investigate the possibility of better answers.

TABLE 8.14

Calculations for adjusted semilog least squares fit, Case Study #9.

x	y' = y − c	log y'*	x log y'*	(log y')²*
1	15	1.176	1.176	1.383
11	25	1.398	15.377	1.954
16	31	1.491	23.862	2.224
21	36	1.556	32.682	2.422
26	45	1.653	42.984	2.733
27	48	1.681	45.394	2.827
28	50	1.699	47.571	2.886
29	53	1.724	50.004	2.973
30	57	1.756	52.676	3.083
31	60	1.778	55.123	3.162
Totals	220	15.912	366.849	25.647

* Values rounded from 7 digit calculations.

TABLE 8.15

Residual calculations for adjusted semilog least squares fit, Case Study #9.

x	y'	$y_i' = 18.80(1.039)^x$ **	$\Delta y = y_i' - y'$	$(\Delta y)^2$ *
1	15	15.34	0.34	0.12
11	25	23.82	−1.18	1.38
16	31	29.69	−1.31	1.72
21	36	37.00	1.00	0.99
26	45	46.11	1.11	1.22
27	48	48.18	0.18	0.03
28	50	50.35	0.35	0.12
29	53	52.61	−0.39	0.15
30	57	54.98	−2.02	4.07
31	60	57.46	−2.54	6.47
Totals			−4.46	16.27

* Values rounded from 7 digit calculations.

Table 8.15 lists the residuals for this solution, and Figure 8.10 shows the resulting graph. This adjustment does produce a smaller sum of the squares of the residuals, a smaller sum of the residuals, and a better coefficient of correlation. Whether or not this is the "best" fit of the data is still a decision to make. Is the sum of the residuals or the correlation coefficient most important? A computer iteration program could investigate other values in the neighborhood of 9.0 and perhaps find a "better" answer.

FIGURE 8.10

Adjusted semilog graph for Case Study #9.

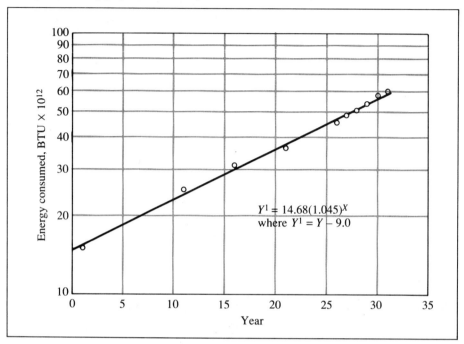

8.7 | Experimentation

Experiments are run when information is needed that cannot be obtained using exact mathematical modeling. Either no theory exists to explain the situation being studied or the theory uses assumptions and boundary conditions that cannot be matched. Experiments may also be used to establish base line information such as material strength properties, or they may be run to find a relationship between variable quantities such as the electrical conductivity as a function of temperature. Questions often lead to experiments.

1. What happens to the shear strength of a material as the rate of loading changes?
2. How does the percent yield from a chemical process change as the temperature changes?
3. How does the traffic flow rate change as the red-green cycle time changes on a traffic light?
4. How does the life of a forging die change when a different lubricant is used?
5. How does the life of a light bulb change as the ambient temperature changes?

Engineers need to know what will happen when changes are made. Sometimes the answer to a question can be found by testing a sample or scale model of the real thing, such as the heat conductivity of a metal. Other times only a real product test will be valid, such as determining the life of a car tire. The purpose of an experiment is to control the variables, whatever they are, establish relationships that can be verified by others doing the same experiment, and obtain relationships to use to predict future events. One of the difficulties with experiments is knowing which variables are affecting the results, and then planning a scheme to keep those variables under control.

After the variables are under control, their values must be randomized so an experiment can be done without equipment or operator bias. If two products, A and B, are to be tested, the person running the experiment should not test all of product A and then test all of product B. This can result in bias and unreliable results. Uncontrolled variables may include the temperature of the test samples, the operating speed of the test machine, or the operator's skill in running the test machine. The instruments showing data values may change calibration during the experiment. Any number of events can occur over the course of the experiment and not be noticeable because of the slow change. Time differences can also affect the hardness of samples if they are subject to precipitation hardening. Time can also allow residual stress relaxation, and creep can occur. By randomizing the order of the test samples any time dependent, uncontrollable, or unknown variations are averaged out. One method for doing this is to give all the test samples a number and select them for test based on a random number table. Even testing on an alternating A-B-A-B basis is better than testing all of one type in sequence. Another method of reducing the effects of uncontrollable variables is to run all samples simultaneously. This could be done if the life of a light bulb experiment is run. If temperature, voltage, and outside vibrations are not controlled, having all the bulbs on at the same time will reduce the effect of these variations, as all the bulbs will be exposed to the same variation.

Another important experiment criterion is the accurate measuring and recording of the data. One of the first experiments run to measure the speed of light using lanterns could not succeed because of the crude measuring technique used. Experimental data with a large deviation or a large range should be suspect and used with caution. Good data can also be lost or become confusing because of poor recording. A person recording data may become bored with the repetition of the task and become careless, or read dials and meters incorrectly. Automatic recording instruments will help avoid that situation, but the instruments must be calibrated and checked from time to time during the experiment.

The following items should be considered when setting up and running an experiment. Each item will be discussed using a light bulb burn-out life experiment as an example.

1. State the problem or question attempted to be answered. Is the 750 hour rated life as advertised for brand X light bulb an average life?

2. Consider appropriate variables and response scales to use. The variable chosen to record is time-to-burnout, measured in hours. Other life criteria could be measured, such as time to reduce to a given level of illumination. A more elaborate experiment could control and vary voltage, ambient temperature, position of bulb filament, or induced vibration.

    Some experiments may require judgment evaluations of variable factors. The degree of cloudiness during a solar collector experiment, or physical comfort during a temperature-humidity experiment are examples. Always determine ahead of time how such judgments are to be handled.

It is also possible that in some experiments the first choice of measuring units will not be appropriate. Samples may have to be run to determine the best units to use. Seconds would be too small a unit for the light bulb experiment, but minutes might be too large a unit for an experiment measuring the speed of light.

3. Select relevant factors to be varied. For a light bulb experiment, one variation is to have one set of bulbs on continuously, and a second set of bulbs cycling on-off. This can be done if equipment is available and the amount of time needed to run the experiment is short. The voltage and ambient temperature can be monitored if not controlled. Bulb position, horizontal, vertical, or at some angle, must be consistent.

4. Select levels of variation. Select an on-off cycle such as a 10 second interval, 5 seconds on, 5 seconds off. If voltage variation is desired, levels such as 110 ± 2, 115 ± 2, 120 ± 2, 125 ± 2, and 130 ± 2 volts are possibilities. If filament position is chosen as a variable, horizontal, vertical up, vertical down, or other angles can be selected. Ambient temperature can also be selected as a variable, any range of temperatures can be selected. To reduce the effect of the uncontrollable variables, both sets of bulbs must be tested at the same time.

   Be cautious about changing the level of variation during an experiment. It is better to wait out a portion of a test and start over with a different level of a variable. Of course a test could be set up as a changeable pattern on purpose. A load cycle, for example, could be one second, two seconds, three seconds, ...

5. Establish combinations of variable levels. In the light bulb experiment this is not necessary because the only variation selected is the on-off cycle.

   To illustrate how this *can be* done, consider an experiment with three variables, each with a range of values: $X_1 \pm \Delta X_1$, $X_2 \pm \Delta X_2$, and $X_3 \pm \Delta X_3$. Figure 8.11 shows how seven different combinations can be used to obtain a first approximation of the effects of variable changes. Figure 8.12 shows how nine different combinations can be used to obtain a first approximation of the effects of variable changes. If the center value (X_1, X_2, X_3) is not used, then the eight points remaining represent the factorial experiment in 3 variables ($2^3 = 8$ runs). Be certain that measuring instruments are capable of accuracies less than the experimental ranges established. Second approximations for increasing accuracy within any of the octants would be appropriate for a second round of experiments (see Exercise 8.52).

   Past experience and judgment are important when considering variable ranges to investigate. It is also important to lay out and document the plan clearly so efficient use of time and equipment can be made. See Case Study #15, Experimental Optimization, in Chapter 9 for additional discussion on establishing experimental variable combinations.

6. Determine the number of observations to make. This is a very difficult pre-experiment judgment to make. It depends on the number of variables and the degree of accuracy desired. In some cases more data may be required after the preliminary tests are performed. Relationships that change abruptly or frequently require more data for analysis. As the desired accuracy increases, so does the need for additional experiments. Many times this is done with no additional variation but repeated for accuracy only. Additional data improves reliability confidence.

FIGURE 8.11

Variations around a point, scheme #1.

1. X_1, X_2, X_3
2. $X_1 + \Delta X_1, X_2, X_3$
3. $X_1 - \Delta X_1, X_2, X_3$
4. $X_1, X_2 + \Delta X_2, X_3$
5. $X_1, X_2 - \Delta X_2, X_3$
6. $X_1, X_2, X_3 + \Delta X_3$
7. $X_1, X_2, X_3 - \Delta X_3$

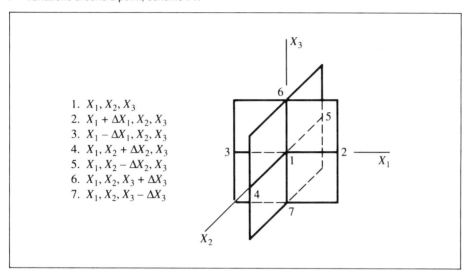

FIGURE 8.12

Variations around a point, scheme #2.

1. X_1, X_2, X_3
2. $X_1 + \Delta X_1, X_2 + \Delta X_2, X_3 + \Delta X_3$
3. $X_1 + \Delta X_1, X_2 + \Delta X_2, X_3 - \Delta X_3$
4. $X_1 + \Delta X_1, X_2 - \Delta X_2, X_3 + \Delta X_3$
5. $X_1 + \Delta X_1, X_2 - \Delta X_2, X_3 - \Delta X_3$
6. $X_1 - \Delta X_1, X_2 + \Delta X_2, X_3 + \Delta X_3$
7. $X_1 - \Delta X_1, X_2 + \Delta X_2, X_3 - \Delta X_3$
8. $X_1 - \Delta X_1, X_2 - \Delta X_2, X_3 + \Delta X_3$
9. $X_1 - \Delta X_1, X_2 - \Delta X_2, X_3 - \Delta X_3$

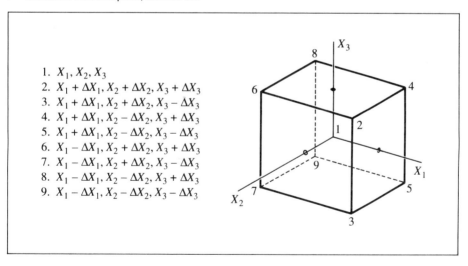

7. Analyze the data collected. In the case of the light bulb experiment the mean bulb life was calculated and compared with the advertised 750 hours. In both cases, steady on and five second on-off, the average life was not 750 hours, but 560.

    If the purpose of an experiment is to determine an empirical relationship then Equations 8.10, 8.11, or 8.12 might be used.

8. Interpret the results. In the case of the light bulb experiment a positive conclusion was difficult to make. This was because the voltage level, which was monitored during the test, was always higher than 120, ranging from 123 to 126 volts. Research on the design of light bulbs indicates that high voltage does reduce life (see Exercise 8.56). The only conclusion made from the samples tested was that at the mean recorded voltage level of 124.5 volts, the mean life was 560 hours with a standard deviation of 115 hours. There was no statistical difference in life to burn out between the bulbs that were on continually and those that were subjected to the on-off cycle.

Occasionally an experiment will be run and no conclusion can be made other than the variables chosen have no effect on each other. If that occurs a different hypothesis must be considered. Do not consider such an experiment a failure.

For all reversible experiments record data going in both directions. This will help uncover warm-up errors or hysteresis phenomenon. In material strength measurements, for example, reverse loading will indicate if the elastic limit was exceeded.

Study a system with and without a component if possible. Then a before and after comparison can be made. This can be done only when a system will function with and without a component. A street intersection, for example, will function with or without a stop sign, and another level of variation would be a yield sign. These three levels of variation are fixed, and provide an example of a step function experiment. Experiments can also be run to optimize a desired output of a process or system, as discussed in Section 9.5.

8.8 Additional Study Topics

1. Absolute measuring uncertainties
2. Relative measuring uncertainties
3. Newton-Raphson root location method
4. Lagrange interpolating polynomial
5. Newton interpolating polynomial
6. Hyperbolic curve fitting
7. Cubic polynomial spline curve fitting

8.9 References

1. Baird, D.C. *An Introduction to Measurement Theory and Experimental Design.* Englewood Cliffs, N.J.: Prentice Hall, 1962.
2. Cheney, Ward and David Kincaid. *Numerical Mathematics and Computing,* 2nd ed. Monterey: Brooks/Cole, 1985.

3. Clothier, Walter. "Design Models Are Naturals For Managing Information." *Design Graphics World,* May 1987, 40-41.
4. Clothier, Walter "Good Models Never Die." *Design Graphics World,* Dec. 1986, 28-30.
5. Dana, Forest C. and L.R. Hillyard. *Engineering Problems Manual*, 5th ed. New York: McGraw-Hill, 1958.
6. David F.W. *Experimental Modeling in Engineering.* Boston: Butterworth, 1982.
7. Dally J. W., W. F. Riley, and K. G. McConnell. *Instrumentation for Engineering Measurement.* New York: John Wiley and Sons, 1984.
8. Gajda, Walter J., Jr. and William E. Biles. *Engineering: Modeling and Computation.* Boston: Houghton Mifflin, 1978.
9. Gallin, Daniel. *Finite Mathematics.* Glencoe, Ill.: Scott Foresman and Co., 1983. Chapters 2 and 5.
10. Hillier, Frederick S. and Gerald J. Lieberman. *Operations Research* , 2nd ed. Oakland, Calif.: Holden-Day, Inc. 1974.
11. John, J.A. *Experiments: Design and Analysis.* New York: Macmillan, 1977.
12. Kapur, J. N. *Mathematical Modeling.* India: Wiley Eastern Limited, New York: John Wiley & Sons 1988.
13. Meredith, Dale D., et al. *Design and Planning of Engineering Systems.* Englewood Cliffs, N.J.: Prentice Hall, 1973.
14. Mischke, Charles R. *Mathematical Model Building*, 2nd ed. Ames, Iowa: Iowa State University Press, 1980.
15. Montgomery, Douglas C. *Design and Analysis of Experiments,* 2nd ed. New York: John Wiley and Sons, 1984.
15. Piascik, Chester. *College Mathematics with Applications.* Columbus: Charles E. Merrill, 1984. Chapter 6.
16. Steidel, Robert F. and Jerald M. Henderson. *The Graphic Language of Engineering.* New York: John Wiley & Sons, 1983. Chapter 8.
17. Tufte, Edward R. *The Visual Display of Quantitative Information.* Cheshire, Conn.: Graphics Press, 1983.

8.10 Exercises

1. Discuss the following models. What kind of a model is each? What is being modeled? What cautions must be observed when using each one?
    a) Strain gage
    b) Flight simulator
    c) Lacquer stress coating
    d) Electricity closed circuit laws (Kirchoff's and others) when analyzing a water system
    e) Current flow equations when analyzing heat flow
2. An alternate method to calculate the radius of a circle when three points on the circle are known is to find the point of intersection of the perpendicular bisectors of lines drawn between the points. Only two bisectors are required. The distance

between the intersection point and any one of the given points is the radius. Determine the models required for this method of analysis.

3. Another mathematical model for determining the radius of a circle when only a portion of it is shown involves the length of an arc chord, L, and the maximum height of the arc above the chord, h.
   a) Derive the formula for the radius of a circle as a function of L and h.
   b) Use the model from (a) on the sample in Figure 8.3.

4. The grooves on a phonograph record are not a series of concentric circles but a tightly wound spiral. Create a model to calculate the length of a groove on a record and discuss how to collect the data needed to use the model.

5. The model $y = 4L^3W/Ebh^3$ determines the maximum deflection of a rectangular cantilevered beam with weight W in pounds on the free end. Identify the other variables including their units and discuss the model limitations.

6. The maximum deflection of a simply supported beam is given as: $y = WL^3/4Ebh^3$ where y = maximum deflection, at the center; W = load in the center; L = length of the beam; E = modulus of elasticity of the beam material; b = width of the beam; h = height of the beam.
   a) Establish a consistent set of units assuming that E will be given as lbf/in^2.
   b) If you made a model of this beam, and made the model geometrically 1/10 as large as the original, used 1/10 of the load, but used the same material, what would you expect the deflection of the model to be compared to the original?

7. Verify Equation 8.3.

8. If the velocity of sound in a gas is considered to be a function of gas density, pressure, and dynamic viscosity, determine the relevant control groups using dimensional analysis.

9. Find the dimensionless groups for the following situation. A hole is drilled in the side of an open tank, a fluid is placed in the container, and the fluid leaves the hole with some velocity.

10. Find a dimensionless groups relationship for a pump using pressure change through the pump, the pipe diameter, and fluid properties density and viscosity.

11. Find the dimensionless groups for the fluid velocity flow over an inclined roof assuming the relative variables are fluid density, roof slope angle, fluid viscosity, surface tension, length of roof in the direction of flow, and gravity.

12. Discuss the following test that was performed on a nuclear fuel rod shipment container. A one-quarter scale model was dropped 30 feet onto concrete. The real tank weighs 50 tons, and holds 200 spent fuel rods and about 200 gallons of water.

13. What is the function of modeling when investigating such concepts as crack growth and stress concentration behavior?

14. Prove the statement, "One inch of circumference change in a circle is associated with a 0.32 inch change in diameter."

15. Prove the statement, "Each half inch of bark will change the circumference of the tree by π inches" mentioned in Case Study #7.

16. Write the models for the coefficient of correlation for

a) semilog line of regression, and
b) log-log line of regression.

17. Occasionally it is efficient to offset X or Y by a fixed amount when the data values are very large or very small. Consider the following data representing oil production in the United States in billions of barrels of oil per year.

Year:	1934	1935	1936	1937	1938	1939	1940	1941
Production:	0.9	1.0	1.1	1.3	1.2	1.3	1.4	1.4

Determine the "best" fit of the data by using the axis translation models $P = m(Y - 1934) + b$ and $P = B(M)^{Y-1934}$.

18. Extrapolate the answer from Exercise 8.17 to current years and determine if the growth of oil production has actually followed the trend established in the years 1934–1941. Library research is required.

19. Repeat the analysis method of Exercises 8.17 and 8.18 for the following data representing the population of the United States in millions of people.

Year:	1880	1900	1910	1920	1930	1940
Population:	62.9	76.0	92.0	105.7	122.8	131.7

For Exercises 8.20–8.24, determine an appropriate empirical equation for the data given. Also determine a coefficient of determination, the sum of the residuals, and the sum of the squares of the residuals.

20. Temperature and resistance of an electronic resistor.

Temperature, °F:	73.2	83.3	87.8	95.0	98.8	105.3	106.2
Resistance, ohms:	24.3	24.9	25.1	25.5	25.9	26.1	26.4

21. Pressure and volume of a gas.

Pressure, psi:	106.0	89.0	70.8	49.0	40.0	30.0	24.5	19.5
Volume, in^3:	1.5	1.7	2.0	2.6	3.0	3.6	4.2	5.0

22. The length of a pendulum and its period of oscillation.

Length, inches:	25.6	31.5	39.4	49.2	59.1	78.7
Period, seconds:	1.60	1.80	2.00	2.24	2.44	2.84

23. Fuel consumption of a particular vehicle.

Speed, mph:	20	22	24	26	28	30	32	34
Gal/mile:	0.183	0.188	0.191	0.193	0.195	0.197	0.198	0.200

24. Altitude and pressure above the earth's surface.

Altitude above sea level, meters:	0	500	1000	1500	2000	2500
Pressure in mm of mercury:	762	716	672	632	593	561

25. Determine an appropriate empirical equation and related coefficient of correlation for the following data.

X:	1.19	2.00	2.32	3.10	3.42	5.30	5.47	6.91	7.33	7.54	7.98	8.26	8.72
Y:	7.50	12.87	15.09	18.63	19.16	21.37	23.89	17.52	17.42	14.75	14.29	12.52	9.83

26. Determine an appropriate empirical equation and related coefficient of correlation for the following data.

X:	1.27	2.21	2.38	3.21	4.32	5.39	6.09	7.30	7.35	7.69	8.18	8.41
Y:	10.75	14.97	17.61	19.06	22.69	22.41	22.33	17.36	16.63	16.77	11.15	9.40

27. Compare the data from Exercises 8.25 and 8.26. Is it probable that the data came from the same universe?

28. An analysis of a set of n data pairs resulted in the equation $Y = 1.1X + 4.2$ with an $r = 0.80$. What is the minimum number of data pairs required so that the equation is a good representation of the data at least 95 percent of the time? At least 99 percent of the time?

29. An analysis of 10 pair of data resulted in the equation $Y = -1.2X + 15.1$ and $r = 0.70$. Is this reasonable?

30. An analysis of 30 pair of data resulted in the equation $Y = -0.9X + 28.2$ and $r = 0.400$. Is this a good representation for the data?

31. Show that a reduction in the degree of the polynomial model by one in Equation 8.22, leads to Equations 8.13 and 8.14.

32. Expand Equation 8.22 used to determine a, b, and c, for the second degree equation

    $Y = aX^2 + bX + c$, and determine a, b, c, and d for the third degree equation $Y = aX^3 + bX^2 + cX + d$.

33. The following model is used to evaluate flow capacity of a gas valve. $F = GA^{0.82}$, where

    F = required flow rate in m^3/minute at standard pressure and temperature

    A = total outside surface area of the storage tank in m^2

    G = 10.66 for propane, and G = 8.98 for butane.

    a) What kind of graph paper will show this model as a straight line?
    b) For a given value of flow capacity, which tank will be the largest, one for propane or butane?

34. What is the significance of the 2.35 in the following model used for a design stress calculation? Maximum stress = $S_{ut}/2.35$.

35. You are going to drive a friend home who lives 9 miles away if you drive the freeway at 55 miles per hour, but who lives only 6.5 miles away if you drive the country road at 35 miles per hour. The cost of driving is estimated at $0.01V^2$ cents per mile, where V is the car speed in miles per hour. Which route will you take under each of the following circumstances?

    a) Your time is worth $0.15 per minute.
    b) Your friend is already late getting home.
    c) Your time is worth $3.00 per minute.
    d) Your friend is pleasant, a good listener, and you enjoy talking.
    e) What additional information would you like to have before answering the questions and making a choice?

36. Suppose you are working on a project designing a control device that could be analyzed with the model: $V = S^{0.5}D^2/WC^3$, where

S = device size
D = the distance between the control device and the object being controlled
W = the weight of the device
C = the cost of the device, and
V = an indicator that relates to product acceptability and profitability, high numbers are best.

   a) Which variable of the project seems to be the most important to analyze, and would you want it to have a high or low value? Why?
   b) Which variable would you begin to work with to change if the product is already in production and you have the job of making it more profitable for your company? Why?
   c) Does this model seem realistic? Why?

37. A house has windows in it for the convenience of the people who live in it. Windows allow outside light to shine inside, allow outside visibility, provide emergency exits, and provide solar heat collecting. A single thickness of regular window glass has an R-value of 1.0, adding a storm window doubles the value to 2.0, and a triple glazed window may increase the value to 3.5.

    Various products are sold to cover windows that increase the effective R-value. Make a list of the items that need to be considered to evaluate whether or not a product should be purchased and added to the windows of a house.

38. The following table lists Iowa health care expenditures per capita per year (Source: *Des Moines Register*, October 26, 1986). The data was extrapolated to arrive at the 1988 and 1990 figures.

Year:	1966	1968	1970	1972	1974	1976	1978	1980	1982	1984	1986	1988	1990
Cost:	$196	242	293	351	453	556	724	931	1182	1358	1588	1877	2206

    a) Determine a probable method of extrapolation and judge its validity.
    b) Consider alternate methods of extrapolation and compare the resulting figures for 1988 and 1990 with those given.

39. Under what circumstances would it be important to know the sequence of data recordings as well as the values, when performing an experiment?

40. List some experiments that contrast the difference between step function and continuous function control.

41. A particular liquid was tested for its boiling temperature in °C at five different pressures. Analyze the 21 data values collected and listed below.

Pressure, psig	67	71	75	79	83
Boiling temperature @ given pressure	90	89	86	84	83
	92	90	87	85	85
		90	87	85	
		91	87	86	
			88	86	
			88	87	
			88		

42. Refer to Figure 8.11 and determine the seven values of the three variables for the octant bounded by 4-1-2-7 using the same system of point selection. How many new points are needed?

43. If it is not known that the deflection of a beam is a function of the section modulus (moment of inertia), a guess might be made that deflection is a function of cross-sectional area.
    a) Use this assumption and determine the Π_i groups in Example 8.9.
    b) What would happen when the value of k is being determined experimentally if this Π group is used?

44. Model a simply supported uniformly loaded beam with a simply supported beam and a center load, assuming equal maximum deflection.

45. Repeat Exercise 8.44 assuming equal end of beam slope angle. Would this model have more or less deflection than the answer to Exercise 8.44?

46. What happens to maximum bending moment in Exercise 8.45 after the model substitution?

47. What happens to the maximum shear stress in Exercise 8.45 after the model substitution?

48. A cylindrical storage tank is being designed to hold 10,000 gallons. The design guideline calls for a diameter: height ratio of 1 : 2.
    a) Determine the dimensions of a scale model of the tank built to a scale of 1/10.
    b) How many gallons will the model hold?

49. A concrete or cement block wall is thin compared to its length and height. A freestanding wall loaded from above can be modeled as a column. Mathematical modeling is difficult when the effects of openings for doors and windows are considered. Discuss the variables and experimental control you would use to empirically determine the effect of wall openings on P_c, the load before buckling.

50. Attached walls also have an effect on the stability of walls and the value of P_c. Discuss the variables and experimental control you would use to empirically determine the effect of attached walls on P_c.

51. An engineer measured the diameter and length of 12 manufactured bolts and calculated a coefficient of correlation between the data of 0.945. The conclusion made was that by controlling the manufacture of only diameter or length, the length or diameter would also be controlled. The engineer is 99 percent confident of this conclusion. Do you agree? Explain.
    Data, measured in inches, (Diameter, length): (0.990, 2.988), (1.002, 3.005), (1.010, 3.012), (0.995, 2.993), (0.998, 2.997), (1.005, 3.008), (1.003, 2.999), (0.997, 2.994), (1.001, 3.001), (1.003, 3.004), (0.997, 2.999), (0.999, 3.000).

52. Consider a light bulb experiment with three bulb positions: vertical up, vertical down, and horizontal; three voltages: 120 ± 1, 125 ± 1, and 115 ± 1; and three ambient temperatures: 70°F, 100°F, and 150°F. Discuss a method of experimentation considering the methods illustrated in Figures 8.12 and 8.13.

53. Prove that Equation 8.10 can also be expressed as $Y - \mathbf{y} = m(X - \mathbf{x})$ where $\mathbf{y}$ is the mean of the y data, and $\mathbf{x}$ is the mean of the x data.

54. The data in Table 8.16 represents federal guidelines for maximum noise exposure to avoid hearing loss.
    a) Explain the rapid percent decline in the time exposure with a corresponding small percent increase in noise level.
    b) Find an empirical equation that best represents the data.

55. While doing research on sewage particulate settling, an engineer developed the model $V_s = \sqrt{[4gd/3C_D][(\rho_s - \rho_1)/\rho_1]}$, where
    V_s = terminal settling velocity, cm/sec
    g = gravitational constant, cm/sec^2
    d = particle diameter, cm
    ρ_s = particle density, g/cm^3
    ρ_1 = liquid density, g/cm^3,
    C_D = coefficient of drag.

    Discuss problems that you might encounter if you were to run experiments to determine the settling velocity of various particulate matter.

56. Experimental work with light bulbs has resulted in a predictor of actual life based on actual voltage: actual life = rated life(rated voltage/actual voltage)$^{13.1}$. Does the voltage level recorded in the experiment discussed in Section 8.7 account for all the reduction in bulb life?

57. The range of data collected in the experiment of Section 8.7 was 338 to 829 hours. What actual voltage existed for the data collected if voltage was the only reason for reduced life?

58. Repeat Exercise 8.57 for data with a range of 441 to 798 hours and a mean of 605 hours.

TABLE 8.16

OSHA sound level exposure limits.

Noise level, dbA	Maximum exposure, hours per day.
90	8.00
92	6.00
95	4.00
97	3.00
100	2.00
102	1.50
105	1.00
110	0.50
115	0.25

9 Optimization

"Why is it we never have time to optimize the design the first time, but always have time to revise the design when something fails?"
♦ AN ANONYMOUS ENGINEER

Optimization methods and analysis techniques give engineers the capacity to manipulate and fine tune their design development work so the results are as efficient, functional, and cost effective as possible. This does not mean that the final design will be the most efficient and most functional and have the lowest cost, because all constraints usually cannot be optimized at the same time. The engineer must select the criteria that are most important, maximize or minimize as desired, but without forcing other constraints to reach unacceptable levels. All discussions of problem solving and specific design considerations in this book have optimization as a primary objective, to select a value for a primary variable that does not create an unacceptable value for a related variable. The engineer's goal is to obtain the best results for a given problem situation within the imposed restrictions. The physically imposed restrictions, including size, weight, material available, processes available, cost, and a due date, are real constraints that must be satisfied. Engineers must also understand that an optimum determined today may not be the optimum of tomorrow, because physical constraints and technology may change as time passes. Another reason for the changing status of selected optimum values is the result of system variances. System variations include changes in material prices, interest rates, future scrap value, available material compositions and sizes, shipping costs, ability of workers, condition of machines used to manufacture products, and consistency of material purchased; as well as changes in the environment such as temperature, humidity, composition of air, and atmospheric pressure. Because of the range of variable values that can occur, assumptions about the status of the variables are made prior to using optimization techniques. A temperature range may be assumed; market potential and sales price estimated; and material availability and cost established. It is important to establish values (or ranges of values) for variables prior to searching for optimum solutions, because solutions are only optimum within the restrictions and the assumptions made. If the assumptions are changed, a different optimum may occur. This is one of the reasons why engineers run many configurations (different sets of assumptions) while developing a product. The goal is to find an optimum that has low sensitivity to changes in boundary conditions. Low slopes on function curves, when a change in an independent variable causes a small change in the dependent variable, indicate low sensitivity.

Even though many optimization techniques are available to an engineer, some are more general in nature and their principles can be extended to a wide variety of problem situations. This chapter will concentrate on several of the more general techniques, both calculus and non-calculus, and their applications.

9.1 Calculus Techniques

Engineers learn early in their education that calculus provides methods of optimization if a mathematical model can be established and is differentiable. This is not always possible, particularly in a totally new situation. Often, however, a mathematical model can be estimated from experimental or observational data as was done in Chapter 8. The specific mathematical model depends on the nature of the data collected. Fitting data to known models (such as Equations 8.10, 8.11, 8.12, 8.20, and 8.21) to determine the best empirical model can itself be an optimization process, as illustrated in Case Study #9, Empirical Formula Refinement.

9.1.1 First Derivative Analysis

If a differentiable mathematical model can be developed, then standard derivative techniques are available to find extreme (maximum or minimum) points. If a function $f(x)$ has a first derivative such that $f'(x_i) = 0$, three possibilities exist.

1. If $f^{II}(x_i) < 0$, or if $f^{II}(x_i) = 0$ and $f^{III}(x_i) = 0$ and $f^{IV}(x_i) < 0$, then $(x_i, f(x_i))$ represents a local maximum extreme value.
2. If $f^{II}(x_i) > 0$, or if $f^{II}(x_i) = 0$ and $f^{III}(x_i) = 0$ and $f^{IV}(x_i) > 0$, then $(x_i, f(x_i))$ represents a local minimum extreme value.
3. If $f^{II}(x_i) = 0$ and $f^{III}(x_i) \neq 0$, then $(x_i, f(x_i))$ represents a point of inflection.

If more than one x_i exists within an interval $[x_a, x_b]$ such that $(x_i, f(x_i))$ are local extreme values, then the global extreme within the interval is determined by comparing the values of all local extreme values. The function may also have a value at the end of an interval that is larger or smaller than local extreme values located using $f'(x_i) = 0$. Be sure to examine the value of the function at the interval endpoints.

Keep in mind that the extreme values found are valid only if the assumptions, usually the constant values in the function, are stable. An engineer should always investigate the effect of changes in the assumptions. The following example illustrates the first derivative method and a variable change sensitivity check.

EXAMPLE 9.1:
Optimize a one variable function.

The cost, in dollars per hour, of running a particular vehicle is expressed as $C_{hour}(v) = 0.05v^2 + 72$, where v = the speed of the vehicle in kilometers/hour. The 0.05 is a cost associated with the speed, and 72 is a fixed cost that applies even if the vehicle is stationary. Determine the speed for minimum cost per kilometer, and determine whether or not the relationship is more sensitive to a change in fixed cost or variable cost.

The cost per kilometer is obtained by dividing the cost per hour by the speed in kilometers per hour: $C_{dist}(v) = 0.05v + 72/v$

Take the first derivative with respect to v, set it equal to zero, and solve for v:

$$dC_{dist}/dv = 0.05 - 72/v^2 = 0$$

$v^2 = 72/0.05$, so $v = 37.94$ kilometers per hour for minimum cost, and the minimum operating cost = \$3.79 per kilometer.

Intuitively, this answer represents a minimum rather than a maximum, because maximum cost per kilometer will occur when v approaches zero and $72/v$ become increasingly large; however, checking the second derivative will verify this. The second derivative is $d^2C_{dist}/dv^2 = 216/v^3$, which is greater than zero for all positive velocities and so a minimum is assured. The graph of $C_{dist}(v) = 0.05v + 72/v$, Figure 9.1, also shows that a minimum occurs.

If a percentage variation in the values of 0.05 and 72 is assumed (±1 percent, ±5 percent, or ±10 percent), then the sensitivity of the 0.05 and 72 can be studied. If

FIGURE 9.1

Cost as a function of velocity, Example 9.1.

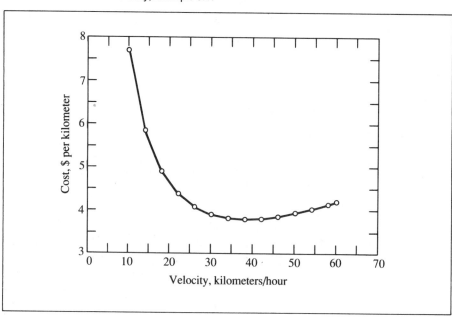

the 72 varies ±10 percent then the minimum speed varies ±5 percent, and if the 0.05 varies ±10 percent the minimum speed also varies ±5 percent, so the relationship has low sensitivity to variations in both constants. The reader should verify these results.

9.1.2 Partial Derivative Analysis

If it is necessary to optimize functions in more than one variable, derivative techniques can be used, but partial derivatives are required, and the tests for extreme values are different. If a function $Z = f(x, y)$ is a model of the problem at hand, take partial derivatives and set them equal to zero, $\partial Z/\partial x = 0$ and $\partial Z/\partial y = 0$. Solve the derivative equations simultaneously, obtain pairs of values (x_i, y_i) that satisfy the equations, and test for extreme values.

The test for extremes requires taking three, second partial derivatives, $\partial^2 Z/\partial x^2$, $\partial^2 Z/\partial y^2$, and $\partial^2 Z/\partial x \partial y$, and defines $\Delta(x, y) = (\partial^2 Z/\partial x^2)(\partial^2 Z/\partial y^2) - (\partial^2 Z/\partial x \partial y)^2$. The value of $\Delta(x, y)$ is calculated at each of the (x_i, y_i) values which are determined by solving the first derivative equations simultaneously. If $\Delta(x, y) > 0$ then $[(x_i, y_i), f(x_i, y_i)]$ represents a maximum if $\partial^2 Z/\partial x^2 < 0$ or if $\partial^2 Z/\partial y^2 < 0$, or a minimum if $\partial^2 Z/\partial x^2 > 0$ or if $\partial^2 Z/\partial y^2 > 0$. If $\Delta(x, y) < 0$ then a saddle point (neither a maximum or minimum) exists, and if $\Delta(x, y) = 0$ the test does not allow a conclusion about an extreme to be made. If $\Delta(x, y) = 0$ then other methods, such as a computer search, may be appropriate. The following example illustrates the partial derivative technique.

EXAMPLE 9.2:

Optimize a two variable function.

Find any extreme values of the function $Z = f(x, y)$, where Z is defined as:

$$Z = xy^2 - 6y^2 + 72y + x^2y - 18xy - 6x^2 + 72x - 216$$

Take partial derivatives, set them equal to zero, and solve the resulting equations simultaneously:

$$\partial Z/\partial x = y^2 + 2xy - 18y - 12x + 72 = 0$$
$$\partial Z/\partial y = 2xy - 12y + 72 + x^2 - 18x = 0$$

The derivatives are the equations of parabolas and can have 0, 1, 2, 3, or 4 distinct pairs of answers. Exercise 9.2 asks for verification of the solutions (6, 6), (6, 0), (0, 6), and (4, 4).

Take second derivatives to check for minimum and maximum: $\partial^2 Z/\partial x^2 = 2y - 12$, $\partial^2 Z/\partial y^2 = 2x - 12$, and $\partial^2 Z/\partial x \partial y = 2y + 2x - 18$, so $\Delta(x, y) = (2y - 12)(2x - 12) - (2y + 2x - 18)^2$:

$\Delta(6, 6) = -36$, so (6, 6) is a saddle point
$\Delta(6, 0) = -36$, so (6, 0) is a saddle point
$\Delta(4, 4) = 12$, $\partial^2 Z/\partial x^2 = -4$, so (4, 4) is a maximum and $f(4, 4) = 8$.

Some extreme value optimization problems can only be solved when restrictions are placed on the variables. Restrictions are necessary when the preliminary partial derivative equations are solved and the only solution is a zero value for all variables, a point without much significance. The following example illustrates this predicament.

EXAMPLE 9.3:
Optimize a two variable function with a restriction on the variables.

Maximize the moment of inertia of a rectangular cross section. The moment of inertia of a rectangle is a function of the height, h, and base, b; $I(b, h) = bh^3/12$. The partial derivatives are $\partial I/\partial b = h^3/12$ and $\partial I/\partial h = 3bh^2/12$. If they are set equal to zero the only solution is (0,0), which results in $I = 0$, a trivial minimum, and no information of

FIGURE 9.2

Graph of $I = bh^3/12$.

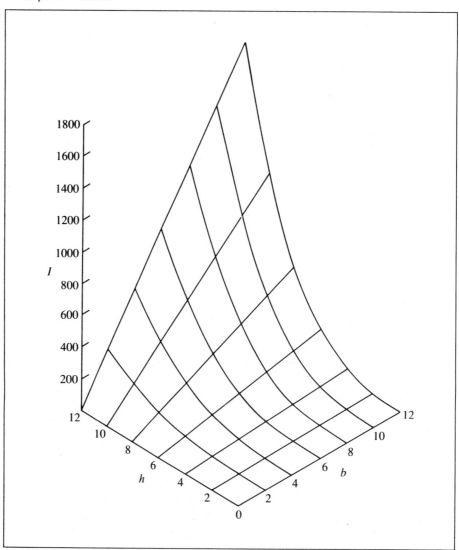

a maximum is found. Figure 9.2 is a graph of $I(b, h) = bh^3/12$ with the domains $0 < b < 12$, and $0 < h < 12$.

If a restriction is placed on the variables the triviality of a zero extreme value can be overcome and a useful extreme determined. Continue the example with the restriction added that the beam is to be cut from a 12 inch (0.30 meter) diameter log (see Figure 9.3). This changes the problem to one of maximizing $I = bh^3/12$ subject to the restrictions $b^2 + h^2 = 12^2$, $0 < h < 12$, and $0 < b < 12$.

The problem can be reduced to a one variable problem by solving $b^2 + h^2 = 12^2$ for b, $b = \sqrt{144 - h^2}$, and substituting in $I(b, h)$. Then $I(b, h)$ becomes $I(h) = h^3[\sqrt{144 - h^2}]/12$.

Now take a non-partial first derivative, set it equal to zero, and solve for h.

$dI/dh = \{h^3(-2h)/[(12)(2)\sqrt{144 - h^2}]\} + \{(3h^2)\sqrt{144 - h^2}/12\}$
$dI/dh = [-h^4 + 432h^2 - 3h^4]/[12\sqrt{144 - h^2}]$
$= [-4h^4 + 432h^2]/[12\sqrt{144 - h^2}] = 0$
Multiply by $[12\sqrt{144 - h^2}]$ (permissible as long as $h \neq 12$),
and $-4h^4 + 432h^2 = 0$
$-4h^2(h^2 - 108) = 0$, and $h_1 = h_2 = 0$, $h_3 = -10.39$, and $h_4 = 10.39$.

The values of h_1, h_2, and h_3 are outside the physical reality of the problem, so if a maximum exists, it must be when $h = 10.39$. When $h = 10.39$ inches (263.9 mm), then $b = 6.00$ inches (152.4 mm), and $I(6.00, 10.39) = 560.81$ in^4(2.33 · 10^8 mm^4). Exercise 9.3 asks for a second derivative verification that $h = 10.39$ produces a maximum value for I, but again the graph, Figure 9.4, shows that $I = bh^3/12$ has a maximum under the restriction $b^2 + h^2 = 144$.

FIGURE 9.3

Beam section to cut from a 12 inch diameter.

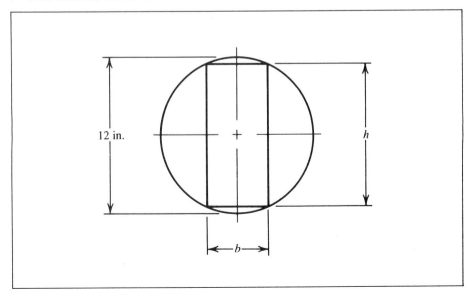

FIGURE 9.4

Graph of $b^2 + h^2 = 144$ superimposed on $I = bh^3/12$.

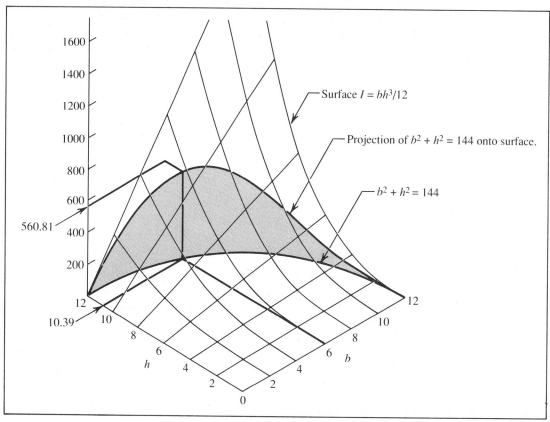

Many times conditions exist such that a single answer is not appropriate even if derivative methods are used. This occurs when the range of limiting values is discrete rather than continuous. The following case study further illustrates the use of derivative analysis, and is an example of a situation where the optimum calculus answer may not be able to be implemented because of physical constraints.

CASE STUDY #10:
Tool Life Optimization

Machine tool speeds and feeds are critical in controlling manufactured part tolerances, surface finish, production rates, and part cost. Total cost to manufacture a part includes the purchase cost of perishable tools such as drills, reamers, lathe turning tools, taps, honing stones, broaches, and milling cutters; and the cost of resharpening them when they become worn through use.

Experiments have been run to determine the effect of cutting speed on tool life so economic decisions can be made about machining processes. A good correlation between cutting speed and tool life has been demonstrated using the model $VT^n = k$ (2,486-490), where: (9.1)

V = Cutting speed at the tool-material interface, feet/minute (meters/minute)

T = Tool life in minutes, and

k and n are parameters determined by such factors as:

1. The material being machined
2. The tool material
3. The surface coating of the cutting tool
4. The tool form and shape
5. The size and shape of the cut, and
6. The cutting fluid used.

The model $VT^n = k$ is a straight line on a log-log graph where n is the slope and k is the value of V when $T = 1$ (see Figure 9.5). This makes k the cutting speed for a tool life of one minute. Combinations of values for n and k can be found in various resources and are available for many common cutting conditions. The relationship $VT^n = k$ can be used to determine the optimum cutting conditions for minimum cost using the following analysis.

FIGURE 9.5

Typical tool-life curves, based on cutting speeds.

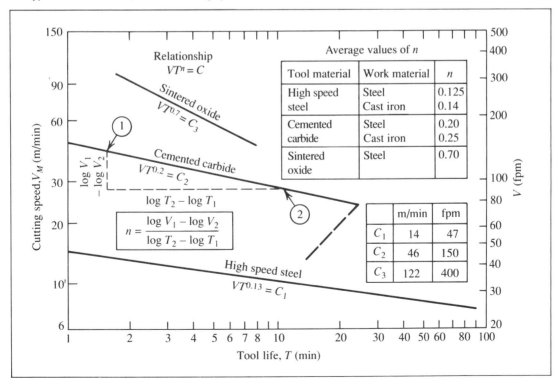

Consider the cost, C_t, to supply a cutting edge to a machining process:

C_t = tool change cost + tool sharpening cost + tool cost

$C_t = (t_1 R_c) + (t_2 R_s/N_1) + [C/(N_1 N_2)]$, where

t_1 = time to change a tool, minutes

t_2 = time to grind a tool, minutes

R_c = machine operator cutting cost including overhead, $/minute

R_s = tool grinding cost including overhead, $/minute

N_1 = number of cutting edges on the tool

C = original tool cost, $, and

N_2 = number of times a tool can be sharpened, $N_2 \geq 1$. If a tool is purchased sharp and cannot be resharpened, then $N_2 = 1$.

The cost added to a part while it is being machined is $C_m = TR_c$, and the material removed during machining time is $Q = KVTfd$, where

Q = volume of material removed, in³ (mm³)

f = width of cut, inches (millimeters)

d = depth of cut, inches (millimeters)

K = 12 in./ft (1 mm/mm), and

T = tool life, minutes.

From Equation 9.1, $V = k/T^n$, so $Q = Kfdk/T^{n-1}$.

The total cost, C_T, of removing material in $/unit³ is:

$$C_T = \frac{C_t + C_m}{Q} = \frac{C_t + TR_c}{Q} = \frac{C_t T^{n-1} + R_c T^n}{Kfdk} \qquad (9.2)$$

Take the first derivative of C_T relative to tool life T, set it equal to zero, and solve for the value of tool life for minimum material removal cost, T_m. (Exercise 9.4 asks for this calculation.) The result is:

$$T_m = (C_t/R_c)(1 - n)/n. \qquad (9.3)$$

Substitute this value of T_m into Equation 9.1 to obtain the velocity required for minimum material removal cost:

$$V = k/T_m^n = k/[(C_t/R_c)(1 - n)/n]^n$$

The cutting speed for maximum production rate can also be determined. The production rate, P, is defined as material removed per minute, and is expressed as:

$$P = Q/(T + t_1) \text{ in}^3/\text{minute (mm}^3/\text{minute)} \qquad (9.4)$$

Substitute $Kfdk/T^{n-1}$ for Q, and $P = (KfdkT^{1-n})/(T + t_1)$.

Take the first derivative relative to tool life T, set the derivative equal to zero, and solve for the value of tool life that results in maximum production rate, T_M. (Exercise 9.6 asks for this calculation.) The result is Equation 9.5, which holds as long as $T \neq -t_1$.

$$T_M = t_1(1 - n)/n \qquad (9.5)$$

Substitute this value of T_M into Equation 9.1 to obtain the velocity required for maximum material removal (maximum production):

$$V = k/T_M^n = k/[t_1(1-n)/n]^n$$

The comparison of T_M with T_m shows that the tool life for maximum production will be less than the tool life for minimum material removal cost if $t_1 < C_t/R_c$.

Note that an implied condition in this analysis is that the supply of cutting tools is unlimited. There is no point in getting high production for six hours and then sitting and waiting for two hours for tools to be resharpened.

Equations 9.3 and 9.5 give different answers for tool life and cutting speed, and in general the practical machine setting is between the two values, because both optimums cannot be satisfied at the same time. The following numeric details illustrate this situation.

$t_1 = 1$ minute $\qquad R_c = \$0.20$/minute
$t_2 = 20$ minutes $\qquad R_s = \$0.25$/minute
$C = \$12.00$ $\qquad N_2 = 10$ times
$N_1 = 8$ edges $\qquad VT^{0.2} = 180$

$C_t = (1)(0.20) + (20)(0.25/8) + 12.00/(10)(8) = \$0.98$

$T_m = (0.98/0.20)(1 - 0.2)/0.2 = 20$ minutes

V for minimum tool cost $= 180/20^{0.2} = 99$ feet/minute (30.2 m/minute)

$T_M = 1(1 - 0.2)/0.2 = 4$ minutes

V for maximum production $= 180/4^{0.2} = 136$ feet/minute (41.5 m/minute)

Now what? An optimum range from 99 to 136 feet/minute (30.2 to 41.5 m/minute) has been determined. When the machine selected for the operation is set up, only one machine setting may exist in this range. In that case no special decision is required; use the setting available. If there is more than one setting available in the range, choose the one closest to the preferred results, minimum material removal cost or maximum production rate. If there are no settings available in the range calculated, then the decision must be made if low tool cost or high production is most important. For higher production use high values of cutting speed, for longer tool life use lower values. Some modern machines are designed with speeds and feeds infinitely variable over a fixed range so that settings at a calculated value are possible. This setting flexibility is made possible by using electronic control, programmable controllers, or direct computer control. Keep in mind that this analysis requires several pieces of data, the most elusive being the values of n and k to use in Equation 9.1.

The results of the maximum tool life/maximum production rate analysis may not result in the tolerances or surface finish specified on the part drawing, or the machine may run out of power when combined with the recommended feed rates. Compromise (there's that word again) is usually necessary. This analysis also does not account for broken tools as the result of improper setup or improper sharpening; all variables are assumed to be under control. Determining cutting speeds and feeds, selecting cutting fluid, and the design of cutting tools is the job of another engineering specialist, the tool engineer.

9.1.3 Lagrange Multipliers

The use of Lagrange multipliers is an alternate calculus method for determining an extreme value of a function with *any* number of variables when the function is subjected to equality constraints. The object is to find an extreme of a function $F(x_i)$, $i = 1, 2, \ldots, n$, subject to restrictions $R_j(x_i)$, $j = 1, 2, \ldots, m$. The procedure is as follows.

Define a new function that is a linear combination of the given function and the restrictions, $F(x_i, \lambda_j) = F(x_i) - \Sigma \lambda_j R_j(x_i)$, where the λ_j are unknown Lagrange multipliers. Take partial derivatives $\partial F/\partial x_i$ and $\partial F/\partial \lambda_j$ and set each of them equal to zero. If the resulting $(m + n)$ simultaneous equations are independent, they will yield values of the variables for extreme function evaluation. If only one extreme is located, then deciding if it is a maximum or minimum is based on intuition, or the geometry, or physical restrictions of the problem. A single extreme, calculated purely mathematically, has no interpretation as a maximum or minimum. If several extreme values are found, then direct comparison of function values will determine if the extremes are maximum or minimum. The extreme values found using Lagrange multipliers are not likely to be the extreme of the function without the constraints. The next example illustrates the technique of Lagrange multipliers using the same situation as Example 9.3.

EXAMPLE 9.4:

Use of the Lagrange multiplier method.

Find the extreme value of $I(b, h) = bh^3/12$ subject to the constraint $R_1 = 12^2 - b^2 - h^2$. The new function is $F(b, h, \lambda_1) = bh^3/12 - \lambda_1(12^2 - b^2 - h^2)$. The three derivatives are:

$\partial F/\partial b = h^3/12 + 2b\lambda_1 = 0$ (1)
$\partial F/\partial h = bh^2/4 + 2h\lambda_1 = 0$ (2)
$\partial F/\partial \lambda_1 = -12^2 + b^2 + h^2 = 0$ (3)

Subtract b times (2) from h times (1) and

$h^4/12 - b^2 h^2/4 = 0$ (4)

Solve (3) for b^2 and substitute in (4), eliminate fractions, combine terms and

$4h^4 - 432h^2 = 0$

Factor, set each factor equal to zero, and $h_1 = h_2 = 0$, $h_3 = -10.39$, and $h_4 = 10.39$. Note that the value of λ_1 did not have to be determined in this case.

The method of Lagrange multipliers can be extended to problems with any number of constraints. The next example illustrates how additional restrictions are included in the procedure.

EXAMPLE 9.5:
Use of the Lagrange multiplier method with more than one constraint.

Find any extreme values of $G(x, y, z) = -2x^2 - y^2 - 3z^2$ subject to the constraints $R_1 = x + 2y + z - 1 = 0$, and $R_2 = 4x + 3y + 2z - 2 = 0$. Note that without restrictions $G(x, y, z)_{max} = 0$ when $x = y = z = 0$, and that $F(x, y, z)_{min}$ does not exist because there is no limit to how small $G(x, y, z)$ can be, as each term of the function is always negative.

Establish $F(x, y, z, \lambda_1, \lambda_2) = -2x^2 - y^2 - 3z^2 - \lambda_1(x + 2y + z - 1) - \lambda_2 (4x + 3y + 2z - 2)$ and take partial derivatives:

$\partial F/\partial x = -4x - \lambda_1 - 4\lambda_2 = 0$

$\partial F/\partial y = -2y - 2\lambda_1 - 3\lambda_2 = 0$

$\partial F/\partial z = -6z - \lambda_1 - 2\lambda_2 = 0$

$\partial F/\partial \lambda_1 = -(x + 2y + z - 1) = 0$

$\partial F/\partial \lambda_2 = -(4x + 3y + 2z - 2) = 0$

Solve simultaneously, Exercise 9.8, and $x = 5/27$, $y = 10/27$, $z = 2/27$, $\lambda_1 = -4/27$, and $\lambda_2 = -4/27$, resulting in the extreme value of $G(5/27, 10/27, 2/27) = -2/9$. Depending on the application to a physical problem, the value $-2/9$ could be a maximum or minimum.

An engineer must never assume that mathematical or computer optimization techniques give answers that can be implemented, as Case Study #10, Tool Life Optimization, showed. Numerical solutions that come from mathematical analysis may not be realistic in ways other than the lack of machine settings. Mathematical solutions of 2.3 people, 4.1 lanes of traffic, or 7.1 floors of a building are also not meaningful. The following case study illustrates a situation where the mathematics is straightforward enough, but the mathematical answer has several physical implementation problems.

CASE STUDY #11:
Pipe Packing Optimization

In the design of fluid flow heat exchangers, the length of pipe in the exchanger influences the rate of heat exchange. The cost of the pipe is a governing factor in the optimization process, as well as the cost of the tank shell within which the pipes are located and the temperature-altering fluid flow. The shell can be fabricated in a variety of shapes: a cylinder, which encloses the most area (most pipes) for a given perimeter; a rectangle, which may occupy less floor space than a cylinder, or some other shape that is easy to fabricate. The manufacture of a cylindrical tank, for example, requires a roll machine to form the metal into shape, and the corners of a rectangular tank require hydraulic or mechanical forming machines.

Consider the cost of a cylindrical heat exchanger tank of diameter D in feet (meters) and length L in feet/pipe-length (meters/pipe-length), that requires a minimum of 300 feet (91.5 m) of pipe to meet heat exchange requirements. The requirement of 300 feet (91.5 m) is itself an optimization process dependent upon heat flow rates, fluid volume flow rates, material heat capacities, incoming fluid temperature, required outgoing temperature, and cost and availability of pipe material. Assume that the cost of the tank shell has been estimated as the sum of the pipe cost of $700, the shell cost (based on other cylindrical tanks that have been built) of $25D^{2.5}L$, and the floor space cost of $20DL$; $C = 700 + 25D^{2.5}L + 20DL$. Additional data given with the tank design is that 20 pipes will fit in 1 ft^2 (0.093 m^2) of tank cross section. The 20 pipes/ft^2 translates into the following analytic restriction of the problem:

$$[\pi D^2/4][(L)(20 \text{ pipes/ft}^2)] \geq 300 \text{ ft}([\pi D^2/4][(L)(20 \text{ pipes}/0.093 \text{ m}^2)]$$
$$\geq 91.5 \text{ m}), \text{ or } \pi D^2 L \geq 60 \text{ ft } (\pi D^2 L \geq 1.70 \text{ m})$$

A minimum cost of $1538 is obtained using the information given, and occurs when $D = 1.37$ ft (0.418 m) and $L = 10.18$ ft/pipe-length (3.104 m/pipe-length) (see Exercises 9.15 and 9.16).

Now analyze the results and determine if the calculations result in a realistic solution. If the length of each pipe is 10.18 ft (3.104 m), and 300 ft (91.5 m) are required, then 300 ft/10.18 ft/pipe = 29.47 pipes are required. That is impossible. Either 29 or 30 pipes must be used. Twenty-nine pipes require that $L = 300/29 = 10.34$ ft (3.152 m), $D = \sqrt{[60/(\pi)(10.34)]} = 1.36$ ft (0.415 m), and $C = \$1539$. Thirty pipes require that $L = 300/30 = 10.00$ ft (3.049 m), $D = \sqrt{60/(\pi)(10)} = 1.38$ ft (0.421 m) and $C = \$1535$. The three costs are approximately the same and reflect a system that is insensitive to small variations in the value of the variables. Keep in mind that no variable can be consistently controlled to an exact value, and the small variation range ($1535–$1538) is likely within the accuracies of calculation and tolerances of the rest of the system.

The commercial availability of the length of pipe is also a factor affecting the design. If material comes in 20.00 ft (6.097 m) lengths, then $L = 10$ ft (3.049 m) is a good choice. If pipe comes in 24 ft (7.317 m) lengths, then additional analysis is required to determine the overall effect if $L = 12.00$ ft (3.659 m). The length of each pipe also affects the design of the tank shell, for its length is the same as the pipe lengths. If the material for the shell comes 10 ft (3.049 m) wide, then again $L = 10$ ft (3.049 m) is a good choice. The widest material the roll machine can handle, perhaps only 8 ft (2.44 m), may also be the limiting factor. But there is a larger difficulty with the heat exchanger analysis than available material and machines. Did you recognize it?

The larger problem is the assumption that the ratio of the number of pipes per square unit is constant regardless of the area. That assumption requires closer analysis. Consider enclosing a number of pipes of diameter two units, in a cylindrical shell, sample cross sections of which are shown in Figure 9.6. If one pipe is enclosed in a cylindrical shell, the area of the cross section of the shell is π unit^2, and the pipe density is 1 pipe/π unit^2 = 0.32 pipes/unit^2. The density of 0.32 pipes/unit^2 is the largest pipe density possible because there is no wasted space with one pipe. If two pipes are enclosed in a cylindrical shell the shell cross section area is 4π units^2, and the pipe density is 2 pipes/4π units^2 = 0.16 pipes/unit^2, half as

FIGURE 9.6

Typical pipe shell cross sections.

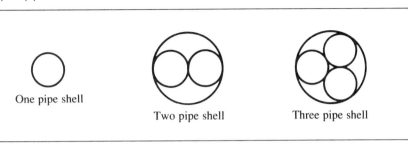

One pipe shell   Two pipe shell   Three pipe shell

efficient as using one pipe. If three pipes are enclosed in a cylindrical shell the shell cross section area is $\{[(2\sqrt{3})/3] + 1\}^2\pi = 4.64\pi$ units^2, and the pipe density = 0.21 pipes/units^2. This analysis can be continued, and in general as the number of pipes increases, the pipe density increases, but never exceeds 0.32. There are, however, some configurations that are more efficient than others on a local level. It is left as an exercise, 9.17, to show that the number of pipes, n, enclosed in a circular cross section is most efficient on a local level when $n = 1, 7, 13, 19, 31, 37,\ldots$. If this is so, then the number of pipes in the shell under study should be 31 and $L = 300/31 = 9.68$ ft (2.95 m). Perhaps the optimization compromise is to use thirty-one 10 ft (3.05 m) pipes and increase the margin of safety, by using 310 ft (94.51 m) of heat transfer pipe rather than 300 ft (91.5 m).

Optimization techniques do not always result in a single answer because of the number of variables connected with the problem and uncertainty about their values. In cases when uncertainty exists, optimization decisions are often made that reflect a combination of variable values. The following case study illustrates such a situation.

CASE STUDY #12:
Cost Ratio Analysis

Consider the cost of a cylindrical food container can of height h, diameter D, volume $V = \pi D^2 h/4$, and t = the material thickness. Then the volume of the material in the can = $\pi Dht + 2(\pi D^2 t/4)$, and the seam joining length = $2\pi D + h$. Assume that a rectangular blank is rolled into a cylinder and two circular blanks are used for the ends.

Let C_1 = the cost of the material in \$/unit^3, C_2 = the cost of seaming in \$/unit, and the total cost (material and seaming) = $C_1 t(\pi Dh + \pi D^2/2) + C_2(2\pi D + h)$. The cost of rolling the cylinder and blanking the ends from sheet metal are considered constant for a wide range of dimensions and are not included in the relationship. Some may prefer to use the volume of material as $t(\pi Dh + 2D^2)$ because the circular ends are probably cut from square blanks. In some manufacturing companies, however, the cost of material includes a credit for the value of scrap sold for recycling. Again a choice must be made (see Exercise 9.24).

Any of the six variables, C_1, C_2, D, h, t, or V, could be considered the main variable, but D and h are probably the two that have the most flexibility. The volume will be governed by the serving size determined by marketing; the material thickness will be determined by required strength and material availability; and the costs C_1 and C_2 will also be governed by market conditions. To reduce the total cost equation to a more manageable relationship, determine the value of h in terms of V and D, substitute in the cost equation, and limit the analysis to D: $h = 4V/\pi D^2$, and $C_t = C_1 t(4V/D + \pi D^2/2) + C_2(2\pi D + 4V/\pi D^2)$.

Take the first derivative relative to D (all other variables are assumed constant) and set the derivative equal to zero in an attempt to find a minimum cost:

$$dC_t/dD = C_1(-4Vt/D^2 + \pi Dt) + C_2(2\pi - 8V/\pi D^3) = 0$$

To reduce the number of parameters let $k = C_2/C_1$, so

$$-4Vt/D^2 + \pi Dt + k(2\pi - 8V/\pi D^3) = 0$$

To eliminate fractions multiply by πD^3, and

$$\pi^2 D^4 t + 2\pi^2 D^3 k - 4\pi VtD - 8Vk = 0 \tag{9.6}$$

This expression can be analyzed for minimum cost in different ways.

1. Determine D based on an assumed volume, material thickness, and values for C_2 and C_1.
2. Consider a range of parameter values that may be desirable and determine their effect on each other. This permits the consideration of variable sensitivity, and may suggest alternatives that have not been considered.

For example, the diameter may be restricted to $2.00 \leq D \leq 4.00$ inches ($0.051 \leq D \leq 0.102$ m) by existing manufacturing capability, and thickness is governed by normally available materials: 30 gage [0.0120 inches (0.3048 mm)], 28 gage [0.0149 inches (0.3785 mm)], 26 gage [0.0179 inches (0.4567 mm)], 24 gage [0.0239 inches (0.6071 mm)], 22 gage [0.0299 inches (0.7595 mm)], and 20 gage [0.0359 inches (0.9119 mm)].

Equation 9.6 has one sign change, so there is one positive real root regardless of the positive values given to t, V, and k. If a value for volume is assumed, then D can be studied as a function of k and t. Figure 9.7 shows the result of letting $V = 20$ in^3 (327.7 cm^3), t = the six gage thickness values listed above, and solving for D. The corresponding value of h is calculated using $h = 4V/\pi D^2$. Figure 9.7 can be used in different ways.

1. Select a value for t, a value for k, and determine the corresponding D and calculate h for minimum cost. Example: If $t = 0.030$ in. and $k = 0.10$, then $D = 2.3$ in. and $h = 4.8$ in.
2. Select a value of t and D and determine the necessary cost ratio for minimum cost. Example: If $t = 0.012$ and $D = 2.15$, then k is approximately equal to 0.11.

Figure 9.7 also shows that as the value of k increases there is less sensitivity in the associated values of D and t. Low sensitivity in a manufacturing operation is good; as k gets larger, however, so does the cost. Total cost can be qualified by changing the expression for C_t to $C_t/C_1 = 4Vt/D + \pi D^2 t/2 + k(2\pi D + 4V/\pi D^2)$. Figure 9.8 illustrates the variance of C_t using the same values of t as Figure 9.7. Compromise is again required, low cost and high sensitivity, high cost and low sensitivity, or someplace in between.

FIGURE 9.7

Analysis summary, Case Study #12.

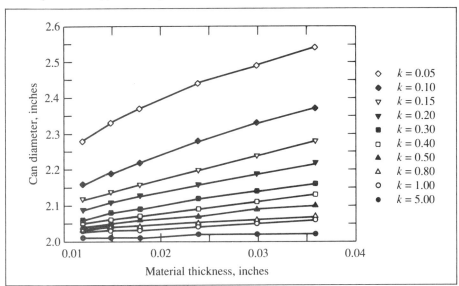

FIGURE 9.8

Cost ratio summary, Case Study #12.

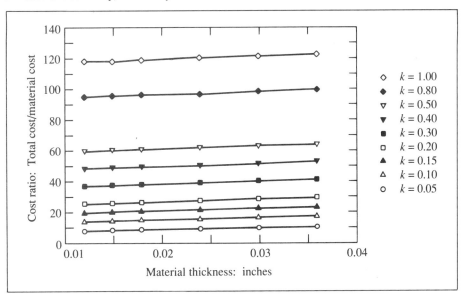

The derivative method of locating extreme values of functions is an easy to use straightforward procedure that can be generalized and extended to problems of more than two variables. The test is to determine the values of the x_i that make $\partial F/\partial x_i = 0$ or nonexistent, for $i = 1, 2, 3, \ldots, n$. The second derivative test to check for maximum, minimum, or saddle point, however, does not generalize to multiple dimensions, and physical interpretations are usually used for making decisions about extremes. This can be done by evaluating the function in the neighborhood of the extreme value in question. Mathematical analysis of second derivatives of functions of more than two variables is possible, but is beyond the scope of this text.

9.2 Non-Calculus Techniques

Non-calculus techniques are required for optimization analysis when an analytical function is difficult or impossible to differentiate, or if an analytic function is not available. When analytic functions are not available, analysis and optimization must be done experimentally. The non-calculus techniques are often called search techniques because the methods search for extreme values using systematic evaluation. Search techniques are valid for analytical or experimental evaluation. The plan for search must be determined prior to performing calculations or experiments so time and resources are used efficiently. This is particularly important when calculations or experiments take a long time and/or are costly in other ways. Four search methods will be discussed here, the equal division search, the Fibonacci search, the golden section search, and the slope search.

9.2.1 Equal Division Search

The most elementary search technique is performed by dividing the domain of the variable(s) into some number of equal segments and evaluating the system at each of the division values. The technique does not place a high premium on finding *the* best, only the best at the ends of the interval selected. The equal division method can be used to fine tune a system beyond the first stage optimization if the domain is narrowed to the segment most likely to contain the desired extreme (minimum or maximum), and additional subdivision and calculation is performed. If the cost of each evaluation is large, as it is when performing a complex experiment, then the number of subdivisions will be small because of cost considerations. If the cost and time to implement each evaluation is small, then a large number of evaluations can be justified, such as the case when computer evaluation is performed.

One use of the equal division search technique is to find the zeros of a function. A function, $f(x)$, is evaluated over a closed interval $[a, b]$, or an open interval (a, b), using a fixed increment Δx, and whenever the values of $f(x_i)$ and $f(x_i + \Delta x)$ have opposite signs, at least one root of the equation is isolated. Further refinement of the root is possible by using an increment smaller than Δx within the interval $[x_i, x_i + \Delta x]$. An exact root of the equation (exact to the accuracy of the calculation) will be found if at some evaluation $f(x_i + \Delta x) = 0$. The equal division search for roots is usually done on a computer because calculations are done rapidly, and good accuracy is possible.

The equal division method can also be used for finding function extremes within an interval. There are two approaches to consider. The first approach determines only the global extreme value within an interval. Evaluation is performed across an interval $[a, b]$ using an increment Δx. The search process is started by letting the location and the value of the local extreme of the function equal $(a, f(a))$. The value at each interval of the domain $f(a + n\Delta x)$, where n = the number of segments used over the domain, is compared to the previous local extreme value. If any $f(a + n\Delta x)$ is larger (or smaller) than the previous extreme, then the new value becomes the local extreme. The local extreme after the interval $[a, b]$ has been searched becomes the global extreme for the interval. If the interval is open, (a, b), then the first local limit is $(a + \Delta x, f(a + \Delta x))$.

A second approach locates *all* local extreme values on an interval, but it requires that the function be differentiated, and the derivative must exist in the interval. The derivative of the function, $f'(x)$, is evaluated over an interval $[a, b]$ using a fixed increment Δx. Whenever the value of $f'(x_i)$ and $f'(x_i + \Delta x)$ have opposite signs, an extreme of the function is isolated. If the value of $f'(x_i) > 0$, then when $f'(x_i)$ and $f'(x_i + \Delta x)$ have opposite signs a local maximum is found. If the value of $f'(x_i) < 0$, then when $f'(x_i)$ and $f'(x_i + \Delta x)$ have opposite signs a local minimum is found. Save the value of $[(x_i + x_i + \Delta x)/2, f(x_i + x_i + \Delta x)/2)]$ as a local extreme each time the signs of $f'(x_i)$ and $f'(x_i + \Delta x)$ are different. A local extreme will also be located if $f'(x_i) = 0$ and $f'(x_i - \Delta x)$ and $f'(x_i + \Delta x)$ have the opposite sign. The global extreme will be the largest (or smallest) of the individual values saved.

The conditions of Example 9.3 will be used to illustrate the equal division search so comparison of the methods can be made.

EXAMPLE 9.6:
The equal division search used on Example 9.3.

Find the maximum value of $I = bh^3/12$ subject to the restrictions $b^2 + h^2 = 12^2$, $0 < h < 12$, and $0 < b < 12$. Divide the range $0 < h < 12$ into 7 equal segments and make evaluations at 6 points. The segments are: (0, 1.71], (1.71, 3.43], (3.43, 5.14], (5.14, 6.86], (6.86, 8.57], (8.57, 10.29], (10.29, 12.00). Table 9.1 summarizes the calculations, and Figure 9.9 illustrates the equal division search in graph form. The maximum value of I is 560.21, only slightly less than the derivative value of 560.81.

TABLE 9.1

Summary of equal division search for I_{max}.

Evaluation #	h	b	I	I_{max}
1	1.71	11.88	4.95	4.95
2	3.43	11.50	38.67	38.67
3	5.14	10.84	122.67	122.67
4	6.86	9.85	264.99	264.99
5	8.57	8.40	440.60	440.60
6	10.29	6.17	560.21	560.21

FIGURE 9.9

I_{max} for the equal division search method. Example 9.6.

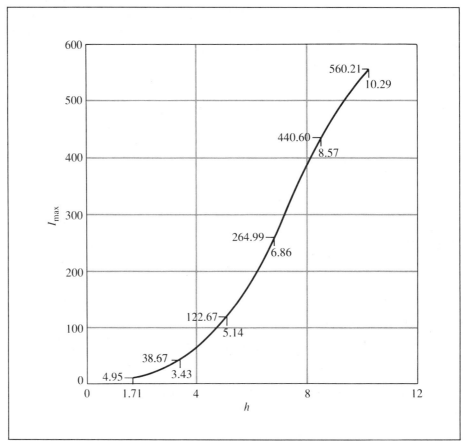

9.2.2 Fibonacci Search

The steps of the Fibonacci search method are similar to those of the equal division search except the segments of the search domain are not equal. The segment length is a function of the number of intervals and the ratio of numbers in the Fibonacci series, F_n, and gets smaller at each successive calculation. The Fibonacci series is an infinite sequence of integers that starts with $F_0 = 1$, $F_1 = 1$ and continues with each successive term equal to the sum of the preceding two terms; $F_2 = 2, F_3 = 3, F_4 = 5, F_5 = 8,\ldots$.

The Fibonacci search is a procedure that discards domain segments that are least likely to contain an extreme value. For computer Fibonacci searches, the number of evaluations can be high with little change in cost. For physical experiments, however, the number of evaluations may be small because of cost or time constraints. To use the Fibonacci search method on an interval $[a, b]$ (or (a, b)), the number of evaluations to make must be specified before the search is started.

The subdivision segment determination, and the evaluation procedure for an open, (a, b), or closed interval, $[a, b]$, of length $(b - a)$ is performed as follows. Specify n, the number of evaluations to make. The first two evaluations are made at a distance $(F_{n-2}/F_n)(b - a)$ from each end of the interval; $x_1 = a + (F_{n-2}/F_n)(b - a)$ and $x_2 = b - (F_{n-2}/F_n)(b - a)$. At each step of the search process the domain is divided into three parts (left, center, and right) and the left or right segment is discarded.

If a maximum function value is being sought and $f(x_1) > f(x_2)$, discard the region of the domain $[x_2, b]$, redefine the search domain $[a, b] = [a, x_2]$, and make the next evaluation at $x_3 = a + (F_{n-3}/F_{n-1})(b - a)$. If a minimum function value is being sought and $f(x_1) > f(x_2)$, discard the region of the domain $[a, x_1]$, redefine the search domain $[a, b] = [x_1, b]$, and make the next evaluation at $x_3 = b - (F_{n-3}/F_{n-1})(b - a)$. Continue to eliminate the right or left segment of the shortened interval, redefine the interval length $[a, b]$, and make another calculation at a distance $(F_{i-2}/F_i)(b - a)$ from the end of the interval that was not changed. When $(n - 1)$ calculations have been made, select a domain value slightly greater or slightly less than the domain value that is presently giving the extreme function value. Evaluate the function at this selected domain value to determine on which side of the present extreme the global extreme lies. This identifies the interval where the global extreme is likely to occur. The next example uses the Fibonacci search technique on the problem from Example 9.3.

EXAMPLE 9.7:
The Fibonacci search used on Example 9.3.

Find the maximum value of $I = bh^3/12$ subject to the restrictions $b^2 + h^2 = 12^2, 0 < h < 12$, and $0 < b < 12$. Set the number of evaluations to six to match the number used in the equal division search. The first two evaluations performed on the interval $(0, 12)$ are performed at $h_1 = 0 + (5/13)(12) = 4.62$ and $h_2 = 12 - (5/13)(12) = 7.38$ (see Figure 9.10).

For $h_1 = 4.62$, $b_1 = 11.07$, and $I_1 = 90.97$.

For $h_2 = 7.38$, $b_2 = 9.46$, and $I_2 = 316.87$.

The second interval is $(4.62, 12)$; $h_3 = 12 - (3/8)(12 - 4.62) = 9.23$, $b_3 = 7.67$, and $I_3 = 502.60$.
The third interval is $(7.38, 12)$; $h_4 = 12 - (2/5)(12 - 7.38) = 10.15$, $b_4 = 6.40$, and $I_4 = 557.70$.
The fourth interval is $(9.23, 12)$; $h_5 = 12 - (1/3)(12 - 9.23) = 11.08$, $b_5 = 4.61$, and $I_5 = 522.56$.
The sixth evaluation is done on one side of 11.08. Choose $I_6 < 11.08$ or $I_6 > 11.08$. If $h_6 = 11.00$, $b_6 = 4.79$, and $I_6 = 531.29$. This indicates that h for maximum I lies in the interval $(10.15, 11.08)$. Notice that this method did not produce a maximum as large as the equal division search. In general it is not predictable which method will result in the value closest to the global extreme; however, the Fibonacci search will narrow to the segment that contains the solution faster than the equal segment search unless the extreme value lies on an equal segment boundary.

FIGURE 9.10

I_{max} for the Fibonacci search method, Example 9.7.

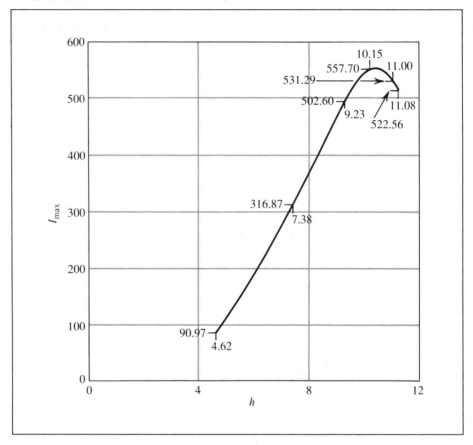

9.2.3 Golden Section Search

The golden section search technique is similar to the equal division and the Fibonacci search methods, with one important difference, it is not necessary to determine the number of divisions and calculations to use before starting the search. The method can be continued until either the size of the division reaches a preset value, or the difference in the value of the function reaches a preset value. This is particularly advantageous when using the computer because a search loop can be continued until a predetermined level of accuracy is reached. The interval divisions are calculated using the golden section ratio value 0.618. The value of 0.618 is the ratio of the width of a rectangle to its length which is considered most pleasing to geometricians and architects. The rectangle ratio of 0.618 is equivalent to dividing a line into a golden section. This is done by starting with a line AB and locating a point C on it such that $AC/AB = BC/AC$ (see Exercise 9.28).

The golden section search for the extreme of a function, $f(x)$, on an interval $[a, b]$ (or (a, b)) begins in a manner similar to the Fibonacci search. Two evaluations are performed, one at $x_1 = a + (0.618)(b - a)$ and one at $x_2 = b - (0.618)(b - a)$. The left or right interval is discarded as was done in the Fibonacci search and a new interval established, either (a, x_2) or (x_1, b). At this point the second difference between the Fibonacci and the golden section searches becomes important. In the Fibonacci search the new evaluation was performed at a calculated distance measured from the end of the interval that did not change. In the golden section search the next evaluation is performed at a distance from the end of the interval that did change; so the next evaluation is performed at either $[x_2 - (0.618)(x_2 - a)]$ or $[x_1 + (0.618)(b - x_1)]$. The internal increment selection-evaluation-discard-new interval iteration is continued until one of three preset conditions is reached:

1. Until a fixed number of evaluations are performed. This is still a real restriction if the cost of each evaluation is high.
2. Until the interval reaches a predetermined small value.
3. Until the difference in function value, $f(x_{n+1}) - f(x_n)$, reaches a predetermined value.

The next example uses the golden section search on the conditions from Example 9.3 so the methods can be compared.

EXAMPLE 9.8:
Golden section search used on Example 9.3.

Find the maximum value of $I = bh^3/12$ subject to the restrictions $b^2 + h^2 = 12^2$, $0 < h < 12$, and $0 < b < 12$. On the interval $(0, 12)$ the first two calculations are performed at $h_1 = 12 - (0.618)(12) = 4.58$, and $h_2 = 0 + (0.618)(12) = 7.42$ (see Figure 9.11).

For $h_1 = 4.58$, $b_1 = 11.09$, and $I_1 = 88.79$.
For $h_2 = 7.42$, $b_2 = 9.43$, and $I_2 = 321.03$.

The second interval is $(4.58, 12)$:

$h_3 = 4.58 + (0.618)(12 - 4.58) = 9.17$, $b_3 = 7.74$, and $I_3 = 497.36$.

The third interval is $(7.42, 12)$:

$h_4 = 7.42 + (0.618)(12 - 7.42) = 10.25$, $b_4 = 6.24$, and $I_4 = 559.98$.

The fourth interval is $(9.17, 12)$:

$h_5 = 9.17 + (0.618)(12 - 9.17) = 10.92$, $b_5 = 4.98$, and $I_5 = 540.40$.

The fifth interval is $(9.17, 10.92)$:

$h_6 = 10.92 - (0.618)(10.92 - 9.17) = 9.84$, $b_6 = 6.87$, and $I_6 = 545.46$.

The maximum value of I found with the golden section search, using the same number of evaluations, is nearly equal to the maximum found using the equal division search. The optimum values determined in general depend on the nature of the function and the number of calculations performed.

FIGURE 9.11

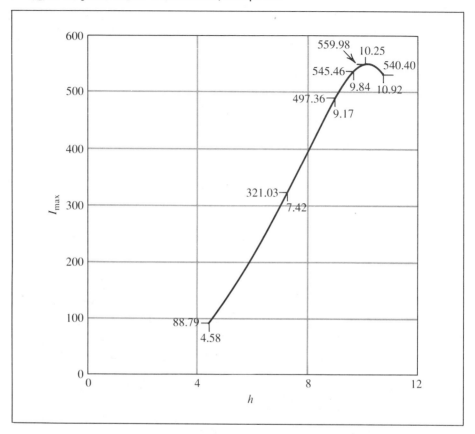

I_{max} for the golden section search method, Example 9.8.

Using the equal division, Fibonacci, or golden section search techniques is no guarantee that the global extreme within the interval will be located. There is also no reason to believe that any one of the methods is more likely than the others to produce the highest maximum, or lowest minimum. Any number of things can happen to a function within a sub-interval, and each method uses a slightly different sub-interval length. High slope values cause rapid changes in function values, and a spike condition may occur within an interval and be missed if the interval is too large.

A second problem arises with the Fibonacci and golden section searches if the evaluation at any two consecutive calculations results in the same function value. If this happens it is not obvious which sub-interval to discard. The usual solution to this problem is to make an evaluation assuming the left segment is discarded, then make an evaluation assuming the right segment is discarded, and select the best of the two choices. If neither calculation improves the extreme value already located, then the midpoint of the interval indicating equal extreme values can be used in a final attempt at improving the extreme value, or the search can be discontinued. An alternative solution to the problem is to reduce the size of the segments and use a larger number of calculations. For function evaluation this solution is satisfactory, particularly if a computer is available; but for experimentation it is not usually practical to increase the number of experiments arbitrarily.

If a function is being evaluated, and a computer is available, then the increment used on the independent variable(s) can be small enough to avoid the problem of missing narrow spikes (or valleys). A second advantage of using the computer is that all local extreme values can be found if a sequential equal division search method is used.

9.2.4 Slope Search

The slope search method is similar to the equal division search method but it permits beginning the search at any value within the interval under study. The equal division search started evaluation at the low end of an interval and progressed incrementally toward the upper end. By selecting a starting point within the interval, at the midpoint for example, it is likely that a local extreme will be located faster. This increase in speed and reduction in the number of evaluations is particularly important if each evaluation requires an experiment rather than analytical evaluation. The speed of analytical evaluation, by hand or by computer, can also be increased by the slope search, but only local extreme values will be located.

Start the slope search method by selecting the midpoint of the interval, $x_1 = (b - a)/2$, or a point where it is believed the extreme to be located, and a search increment, Δx. The function is evaluated at three points: the base point x_1, $x_1 + \Delta x$, and $x_1 - \Delta x$. If a maximum is being sought and $f(x_1 + \Delta x) > f(x_1 - \Delta x)$ the second base point $= x_2 = x_1 + \Delta x$; if $f(x_1 - \Delta x) > f(x_1 + \Delta x)$, then $x_2 = x_1 - \Delta x$. If the search is for a minimum, then the direction of the inequalities is reversed. Once a direction of function increase is determined, then advancement toward a maximum is continued for a fixed number of evaluations, or until $f(x_{i+1}) < f(x_i)$ at which time $(x_i, f(x_i))$ is a local maximum. In a similar manner, advancement toward a minimum is continued for a fixed number of evaluations, or until $f(x_{i+1}) > f(x_i)$ at which time $(x_i, f(x_i))$ is a local minimum. If the function is analytic, and a computer is available, then Δx can be small and the answer can be made quite accurate. If the problem is not analytic and experimentation is required, then the economics of the size of Δx will limit the accuracy. The method does not give all extreme values if the search is stopped at the first maximum or minimum located. The slope search method is illustrated in the next example using the conditions from Example 9.3 so comparison with the other methods can be made.

EXAMPLE 9.9:
Slope search method used on Example 9.3.

Find the maximum value of $I = bh^3/12$ subject to the restrictions $b^2 + h^2 = 12^2$, $0 < h < 12$, and $0 < b < 12$. To be consistent with the other methods, six evaluations will be used. If the midpoint of the interval is selected as the starting value, then $h_1 = (12 - 0)/2 = 6$, and by letting $\Delta h = 1.2$, then five evaluations will divide the half interval equally. One evaluation is necessary in the other half of the interval to determine in which direction to increment. All six evaluations will not be required if the maximum is found early in the search.

FIGURE 9.12

I_{max} for the slope search method, Example 9.9.

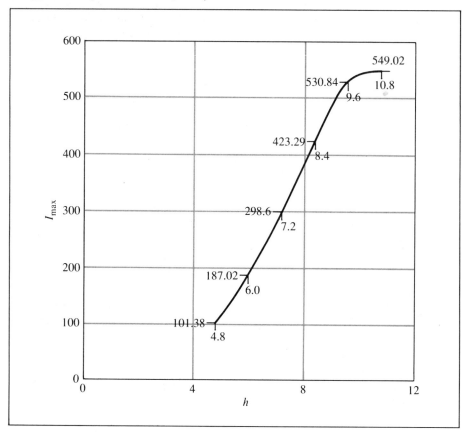

Start evaluation at the base point $h_1 = 6$, then $b_1 = 10.39$, $I_1 = 187.02$. The next two calculations are done on either side of the base point (see Figure 9.12).

$h_2 = h_1 + \Delta h = 7.20$, $b_2 = 9.60$, $I_2 = 298.60$, and

$h_3 = h_1 - \Delta h = 4.80$, $b_3 = 11.00$, $I_3 = 101.38$

The new base point becomes $h = 7.20$ because it resulted in the largest output, and the next evaluation is made at an increment larger than the base point because that is the direction of increasing output:

$h_4 = 7.20 + 1.20 = 8.40$, $b_4 = 8.57$, and $I_4 = 423.29$.

Continue increasing h in that direction because $423.29 > 298.60$:

$h_5 = 8.40 + 1.20 = 9.60$, $b_5 = 7.20$, and $I_5 = 530.84$.

Continue increasing h in that direction because $530.84 > 423.29$:

$h_6 = 9.60 + 1.20 = 10.80$, $b_6 = 5.23$, and $I_6 = 549.02$.

9.2.5 Method Comparison

Two factors are important when comparing the efficiency of search methods.

1. The speed of the method to reach a given limit of accuracy, evaluated by noting the interval size that likely contains the desired extreme;
2. The extreme value obtained.

The final interval of the search methods, and the numerical value obtained, taken from Examples 9.3, 9.6, 9.7, 9.8, and 9.9 are summarized in Table 9.2.

The Fibonacci search ends with the smallest interval when the methods all use the same number of evaluations. This is always true; the Fibonacci search converges to an extreme value faster than the other search methods. There is no reason to believe, however, that there is a correlation between the final interval size and the maximum (or minimum) value obtained. Any of the search methods may result in the most extreme value. In this example the smallest final interval size produced the third largest extreme.

9.2.6 Multi-dimensional Search

All of the search methods can be extended and used on multi-variable problems, but the number of evaluations increases rapidly. If six evaluations per variable are being considered then an n variable problem requires 6^n evaluations. This is not much of a problem for analytic function evaluation, particularly if a computer is used, but for experimental evaluation it can be an important factor as it increases the time and cost of data collection. The following example illustrates the equal division search on a multi-variable problem.

EXAMPLE 9.10:
Multi-variable, equal division search.

Consider the total heating cost per year for a special purpose, 1000 cubic unit rectangular storage cubicle, which is estimated at $4/year per square unit of ceiling area, $3/year per square unit for the east, north, and west walls, and $2/year per square unit for the south wall. Find the dimensions of the cubicle for minimum heating cost, and determine the annual heating cost.

TABLE 9.2

Comparison of optimization methods.

Method	Example #	Final interval size	I_{max}
First derivative	9.3	0	560.81
Fibonacci search	9.7	11.08 – 10.15 = 0.93	557.20
Equal division search	9.6	12.00 – 10.29 = 1.71	560.21
Golden section search	9.8	10.92 – 9.17 = 1.75	559.98
Slope search	9.9	12.00 – 9.60 = 2.40	549.02

Let x = the south and north wall length dimension, y = the east and west wall length dimension, and z = the height of the building. Then the annual heating cost is $C = 4xy + 3[2(zy) + zx] + 2(zx) = 4xy + 6zy + 5zx$, and $xyz = 1000$ is the volume restriction.

Solve $1000 = xyz$ for any one of the variables, substitute in the cost equation, determine a reasonable domain for the other two variables, select a sub-interval for each domain, substitute sequentially in the cost equation, and select the minimum value calculated.

Solve $xyz = 1000$ for x, $x = 1000/yz$, substitute for x in $C = 4xy + 6zy + 5zx$, and $C = 4000/z + 6zy + 5000/y$. The domain for y and z, and the increment to use are arbitrary, but it seems logical to start with small values and work up rather than trying to determine an upper limit and work down; let $\Delta y = 3$ and $\Delta z = 3$. Table 9.3 summarizes the search calculations, and Figure 9.13 shows the graph of the data.

The results of the search for minimum cost in the intervals $0 < y < 18$ and $0 < z < 18$ are $y = 9$, $z = 9$, and cost = \$1486. Substitute $y = 9$, and $z = 9$ into $x = 1000/yz$, and $x = 12.3$. If it is desirable to find a more accurate solution, then the intervals $6 < y < 12$ and $6 < z < 12$ can be divided into smaller increments and calculations continued, for example let $\Delta y = \Delta z = 1$.

The Fibonacci search and the golden section search are applied to multi-variable functions in a manner similar to the equal division search. Each variable is assigned values based on the interval division criteria of the method selected, and the function is evaluated at each variable combination.

The slope search method can also be extended to n dimensions, but the number of slopes to calculate increases to at least 2^n at each point selected. In a two-variable problem ($n = 2$) the four incremental slopes from a base point (x_1, y_1) are calculated to $(x_1, y_1 + \Delta y)$, $(x_1, y_1 - \Delta y)$, $(x_1 + \Delta x, y_1)$, and $(x_1 - \Delta x, y_1)$. If diagonal slopes are considered, then the number of calculations is 2^{n+1}, and additional slopes are calculated from (x_1, y_1) to $(x_1 + \Delta x, y_1 + \Delta y)$, $(x_1 + \Delta x, y_1 - \Delta y)$, $(x_1 - \Delta x, y_1 + \Delta y)$, and $(x_1 - \Delta x, y_1 - \Delta y)$. After the incremental evaluations are made from the first base point, the base point moves to the location where the function has the largest value if a maximum is being sought, or the smallest value if a minimum is being sought. The procedure is continued until the four calculations from the current base point all result in smaller (or larger) values of the function.

If the slope search method is used for experimental situations, then often the base point is moved more than one increment in the direction of largest (or smallest) output. The hope is that fewer experiments can be run, and that the global extreme will not be skipped over in the process.

9.3 Linear Programming

Linear programming is a non-calculus optimization tool used to maximize or minimize a function, $F(x_i)$, which is dependent upon a variety of *inequality* constraints, $R_j(x_i)$, rather than *equality* constraints. The inequalities may be < 0, ≤ 0, > 0, or ≥ 0. The function to be optimized, and the constraints are usually continuous on the interval under consideration. If the assumption of continuity is not valid, then the optimum solution

TABLE 9.3

Cost in dollars for incremental values of y and z for Example 9.10.

		y				
		3	6	9	12	15
	3	3036	2275	2051	1966	1937
	6	2441	1716	1546	1515	1540
z	9	2273	1602	**1486**	1509	1588
	12	2216	1599	1537	1614	1747
	15	2203	1640	1632	1763	1950

FIGURE 9.13

Graph of multi-variable search, Example 9.10.

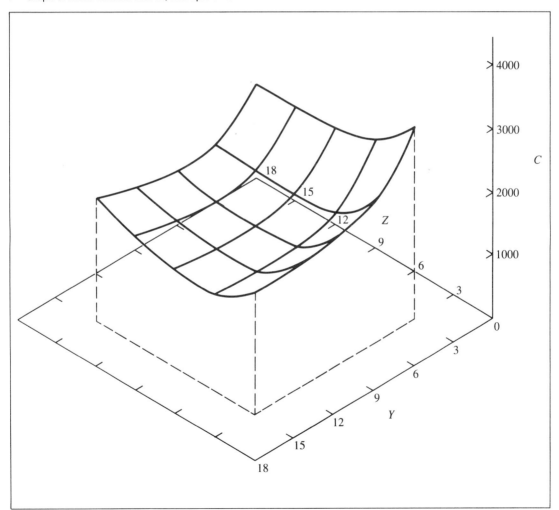

may have to be modified to fit the physical environment as was the situation in Case Study #10, Tool Life Optimization.

Linear programming is an analysis method used in business and industry to help make decisions on questions of product mix or resource distribution planning. As with other analysis methods, linear programming results must be adjusted for the practicalities of using resources such as time, money, and personnel. The following examples illustrate the general technique of linear programming, and practical implementation concerns that should follow.

EXAMPLE 9.11:

Optimize the output of a chemical process.

The maximum production of a particular chemical process is 1000 units per day. The chemical process is capable of making two product grades, A and B. A minimum of 500 units of grade A and 200 units of grade B are required each day to meet customer commitments. The profit on the sale of grade A is $2 per unit and $1 per unit on grade B. If all the output from the process can be sold, what is the maximum profit possible from the process and what output ratio will produce this result?

Let A = Units of grade A to be produced per day

Let B = Units of grade B to be produced per day

Let P = Total profit per day, dollars

The objective is to maximize the profit function $P(A, B) = 1B + 2A$, subject to the constraints: $A \geq 500$ units, $B \geq 200$ units, and $A + B \leq 1000$ units

The shaded area of Figure 9.14 shows the region of the A-B plane that satisfies all the constraints. The coordinates of the vertices of the shaded area can be found by carefully reading the graph or by solving simultaneous equations. The object is to determine the location of the profit function on the graph that results in highest profit. Because the profit function is linear, only the vertices of the shaded area need be checked in the profit function to determine a maximum profit. When the function to maximize or minimize is not linear, then its location on the graph must be determined using the equations of the lines between vertex points as well as the vertices.

$P(200, 500) = 200 + 1000 = \$1200$

$P(500, 500) = 500 + 1000 = \$1500$

$P(200, 800) = 200 + 1600 = \$1800$, which is maximum. The graph shows this profit line as a dashed line.

The required production for a maximum profit of $1800 per day is 200 units of grade B per day and 800 units of grade A per day. The solution to this problem would not have to have been integral values; decimal parts of a unit are acceptable because production is continuous.

This is a simple straightforward example of the linear programming procedure. Though many students feel at this point they are done, the analytic solution is only the start of a complete analysis. In a real example additional questions should be asked and further analysis conducted considering such items as included in the following list.

FIGURE 9.14

Graph for Example 9.11.

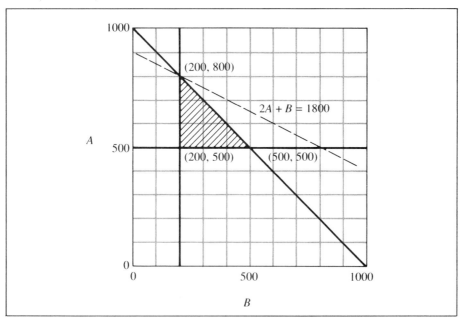

1. The difference in profit between grade A and B indicates a preference to sell grade A. Why not make only grade A?
2. Perhaps there is too much capacity if just grade A is produced. Was grade B added to the product line just to use up excess capacity?
3. Perhaps the difference in profit indicates inefficiencies in the production of grade B. What can be done to improve the profitability of grade B?
4. Perhaps the profit on grade B is low because of market competition and this is a temporary situation. Should grade B be advertised more? Can additional markets be found for grade B?
5. It is assumed that all of the product produced can be sold. Should there be a limit on the daily output of either or both grades?
6. Is this profit worth the investment in the machinery needed to produce the product?
7. Is the market potential great enough to warrant investment in more productive equipment?
8. Are there waste products that occur that need special treatment? Can they be used to improve the profitability of the whole process?
9. Are there economies of production mix ratios that should be investigated?

As an engineer, always go beyond the first question asked and consider the future of your activities in relation to the overall objectives of your company. Proper planning and appraisal of activities can make the difference between a "flash in the pan" and a sound, stable operation.

Notice also in Figure 9.14 that as the profits for grades A and B become more equal, that is, a profit function slope closer to -1, it will make little difference which grade is made, and the limiting factor is the process limit of 1000 units. Notice also that if the minimum commitment for grade B is gone, there would be no reason to make any grade B unless it is done to use personnel and machines efficiently. It is often inefficient for processes to be shut off and started up except for major overhaul or maintenance. Reducing the production of one product may raise the cost of the other beyond economic operation.

EXAMPLE 9.12:
Optimize a manufacturing mix.

A company manufactures two products, each of which is manufactured using the same three production machines. Product A requires one hour, one hour, and two hours on machines #1, #2, and #3, respectively. Product B requires three hours, one hour, and one hour on each of the machines, respectively. Each product A sold yields a profit of $20, and each product B sold yields a profit of $30. Machine #1 is available 24 hours a day, machine #2 is available only 10 hours a day, and machine #3 is available 16 hours a day due to other product line usage. Find the most profitable product mix under these circumstances.

Let A = the number of product A's to make each day

Let B = the number of product B's to make each day

Let P = the profit for a day's production, dollars

Maximize the profit function, $P(A, B) = 20A + 30B$, subject to the following constraints:

$A \geqslant 0, B \geqslant 0,$

$1A + 3B \leqslant 24$, hours, the limit on machine #1

$1A + 1B \leqslant 10$, hours, the limit on machine #2

$2A + 1B \leqslant 16$, hours, the limit on machine #3

Figure 9.15 shows the region that satisfies the constraints. The vertices of the acceptable area are found by carefully reading the graph or by solving simultaneous equations. This example illustrates an error that can occur in linear programming analysis. If the graph is not drawn, an error will occur if the intersection point (4.8, 9.6) is used. This point is the intersection of two of the lines, but it lies outside the region satisfying all the constraints.

Evaluation of the profit function for the vertices of the shaded region results in $P(0, 0) = \$0$, $P(0, 8) = \$240$, $P(3, 7) = \$270$, $P(6, 4) = \$240$, and $P(8, 0) = \$160$.

The answer to the original question is to make three product A's and seven product B's for a daily profit of $270. The optimum answer does not have to be an integer because the process is assumed continuous and products that are partially complete on one day can be completed the next day. Notice that two other vertices give answers close to $270 and a change in profit or customer demand might

FIGURE 9.15

Graph for Example 9.12.

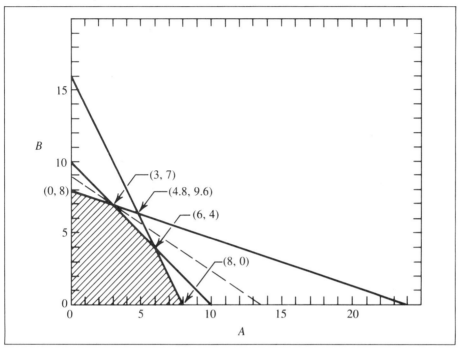

TABLE 9.4

Profit variations for Example 9.12.

P(A, B)	P(0, 8)	P(3, 7)	P(6, 4)
18A + 27B	$216	$243	$216
18A + 33B	264	285	240
22A + 27B	216	255	240
22A + 33B	264	297	264

influence the product mix decision. Suppose, for example, that the estimates for profit could be off by ±10 percent. The profit on A could range from $18 to $22, and the profit on B could range from $27 to $33. Would this make a difference? There are now four profit functions to use with the limits of product profit potential. Table 9.4 lists the four functions and the possibilities to consider. In each case the product mix for maximum profit stays the same, but maximum profit ranges from $243 to $297. This indicates that the system as modeled is "robust" (not sensitive to minor fluctuations in the values of the parameters.) This is good because minor changes are impossible to avoid. Other less robust systems could result in a different product mix as the profit ratio changes.

Linear *integer* programming is required when the values of the variables must be integers, such as the number of pipes in a heat exchanger, the number of people assigned to a task, or the number of vehicles needed to satisfy a transportation problem. Linear integer programming analysis is done using the same techniques as non-integer linear programming; however, the check for optimum may require moving away from the calculated extreme value if the calculation does not produce integers. See Exercise 9.45 for such a situation.

9.4 Additional Concerns

Engineers must use optimization techniques with care during the design process, just like all other methods that are available to them. No one method will always be the best to use, so flexibility and combinations of methods often make the search for the best more efficient. The following two case studies illustrate the flexibility and adaptability of optimization procedures, and they focus attention on additional optimization concerns. The first study illustrates a problem whose solution is facilitated by using the computer.

CASE STUDY #13:
Balcony Seating Optimization

Figure 9.16 is a layout of an older style field house basketball court with a balcony above. The balcony was once a jogging track but is now partially occupied with bleacher seating for basketball games and other activities held in the field house. The main problem with the seating in the balcony is visibility. The seats at the side of the balcony are so close to the edge of the basketball court that only first row spectators can see the whole court, and then only by leaning over a bit. The areas at the ends of the court presently do not have seating, but offer potential for better visibility because the edge of the balcony is farther back from the end of the court. In this design problem the input is not variable, the basic restriction is that no changes can be made to the existing structure, only the addition of seats is permitted.

Optimization situations such as this require a decision to be made on the output criteria, because the input is fixed. The criteria to optimize could be:

1. Place the maximum number of seats to have 100 percent visibility of the basketball court without regard to how many seats are available.
2. Maximize the number of seats that can see a defined percent of the court.
3. Maximize the number of seats without regard to visibility.

In this case one row of seats will be placed for optimum (maximum) visibility, close to the edge of the balcony and as high as practical, and the visibility determined. If each seat has 100 percent visibility, then a second row of seats will be considered. The analysis, therefore, must determine the visibility of the edge of the court from each seat.

FIGURE 9.16

Court layout for Case Study #13.

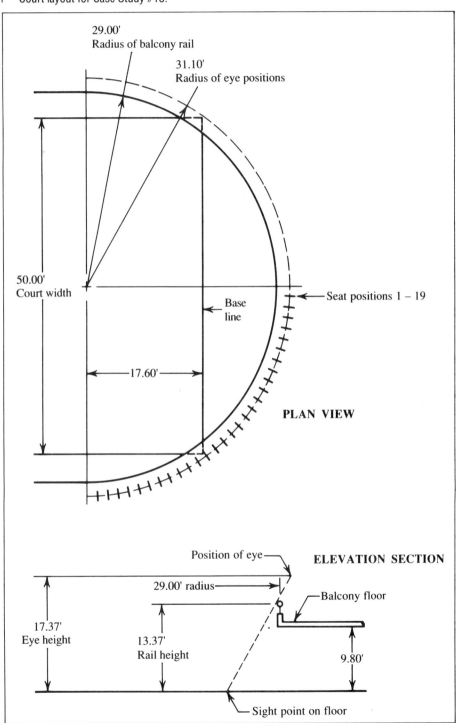

Figure 9.17 shows the details of one-half of the balcony area end zone behind one basket, the other end of the court being the same. Visibility will be improved when each seat is close to the edge of the balcony, and as high as practical. The limit of 17.37 foot (5.30 m) maximum height for the eyes of a seated spectator is controlled by the ceiling height. The spectator must have head clearance when standing up. The 31.10 foot (9.48 m) radius of the eyes of a seated spectator allows 1.30 foot (0.40 m) clearance between the edge of the spectator's seat and the balcony rail. The 1.30 foot (0.40 m) clearance is judged the minimum acceptable for the average spectator. Room between adjacent seats on the 31.10 foot (9.48 m) radius is set at 1.50 feet (0.46 m). This allows 19 seats on the 31.10 foot (9.48 m) arc before reaching the sideline. *Note*: If S = the number of seats, then $1.50S/31.10 \leq$ arctangent $(25.00/17.60)$ and $S \leq 19$.

How much of the floor can be seen from the eye position of the spectator in each of the 19 seats? A layout can be made from each eye position: P_i, $0 \leq i \leq 19$, toward various points on the court baseline; S_j, $0 \leq j \leq 25$ feet, to locate the sight points on the floor. The locus of all such points describes the visibility range for each seat. This locus will be an ellipse because it is the intersection of a cone and a flat plane.

For 19 seats and 25 vision lines, 475 separate layouts need be made to determine the visibility patterns, and that is just for half the baseline; 475 are also required for the other half of the base line. Perhaps there is an easier way.

If an equation can be developed to define the geometry of the situation, then a computer program can be written to solve the problem. Following the graphical method, the computer algorithm is as follows.

1. Determine the coordinates of point P_i using the equations $XP = 31.10 \cos(1.50i/31.10)$, $YP = 31.10 \sin(1.50i/31.10)$, and $ZP = 17.37$. The argument of the sine and cosine are in radians, the center of the balcony radius is the origin of the coordinate system, and $0 \leq i \leq 19$.

2. Determine the equation of the line, $Y = mX + b$, on the floor from directly under $P_i(XP, YP)$ to $S_j(XS, YS)$. This changes for each vision line. For a complete analysis, $-25 \leq j \leq +25$. Only when $i = 0$ is the pattern symmetric and $0 \leq j \leq 25$ adequate.

3. Solve the equation from step #2 with the equation of the projection of the balcony edge on the floor, $XC^2 + YC^2 = 29.00^2$. The results are the coordinates of point $L(XL, YL, 13.37)$.

4. Calculate $AISLE = \sqrt{(XP - XL)^2 + (YP - YL)^2}$. $AISLE$ should always be greater than or equal to 2.1.

5. Calculate $A = 13.37[AISLE/(17.37 - 13.37)]$.

6. Calculate $HY = \sqrt{(XS - XP)^2 + (YS - YP)^2}$, and if $A < HY$, then the baseline is visible.

7. Calculate $YA = YP - A(YP - YS)/HY$.

8. Calculate $XA = (YA - b)/m$.

9. Increment j until $j = 25$ and check again.

10. Reset $j = 0$, increment i, and start again.

11. Quit when $i = 19$.

With graphics added, the vision limits for each seat can be generated. Figure 9.18 illustrates the vision limit for seat #5 ($i = 5$). When the data from the computer for each seat are analyzed, it is discovered that at least 82 percent of the baseline can be seen from each of the 19 seats, but only from the first 5 seats from the court center line can the entire baseline be seen.

The case study solution is a compromise on the mode of optimization, for it allows only one row of seats to be placed around the curved end, and the question was how much of the court could be seen from each seat.

FIGURE 9.17

Court details for computer program.

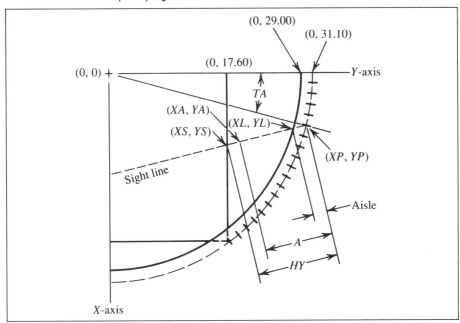

FIGURE 9.18

Line of sight from seat #5.

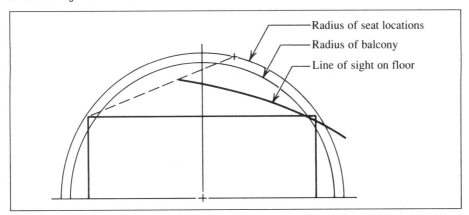

The next case study illustrates how existing material sizes, properties, and costs can influence the selection of an optimum material for a design requiring a minimum capacity, a restricted search problem.

CASE STUDY #14:
Spherical Storage Tank Optimization

Consider a spherical storage tank to be designed to withstand a maximum internal pressure of 100 psi (0.689 MPa).

The stress in the shell, if $t \leq 0.05D/2$, is $\sigma = Dp/4t$, where (9.7)

σ = allowable material tensile stress, psi (MPa)
D = shell internal diameter, inches (meters)
p = internal pressure, psi (MPa), and
t = shell thickness, inches (meters).

The cost of the shell material is: $C = k\rho\pi D^2 t$, where: (9.8)

C = material cost, dollars
ρ = material weight density, lb/in³(kg/m³), and
k = material cost, dollars/lb (dollars/kg).

Choose a material from the list in Table 9.5 that will give lowest cost for this application.

Some algebraic manipulation may help the analysis. Begin by solving Equation 9.7 for D and substituting in Equation 9.8, then

$$C = 16k\rho\pi t^3 \sigma^2/p^2. \qquad (9.9)$$

Some calculus students might take the first derivative of Equation 9.9 and set the result equal to zero to get a minimum cost, but this gives no practical value for t. Try it.

TABLE 9.5

Materials available for Case Study #14.

Material	k $/lb	k $/kg	ρ lb/in³	ρ g/cm³	σ lb/in²	σ MPa
1	0.26	0.57	0.282	7.81	13,000	89.6
2	0.29	0.64	0.282	7.81	14,000	96.5
3	0.31	0.68	0.282	7.81	15,000	103.4
4	0.75	1.65	0.282	7.81	30,000	206.8
5	0.70	1.54	0.098	2.71	10,000	68.9

As a beginning, it might be of value to keep the numerator of the cost function a minimum by making the product $k\sigma^2\rho$ a minimum. Table 9.6 gives these results. This preliminary analysis does not give a conclusive answer to the question because no overall tank size has been given. How big does the tank have to be? More data is required. Assume the tank will have to hold a minimum of 10,000 ft³ (283.2 m³).

The volume of the tank is $V = (4\pi/3)(D/2)^3$ and the corresponding value of D for a minimum volume of 10,000 ft³ (283.2 m³) is $D \geq [(10,000 \text{ ft}^3)(1728 \text{ in}^3/\text{ft}^3)(6/\pi)]^{1/3} = 321$ inches (8.15 m). From $D = 4t\sigma/\rho$, $t\sigma \geq (321 \text{ in.})(100 \text{ lb/in}^2)/4 = 8020$ in.·lb/in², and $t \geq 8020/\sigma$ inches (1.403/σ m). For each material a limiting value of t can be determined (see Table 9.7). Commercially available sizes are also included in Table 9.7, based on the assumption that material is available in 0.0625 inch (1.588 mm) increments.

Table 9.8 lists the tank size available for each preliminary answer from Table 9.7. Volumes are compared by calculating D from $D = 4t\sigma/\rho$ and then the volume from $V = (4\pi/3)(D/2)^3$. The increased thickness over the minimum allows the tank to be built larger than the minimum of 10,000 ft³ (283.2 m³). Material #1 or

TABLE 9.6

Material property products for Equation 9.9 and materials from Table 9.5.

| Material | $k\sigma^2\rho$ | | Rank |
	English	Metric	(1 = smallest)
1	1.24·10⁷	3.57·10⁴	2
2	1.60·10⁷	4.65·10⁴	3
3	1.97·10⁷	5.68·10⁴	4
4	19.04·10⁷	55.11·10⁴	5
5	0.69·10⁷	1.98·10⁴	1

TABLE 9.7

Minimum material thickness, Case Study #14.

| Material | Calculated minimum t | | Standard available t | | Rank |
	inches	mm	inches	mm	(1 = thinnest)
1	0.617	15.67	0.625	15.88	3,4
2	0.573	14.55	0.625	15.88	3,4
3	0.535	13.59	0.5625	14.29	2
4	0.267	6.78	0.3125	7.94	1
5	0.802	20.38	0.8125	20.64	5

TABLE 9.8

Actual tank diameter and volume, Case Study #14.

Material	Purchased thickness, t		Diameter based on purchased t		Tank volume		Rank (1 = closest to objective)
	inches	mm	inches	mm	ft³	m³	
1	0.625	15.88	325	8.26	10,400	294.5	1,2
2	0.625	15.88	350	8.89	12,990	367.9	4
3	0.5625	14.29	338	8.59	11,700	331.4	3
4	0.3125	7.94	375	9.53	15,980	450.0	5
5	0.8125	20.64	325	8.26	10,400	294.5	1,2

TABLE 9.9

Cost of storage tank material, Case Study #14.

Material	Cost based on purchased t	Cost of storage $/ft³	$/m³
1	$15,200	1.46	51.55
2	$19,670	1.51	53.32
3	$17,600	1.50	52.97
4	$29,200	1.83	64.62
5	$18,500	1.78	62.85

#5 appears to be "best," but cost still must be checked. Table 9.9 lists costs based on the value of purchased t listed in Table 9.8.

In this particular case, material #1 has the lowest total cost and the lowest cost per unit volume, and results in the optimum solution. If a safety allowance of 10 percent is used for storage volume, then materials #2 and #3 become more attractive. If the manufacturing processes can only handle a maximum of 0.500 inch (12.7 mm) thickness, then material #4 would be the optimum selection.

This analysis has been simplified and many other factors need to be considered before a tank can be fabricated. Other factors to be considered include:

1. Total weight and the cost of the supporting structure
2. Land area required, its availability, and cost
3. Annual maintenance cost for painting, a function of surface area and type of paint used
4. Delivery schedule of the preferred material compared to the project schedule

5. Weldability of the material selected
6. Fracture toughness of the selected material
7. Anticipated environment (salt water, acid rain).

9.5 Experimental Optimization

Several times in this chapter reference has been made to experimental optimization, and for many engineering problems it is the only solution method available. Experimental techniques are necessary when the physical situation cannot be modeled with an analytic function, or when variation of an input is observed to have an effect on the output, but the mathematical relationships are not known. This is a common occurrence in many engineering design situations. Frequently it is not known which variables have an effect on the outcome, or, if the outcome is affected, by how much. Experiments are conducted to answer these questions. Investigations are conducted to study relationships such as:

1. Material hardness as a function of annealing time with temperature held constant
2. Material hardness as a function of annealing temperature with time held constant
3. Melting temperature of a two element alloy as a function of element ratios.

Perhaps *after* an experiment is run and data collected a mathematical relationship may be developed. Then analytic methods can be used for additional analysis. If only one variable is under investigation, any of the search methods of Section 9.2 may be used to study the experimental input-output relationship.

The basic search methods can also be used if more than one variable is being considered if each variable is assigned incremental values within its domain, but this increases the number of experiments exponentially. If the cost of each experiment is high, or the time to conduct the experiment is long, multiple variable experiments can prolong an investigation.

An alternate to the basic search methods is to select and test various combinations of the variables and carefully study the results. Pre-experiment decisions are made to select the domain of values in the experiment and the increment used to cover the domain. For a two variable experiment four combinations are used to start the process:

1. Assign each variable the value at the low end of its domain
2. Assign each variable the value at the high end of its domain
3. Assign the first variable the value at the low end of its domain and the second variable the value at the high end of its domain
4. Assign the first variable the value at the high end of its domain and the second variable the value at the low end of its domain.

The results of these four experiments will indicate where macro changes are occurring, and indicate where subsequent experiments should be conducted to lead to the desired extreme. The following example illustrates this procedure, and also illustrates the Yates algorithm used to determine which variable(s), if any, is (are) most active. A change in the most active variable affects the output of the experiment the most.

EXAMPLE 9.13:
Experimental optimization for a two variable problem.

Consider an experiment in two variables, a and b.

Experiment #1, when both variables were at their low level, symbolized as the $(--)$ condition, had an output of 210. Experiment #2, when a was kept low and b was high, symbolized $(-+)$, had an output of 180. Experiment #3, when a was at its high value and b was at its low level, symbolized $(+-)$, had an output of 240. Experiment #4, when a and b were both at their high level, symbolized $(++)$, was an output of 200. The conclusion is that a change in a is the best thing to do to increase the output level, because $240 > 210 > 200 > 180$. The conclusion cannot be made that the condition $(+-)$ produces maximum results, but the maximum is probably in the vicinity of $(+-)$, and additional experiments can be conducted near that condition in search of a better optimization. The rectangular domain area created by conditions $(--)$, $(-+)$, $(+-)$, and $(++)$ can be subdivided into quarters, eights, or smaller partitions to narrow the region in which to conduct additional experiments.

The Yates analysis algorithm, used to determine the most active variable(s), requires that the experimental results be listed in special order by experimental code. Table 9.10 shows this order and also the calculations required for the additional analysis. The required sequence of table entries is controlled by column #1, the experimental code.

Column #2 entries are the results of the experiment. The numbers in the column headed (1) are obtained from the experimental response, column #2. Entry #1 in column (1) is the sum of the first two experimental response values; entry #3 is their difference. Entry #2 in column (1) is the sum of the third and fourth experimental response values; entry #4 is their difference. The entries in column (2) are determined using the values from column (1) and performing the same operations as were performed on the experimental response entries to obtain the entries in column (1). Be sure to subtract the number lower in the column from the number higher in the column. Two iterations are required because there are two variables. The second, third, and fourth entries in column (2) indicate the influence of each of the variable combinations on the outcome of the experiment, and the improvement due to a alone is confirmed because $50 > -10 > -10$. The first entry in the effect estimate column is obtained by dividing the first entry from column (2) by 2^2 and the remaining entries are obtained by dividing the column (2) entries by 2^{2-1}. The exponent 2 in these calculations is used because there are two variables under study.

In many engineering situations, it is not only desirable to optimize the situation, such as maximum output or minimum cost, but also to determine the sensitivity of the system to variation and combinations of one or more variables. This condition can be analyzed by performing a graphical analysis, hence the need for the last three columns in Table 9.10.

The graphical analysis is performed as follows. List all entries except the first from the effect estimate column in order of magnitude, along with P, their respective percentage of the whole. The percent of the whole is determined using the formula $P = 100(i - 1/2)/n$, where $i = 1, 2, \ldots, n$, and $n =$ the number of values listed. Table 9.11 shows the results of this listing and calculation for the data from Table 9.10. After the calculations are made, the values are plotted on normal probability paper as shown in Figure 9.19. In this problem the three points lie on a

TABLE 9.10

Calculations for the Yates analysis method for the two variable analysis of Example 9.13.

Experimental code	Experimental response	(1)	(2)	Effect estimate
(− −) 1	210	450	830	208
(+ −) a	240	380	50	25
(− +) b	180	30	−70	−30
(+ +) ab	200	20	−10	−5

TABLE 9.11

Evaluations to use for graphical analysis of variable activity.

i	1	2	3
P	16.7 percent	50 percent	83.3 percent
Effect estimate	−5	−35	25

straight line, so no condition appears more active than any other, and it is not clear which variable to manipulate for maximum response. Had there been a factor more active than the others, the points would not have been on a straight line. If the points are not on a straight line, the variable(s) responsible for the point(s) highest and to the right are the most active.

When three variables are under consideration, the minimum-maximum variable combinations again can be used, and if properly analyzed, can also lead to the variables that are having the greatest effect on the experimental outcome. The following case study illustrates an experiment conducted to determine the most active variable of a three variable analysis, as well as progressing toward a maximum output value.

CASE STUDY #15:

Experimental Optimization Performed on Three Variables.

This problem requires the maximization of the output from an alternating current electric generator being driven by a windmill. A preliminary analysis was conducted and the decision was made to limit the investigation to the following variables:

1. The number of blades on the windmill shaft
2. The shape of the blades
3. The angle that the blades made with the plane of rotation of the windmill.

FIGURE 9.19

Effect estimate plot for Example 9.13.

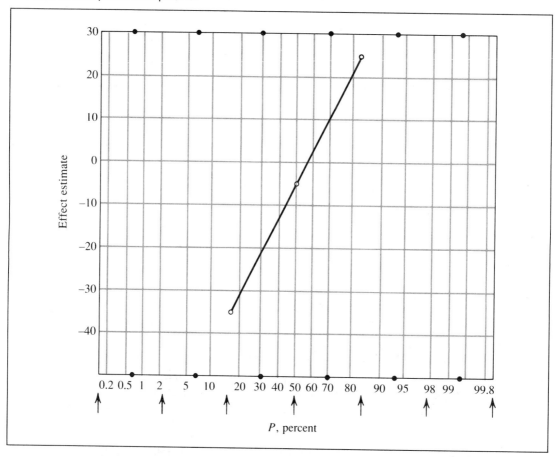

All experiments used a wind speed of 8 miles/hour (12.88 kilometers/hour), and a blade diameter of 15.40 inches (0.391 m). The power output was monitored by measuring the voltage and current flow of a one resistor circuit. Preliminary experiments determined which direction to proceed for additional experiments, and which of the three variables is the most active and influential on maximizing the output.

Each variable was coded as follows. The variable a was assigned to the number of blades, with a low value (−) of two blades and a high value (+) of four blades. The variable b was assigned to the shape of the blades, with a low value (−) representing a flat blade and a high value (+) assigned to an airfoil shape. Variable c was assigned to the angle of the blade with the plane of rotation, with a low value (−) of 55° and a high value (+) of 45°. The decision as to which value is the + and which is the − is arbitrary, final analysis sorts out the variation direction.

A three variable problem requires 8 experiments to be run because there are $2^3 = 8$ combinations possible when three variables each have 2 values. The eight combinations and the experimental order used are shown in Table 9.12, generalized using the (+) (−) notation. The order of the experiments is not critical but the coding is. When all variables are at the low level the condition (− − −) exists and is coded 1. Each subsequent code uses the letter for the variable(s) that are in the high value mode, hence (+ + −) is ab, (+ − +) is ac, and (− + −) is b.

The next step in the procedure requires a special order of the experimental code as listed in Table 9.13. The first column of Table 9.13 is the three variable extension of the first column of Table 9.10. Table 9.13 also includes the experimental response and the necessary Yates analysis calculations.

Experimental response shows that increasing the number of blades from two to four is the greatest single factor affecting higher output. Changing the angle from 55° to 45° is the second most important factor affecting higher output, and

TABLE 9.12

Combination codes and output for eight experiments in a three variable experiment.

Experiment number	Variable A: Number of blades	Variable B: Shape	Variable C: Angle	Experiment code	Output: milliwatts
#1	− (2 blades)	− (flat blades)	+ (45°)	c	7.1
#2	+ (4 blades)	−	+	ac	14.1
#3	−	+ (airfoil blades)	+	bc	5.7
#4	+	+	+	abc	15.8
#5	−	−	− (55°)	1	4.5
#6	+	−	−	a	8.2
#7	−	+	−	b	4.7
#8	+	+	−	ab	11.0

TABLE 9.13

Calculations for the Yates method of experimental data analysis for three variables.

Experimental code	Response: milliwatts	1	2	3	Effect estimate
1	4.5	12.7	28.4	71.1	8.9
a	8.2	15.7	42.7	27.1	6.8
b	4.7	21.1	10.0	3.3	0.8
ab	11.0	21.5	17.1	5.7	1.4
c	7.1	3.7	3.0	14.3	3.6
ac	14.1	6.3	0.3	7.1	1.8
bc	5.7	7.0	2.6	−2.7	−0.7
abc	15.8	10.1	3.1	0.5	0.1

changing the blade shape has the least (but some) effect on increasing output. This observation is based on the data relationships 8.2 > 7.1 > 4.7 > 4.5. The highest output occurred when all three changes (four blades, airfoil shape, and 45°) were made at the same time, 15.8 being higher than any other output.

The additional calculations required for the Yates analysis for a three variable experiment are extensions of the calculations performed for the two variable experiment, but the addition-subtraction iteration calculation is performed three times instead of two. The first four entries in column (1) are formed by adding consecutive entries in the response column; for example, the first entry in column (1) is 12.7, the sum of the first two entries (4.5 and 8.2) in the response column. The second entry in column (1) is 15.7, the sum of the next two entries (4.7 and 11.0) in the response column. The last four entries in column (1) are the differences of the consecutive entries in the response column. The first four entries in column (2) are formed by adding consecutive entries in column (1), and the last four entries in column (2) are formed by subtracting the consecutive entries in column (1). Column (3) is formed from column (2) as column (2) was formed from column (1). The entries in the effect estimate column are obtained by dividing the first entry in column (3) by 2^3 and subsequent entries by 2^{3-1}.

Continue the analysis in search of the most active factors. The effect estimate values (from the last column in Table 9.12) are placed in order, along with the corresponding P value, as shown in Table 9.14; use the formula for P from Example 9.13. The data is graphed on normal probability paper, Figure 9.20, and shows two points that are not on the straight line formed by the other points. These two points are associated with the single factor changes (the angle of the blades, and the number of blades), not combined factor changes. The conclusion is that the number of blades alone, and the blade angle alone, are the most active variables, and should be pursued independently of the other combinations.

The second half of the study focused on the effect of blade angle on power output. Four airfoil blades were used because that combination produced increases in output in the first series of experiments. The time to run each experiment in the next series of experiments was short enough so the domain was set at 0° < angle < 90° with an increment of 10°, even though it became clear the maximum output occurred at small angles. Table 9.15 contains the results of experiment #2, the output with 10° increment changes in blade angle. The highest output of 82.4 milliwatts came at 10° and represents an 1800 percent increase over the starting output value of 4.5 milliwatts. The experiment was repeated at 45° to monitor experimental consistency. The 15.8 milliwatt output compares with a 7.0 milliwatt

TABLE 9.14

Evaluations to use for graphical analysis of variable activity.

i	1	2	3	4	5	6	7
P	7.1%	21.4%	35.7%	50%	64.3%	78.5%	92.9%
Effect Estimate	−0.7	0.1	0.8	1.4	1.8	3.6	6.8

FIGURE 9.20

Effect estimate plot for Case Study #15.

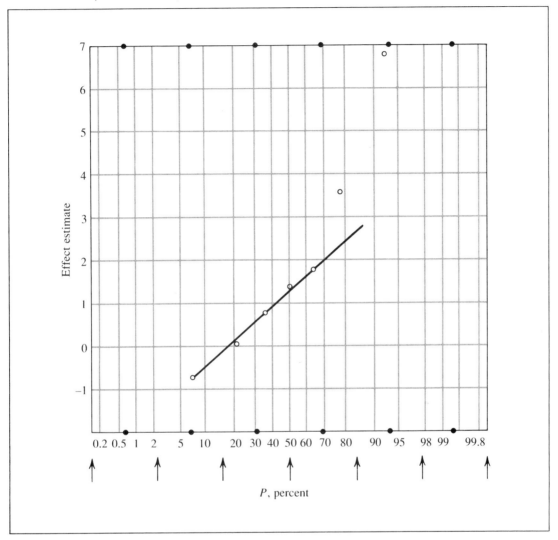

output from the (+ + +) condition from experiment #1. This variation in output is attributed mainly to a non-constant wind source (see Exercises 9.58 and 9.59), variation in bearing tightness on the wind mill shaft, and the rigidity and induced vibration in the test apparatus. The density of the air, which is dependent on atmospheric pressure, temperature, and humidity can also contribute to output variation. The mathematical model in Exercise 9.58 shows how the density of air enters into the analysis.

Table 9.15 also contains a listing of the results of experiment #3, which used an angle increment of 2°. The maximum output from experiment #3 occurred at 8°.

TABLE 9.15

Results of blade angle variation on power output.

Experiment #2, Δblade angle = 10°		Experiment #3, Δblade angle = 2°	
Blade angle	Power output, milliwatts	Blade angle	Power output, milliwatts
0°	0	0°	0
10°	82.4	2°	0
20°	46.7	4°	64.8
30°	18.5	6°	78.3
40°	8.2	8°	103.1
45°	7.1	10°	86.5
50°	5.4	12°	80.8
60°	3.4	14°	64.1
70°	2.0	16°	55.2
80°	1.1	18°	45.8
90°	0.3	20°	41.2

Notice again the experimental variation in power output at the monitoring values of 10° and 20°: 82.4 compared to 86.5, and 46.7 compared to 41.2.

Even though the output has been increased, the study is not complete. The repeatability of the experiment must be improved by controlling wind speed, perhaps by using a wind tunnel. Structural changes in the windmill support and bearing adjustment will also help control the experimental variation. Each experiment did increase the output, but percent changes are not a reliable measure of differences because of the variance in results. Optimizing based on one condition does not guarantee a global optimum, only a local optimum. Additional experiments will show whether other variations will affect the output more. These experiments should investigate:

1. The effect of additional blades
2. The effect of combinations of the number of blades and blade angle
3. The effect of other blade shapes.

To add to the difficulty of this experiment, a problem was created during the last experiment that was not present at the beginning of the study. The 8° blade angle, airfoil shape, four blade configuration, which resulted in maximum power output, would not self start. The configuration did not have enough starting torque to overcome static friction; so at the benefit of higher output, the cost of needing an auxiliary start was created. More problems for the engineer.

Optimization procedures, like all engineering analysis tools, must be selected and used after both their advantages and disadvantages have been considered. The same method will not always be the most appropriate to use. Decision judgment must be tempered with concern for accuracy, cost, personnel and equipment utilization, and the health and safety consequences of unacceptable conclusions. The results of analysis must be balanced with experience, responsibility, and a regard for professional ethics.

9.6 Additional Study Topics

1. Jacobian maximization technique
2. Derivative techniques for functions of more than three dimensions
3. Additional search techniques
    a) Method of Bolzano
    b) Method of Rosenbrock
    c) Method of Powell
    d) Brent's Praxis Method
    e) Internal halving
    f) Exhaustive search
4. Nonlinear programming techniques
5. Function root refinement techniques
    a) Bisection method
    b) Newton method

9.7 References

1. Berkey, Dennis D. *Calculus.* New York: Saunders College Publishing, 1984.
2. Budnick, Frank S. *Applied Mathematics.* New York: McGraw-Hill, 1979. Chapters 6, 7, 12, 13, and 14.
3. Coleman, Hugh W. and W. Glenn Steele, Jr. *Experimentation and Uncertainty Analysis for Engineers.* New York: John Wiley & Sons, 1989.
4. Doyle, Lawrence E. *Manufacturing Processes and Materials for Engineers,* 3rd ed. Englewood Cliffs, N.J.: Prentice-Hall, 1985.
5. Foulds, L. R. *Optimization Techniques.* New York: Springer-Verlag, 1981.
6. Frazer, J. Ronald. *Applied Linear Programming.* Englewood Cliffs, N.J.: Prentice Hall, 1968.
7. Gajda, Walter J., Jr. and William E. Biles. *Engineering: Modeling and Computation.* Boston: Houghton Mifflin, 1978. Chapter 5.
8. Hicks, Charles R. *Fundamental Concepts in the Design of Experiments*, 3rd ed. New York: Holt, Rinehart and Winston, 1982.
9. McDonald, T. Marll. *Mathematical Methods for Social and Management Scientists.* Boston: Houghton Mifflin, 1974. Chapter 8.
10. Piascik, Chester. *College Mathematics: With Applications.* Columbus, Ohio: Charles E. Merrill, 1984. Chapter 6.
11. Woodson, Thomas T. *Introduction to Engineering Design.* New York: McGraw-Hill, 1966. Chapter 15.

9.8 Exercises

1. Expand the discussion of extreme function analysis for the condition when $f^{I}(x_i) = 0$, $f^{II}(x_i) = 0$, $f^{III}(x_i) = 0$, and $f^{IV}(x_i) = 0$.
2. Verify the intersection points of the two parabolas in Example 9.2.
3. Verify that the answer of $h = 10.39$ in Example 9.3 does give a maximum by checking the value of the second derivative.
4. Verify Equation 9.3 for minimum material removal cost: $T_m = (C_t/R_c)\,((1-n)/n)$.
5. Show that $T_m = (C_t/R_c)\,((1-n)/n)$ does give a minimum cost value for removing material by evaluating the second derivative of Equation 9.2 and showing that it is greater than 0.
6. Verify the value of T for maximum production $T_M = (t_1)(1-n)/n$, and that it does give a maximum by evaluating the second derivative of Equation 9.4.
7. The following represent some typical $VT^n = k$ relationships:
   High speed steel cutter on
Steel	$VT^{0.125} = 47$ ft/min
Cast iron	$VT^{0.14} = 47$ ft/min
Nonferrous cast alloys	$VT^{0.16} = 47$ ft/min

   Cemented carbide cutter on
Steel	$VT^{0.2} = 150$ ft/min
Cast iron	$VT^{0.25} = 150$ ft/min

   Sintered carbide on
Steel	$VT^{0.7} = 400$ ft/min

   Discuss the significance of the exponent change and the constant change.
8. Verify with calculations the extreme value of $-2/9$ for Example 9.5.
9. Find any minimum or maximum values of $f(x,y) = xy + 8/x + 8/y$ for real values of x and y.
10. Find any minimum or maximum values of $f(x,y) = (x-2)^2 - y^2$.
11. Find any maximum, minimum, or inflection points of each of the following functions using calculus derivatives. Use a domain of all real numbers.
    a) $f(x) = x^4 + 6$
    b) $f(x) = x^4 - 12x^3 + 52x^2 - 96x + 64$
    c) $f(x) = x^2 - 4.25x + 4$
12. Find any maximum, minimum, or inflection points of the function $f(x) = x + \cos x$ using calculus derivatives. Use the domain $0 \leq x \leq 2\pi$.
13. A wooden I beam is made from a log 12 inches (0.305 m) in diameter shown in Figure 9.21. The center section of the I beam is made from the longest piece, the top and bottom made from the shorter pieces, and all three pieces have the same thickness. Determine the dimension for x that results in maximum area moment.
14. Will changing the thickness of the center section of the I beam in Exercise 9.13 to $x/2$ increase or decrease the maximum moment possible? By how much?

15. Verify the minimum cost of $1538 in Case Study #11 using derivative calculus techniques.

16. Verify the minimum cost of $1538 in Case Study #11 using Lagrange multipliers.

17. Verify that the optimum packing of circles within a larger circle, as used in Case Study #11, occurs on a local level when the number of small circles follows this sequence: $P_1 = 1, P_2 = 7, P_3 = 13, P_4 = 19, P_5 = 31, P_6 = 37,\ldots$ where the general term is $P_k = P_{k-1} + 6$ for each even numbered term in the sequence, and $P_k = P_{k-1} + 3(k-1)$ for each odd numbered term in the sequence, $k > 1$.

18. If the radius of each small circle enclosed in the larger circle is r, then verify that the radius of the large enclosing circle for each term in the sequence in Exercise 9.17 follows this sequence: $r, 3r, 2r\sqrt{3} + r, 5r, \sqrt{3(r\sqrt{3})^2 + r^2} + r, 7r,\ldots, (n+1)r$ when n is even,$\ldots$, $\sqrt{(((n+1)/2)(r\sqrt{3}))^2 + ((n-3)r/2)^2} + r$ when n is odd.

19. Case Study #11 used a cost estimate of a cylindrical shell as $25D^{2.5}L$ based on other shells manufactured. The use of a base model for estimating costs is not unusual, and is a satisfactory procedure within specified boundary conditions. Some of these conditions are the physical limits on material and manufacturing processes, and physical limits on the installation site including ceiling heights, door opening sizes, and available handling equipment. The cost of a cylindrical shell can be more closely estimated if more detail is used, and this allows a better comparison to be made with alternate designs. Determine a cost model for fabricating a welded cylindrical shell, made with a rectangular blank rolled into a cylinder and capped with two circular blanks. Let D = tank diameter, t = material thickness, and L = tank length. Comment on the simplification to the cost model $25D^{2.5}L$ used in the case study. Is the simplification justifiable?

20. Repeat Exercise 9.19 for a rectangular cross section tank and compare results. Compare the total cost of installing the rectangular tank with the total cost of installing the circular tank, $700 + 25D^{2.5}L + 20DL$.

FIGURE 9.21

Sections to cut from a log to construct an I beam, Exercise 9.13.

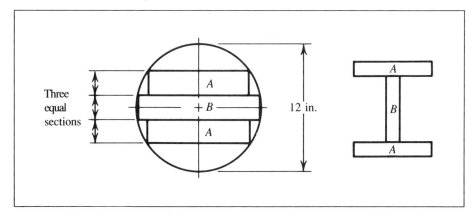

Refer to Case Study #12 for Exercises 21-24.

21. What happens to Figure 9.7 when a larger volume is used, such as 30 in^3?
22. A 20 in^3 can is being made at minimum cost with 0.0239 thick material and a 2.16 diameter.
    a) What is the height of the can?
    b) If seaming costs $0.01/inch, what is material cost?
    c) What is the total material and seaming cost?
23. Suppose 20 in^3 cans are in production with $t = 0.0239$ inches and $D = 2.28$ inches when you discover that material with $t = 0.0179$ inches is strong enough to be used. What action would you take to reduce cost?
24. Repeat the calculation for Case Study #12, Cost Ratio Analysis, using the material volume as $t(\pi Dh + 2D^2)$.
25. Repeat Example 9.6 by using seven equal division search calculations on the variable b instead of h.
26. Illustrate that the equal division method may result in the global maximum, and that fewer divisions may be more efficient than more divisions, by redoing Example 9.6 using six divisions instead of seven.
27. Use the equal division search technique on the functions in Exercise 9.11 on the interval $-2 \leq x \leq 6$. Use eight equal segments and search for a minimum function value.
28. Use the Fibonacci search technique on the functions in Exercise 9.11 on the interval $-2 \leq x \leq 6$. Use nine evaluations and search for a minimum function value.
29. Prove that the golden mean ratio is 0.618 by using the ratio definition $AC/AB = BC/AC$ given in the text. Calculate AC as a percentage of AB.
30. Illustrate that the ratio of consecutive terms of the Fibonacci sequence is the same as the golden section search ratio if $n > 8$.
31. Use the golden section search technique on the functions in Exercise 9.11 on the interval $-2 \leq x \leq 6$. Use nine evaluations and search for a minimum function value.
32. Use the equal division search technique on Exercise 9.9 looking for a minimum function value. Use six intervals on the domains $0 < x < 5$, and $0 < y < 5$.
33. Use the Fibonacci search technique on Exercise 9.9 looking for a minimum function value. Use six intervals on the domains $0 < x < 5$, and $0 < y < 5$.
34. Use the golden section search technique on Exercise 9.9 looking for a minimum function value. Use six intervals on the domains $0 < x < 5$, and $0 < y < 5$.
35. Write an algorithm to search a domain of a function for the minimum and maximum value within the interval.
36. Refine the solution of Example 9.10 in search of a minimum by subdividing the intervals $6 < y < 12$ and $6 < z < 12$ using an increment equal to 1.
37. Solve Example 9.10 using the Fibonacci search and five evaluations on each of the intervals $0 < y < 18$ and $0 < z < 18$.
38. Solve Example 9.10 using the golden section search and five evaluations on each of the intervals $0 < y < 18$ and $0 < z < 18$.

39. Solve Example 9.10 using the four direction slope search method starting at $y = 4$, $z = 6$, and $\Delta y = \Delta z = 1$.

40. There are equations that can be solved algebraically, either for a root or a functional value at a given location, that are often solved by using the computer search method, because it is faster and more consistent. In these cases it is helpful to do a little pre-computer analysis so the search interval is selected wisely. This is particularly helpful when only *real* roots or functional values are meaningful. For example, the function $f(x) = \sqrt{x - 5} - 6$ has real values only if $(x - 5) \geq 0$, so for real values the domain is $x \geq 5$. Analyze each of the following functions in a similar way to determine the acceptable domain for real function values search.

    a) $f(x) = \sqrt[3]{x - 1} + \sqrt{x + 3} - 2$
    b) $f(x) = \sqrt{2x + 1} - \sqrt{4x - 5} + \sqrt{6x - 8}$
    c) $f(x) = \sqrt{x^2 + 2x - 2\sqrt{x^2 - x - 6}} - x$

41. The radial flow of heat between two concentric spheres can be represented as: $H(r, R, t, T) = [(t - T)4\pi k R r]/(R - r)$, where k depends on the material involved, r = the radius of the inner sphere, R = the radius of the outer sphere, t = the temperature of the inner sphere, and T = the temperature of the outer sphere.

    An engineer wants to design a two-sphere system to contain a heat producing experiment, but is concerned that the estimates of the temperature difference from inner sphere to outer sphere may be off by as much as a factor of two. The engineer does not want to change material (k is the material constant), but wants flexibility to increase the outer sphere radius to handle any variations that may arise. Analyze the mathematical model and determine if there are any particular sphere radii combinations that should be avoided if the engineer wants flexibility to change the radius of the outer sphere during the experiment.

42. A five sided rectangular box is to be made by folding the cutout from a single sheet of material as shown in Figure 9.22. Use one of the optimization methods to determine the dimensions L, W, and x for minimum surface area for a box to hold one cubic unit. What are the dimensions of the box?

FIGURE 9.22

Material layout for Exercise 9.42.

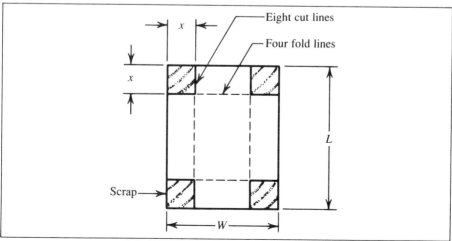

43. Determine the dimensions of a rectangular box so its volume is a maximum. The only restriction is that the perimeter of the smallest right cross section plus the length perpendicular to the cross section is less than or equal to 84 inches.

44. Analyze the relationship between the three dimensions of a six sided box of a given volume for minimum cost, considering that the material for the top and bottom has a cost different from the material for the four sides. Use the cost ratio between the two as a single quantity.

45. A company makes tables and chairs. The tables require three machine hours and one hour of hand labor each. The chairs require one hour on the machine and two hours of hand labor. One machine is available nine hours per day and there are eight hours of hand labor available per day for this operation. If the profit on a chair is $4 and the profit on a table is $10, what product mix should be manufactured per day for maximum profit? Assume there is no market problem and all the products made can be sold. Can the answer legitimately be a fraction? Explain.

46. What happens to the analysis of Exercise 9.45 if the requirement is changed to requiring that one table and four chairs be sold as sets? What recommendation would you make to management concerning machinery and labor to be made available? What effect do you suppose this might have on profit?

47. To make the operation in Exercise 9.46 more profitable would you add machines or hand labor? What quantity would you add?

48. Consider the following situation. A football team has 2.50 minutes left in a game and is behind by 4 points. They have 40 yards to go for a touchdown and have been having success with running and passing plays. Running plays have been averaging 6 yards per play while taking 30 seconds, and pass plays have averaged 12 yards while taking only 20 seconds. What combinations of plays will best suit the team's situation? Ignore time outs. *Hint:* There is no function to maximize or minimize in this problem.

49. Refer to Example 9.12 and answer the following questions:
    a) How many hours per day are each of the three machines used to make maximum profit?
    b) Which of the three machines would run out of capacity first if production needed to be increased? Explain how this affects plant capacity limits.
    c) Is it reasonable to assume that machine #1 will be available 24 hours per day? Discuss your answer.

50. What minimum profit variation (percent or absolute) would change the production mix in Example 9.12 to:
    a) No A's and eight B's?   b) Six A's and four B's?

51. A chair manufacturer has on hand 1000 board feet of hardwood and 750 board feet of softwood. The cost is $0.30 and $0.15 per board foot, respectively. The company can make chairs, which use five board feet of hardwood and ten board feet of softwood, or can make tables, which use 25 board feet of hardwood and 15 board feet of softwood. The chairs can be sold for $75 each and yield a profit of $12, while tables can be sold for $150 each and yield a profit of $30. What combination of chairs and tables should be produced to make a maximum profit? How much wood is left over if this is done? What is the gross income?

52. Reevaluate Exercise 9.51 assuming that a table must be sold with six chairs.

53. A salesman working for the company in Exercise 9.51 suggests selling four chairs with one table. Is that a good suggestion?

54. Problems in linear programming often involve three variables and a three-dimensional graphical analysis and can be completed with the method presented in section 9.3. The graph can be drawn in any of the common three-dimensional methods, such as isometric or oblique. Expand the two-dimensional method to three-dimensions and solve this problem that way. A company wants to make a new alloy containing at least 25 percent of each of three special elements. The alloys do not come separated but as part of available compounds. The available compounds, their percentages of the three elements, and their cost are shown in Table 9.16. Determine how many ounces of each compound should be used to make the new alloy with a minimum percent of each element of 25 percent, and keep cost to a minimum.

55. Select a material to minimize the cost of material for a spherical storage tank, $C = k\rho\pi D^2 t$ subject to the volume constraint $(4\pi/3)(D/2)^3 \geq 1000$ in^3. Select from the materials in Case Study #14.

56. Repeat Case Study #14 assuming the materials are only available in 0.125 inch thickness increments.

57. Expand the Yates algorithm as used in Case Study #15 and outline the table and instructions necessary to carry out a four variable experiment.

58. Derive the model for the power of wind air flow, $P_{in} = (\pi D^2)(\rho)(0.447v)^3/8$, where D = the diameter of the rotating windmill blades in meters, ρ = the density of the air in kg/m^3, v = the air flow speed in miles/hour, and 0.447 converts miles/hour to meters/second. *Hint:* Consider the kinetic energy in a column of moving air the same diameter as that of the rotating blades.

59. Using the model from Exercise 9.58, demonstrate that at the low air speeds used in Case Study #15 a small variation in wind speed results in a large variance in experimental conditions. In particular show that a wind speed of 8 ± 1 miles/hour results in approximately 112 percent variation in power from the windmill experiment.

60. Investigate the effect of temperature and humidity on the density of air and determine if the variation is a significant factor in explaining the experimental variation of Case Study #15.

TABLE 9.16

Data for Exercise 9.54.

Compound	Element 1	Element 2	Element 3	Cost/unit
A	45 percent	20 percent	35 percent	$0.73
B	45 percent	45 percent	10 percent	0.90
C	5 percent	25 percent	35 percent	0.80

FIGURE 9.23

Diagram for Exercises 9.61–9.64.

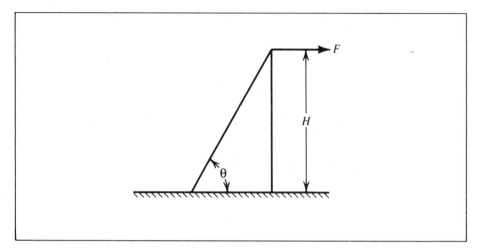

61. A pole on which a horizontal pull of F lbf is applied is to have a wire-rope guy wire attached in the plane of the pull as shown in Figure 9.23. The diameter of the guy wire depends on its strength, S_{ut}, the angle θ, and the load F. If the cost of the guy wire installation is a function of its volume, determine the angle θ for minimum wire cost.

62. Continue Exercise 9.61 by determining general expressions for:
    a) The length of the wire-rope as a function of H, the height of the pole and θ, the angle.
    b) The volume of the rope as a function of H, θ, F, the pull on the pole, and S_{ut}, the strength of the wire rope.
    c) The diameter of the rope as a function of F, θ, and S_{ut}.

63. One grade of steel wire-rope comes in diameter increments of 0.125 inches in the range $0.250 \leq D \leq 1.00$ inches. The wire-rope strength is 76,000 psi based on the nominal area of the rope, $0.38D^2$, rather than $\pi D^2/4$, and its cost is proportional to the volume used, $V = \text{(area)(length)} = 0.38D^2L$. The wire-rope is to be used as guy wire support, so a safety factor of 3.5 is appropriate. Assume the pole in Figure 9.23 is 20 feet high and the pole and guy wire must support a horizontal pull of 1000 lbf. Determine the angle θ, the wire diameter, and the length to use to minimize the cost as analyzed in Exercises 9.61 and 9.62.

64. Analyze the situation and discuss the implications if the load F in Exercise 9.61 increases by 1000 lbf increments.

65. Physical limitations often exist that restrict the location of a guy wire. One restriction is where the cable attaches to the ground. The optimum location may be in the middle of a road, or the location may interfere with some other space use. The

FIGURE 9.24

Diagram for Exercises 9.65–9.68.

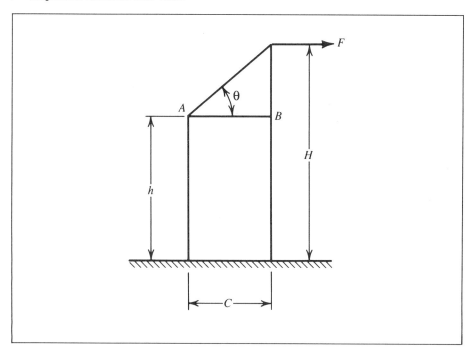

restriction may limit the location to a range of locations, or a single location as identified by C in Figure 9.24. A second restriction may be a minimum vertical clearance next to the pole, identified by h in the figure. One solution to this problem is to add a rigid member AB that creates the necessary clearance. The diameter of the guy wire depends on its strength, S_{ut}, the angle θ, the distance C, and the load F. If the cost of the guy wire installation is a function of its volume, determine the angle θ for minimum cost. Assume the cost of member AB is constant regardless of the force it must withstand, that C is constant, and that the total length of wire must be the same diameter.

66. Continue Exercise 9.65 by determining general expressions for:
    a) The length of the wire-rope as a function of H, the height of the pole, C, the horizontal clearance distance, and θ, the angle.
    b) The volume of the rope as a function of H, θ, F, and S_{ut}.
    c) The diameter of the rope as a function of F, θ, C, and S_{ut}.

67. Using the data from Exercise 9.63, and the restrictions that $C = 5$ feet and $h \geqslant 7$ feet, determine the specifics of the design for minimum cost.

68. Analyze the situation and discuss the implications if the load in Exercise 9.65 increases by 1000 lbf increments.

69. Silos are used to store grain and other foodstuffs. Silos are built round because it is well known that for a given perimeter, a circle encloses the most area, and so a circular silo will contain the most volume for a given surface area. Supposing, however, that two silos are to be built next to each other. Is there any advantage in building the silos together with a common straight wall as shown in Figure 9.25? The question that needs answering is whether or not at some angle θ, the total area of the two partial adjoining circles is greater than the area of two separate circles. To answer this question maximize the combined area of the two partial circles and compare with $2(\pi R_1^2)$ while keeping total perimeter constant.
Note: $R_2 \neq R_1$.

70. Repeat Exercise 9.69, but change the conditions so that the straight common segment between the partial circles costs half as much as the circular segments.

FIGURE 9.25

Circle layouts for Exercises 9.69 and 9.70.

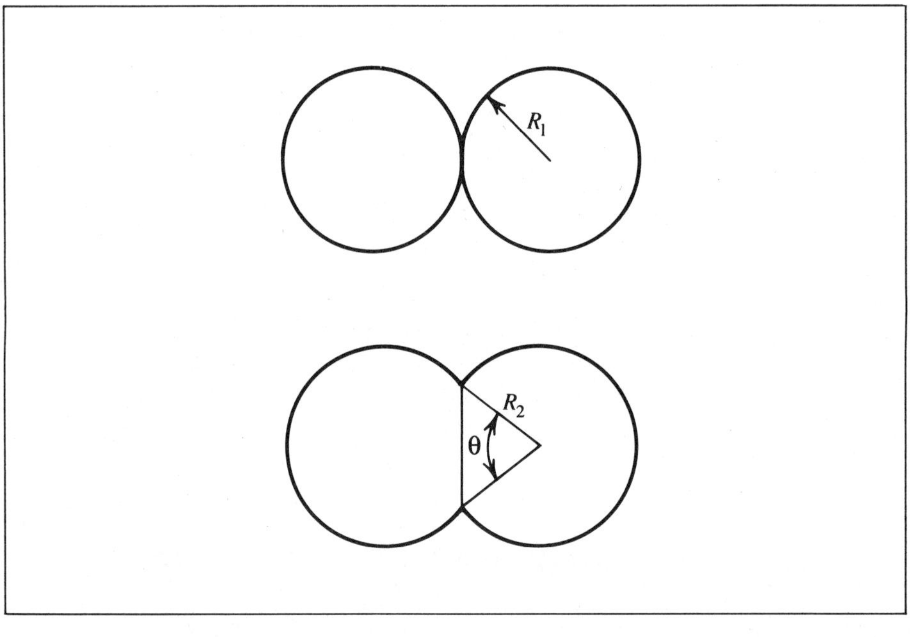

10 Ethics

"Example is not the main thing in influencing others. It is the only thing."
◆ ALBERT SCHWEITZER

The purpose of this chapter is to provide food for thought concerning the professional ethics of an engineer. Ethics are influenced and controlled by individuals. Ethics affect the person and the person's profession, and must be considered globally when performing engineering activities. Even though your basic moral character is probably well ingrained in your life, this chapter asks you to stop and consider if your attitudes are consistent with those expected of a professional engineer. It also raises questions to make you think about specifics that you may never have considered. Additional reading of cases that have occurred is good study material. Concern for ethical matters will prepare you to maintain a proper perspective when an unusual or unexpected situation arises. Planning ahead in ethical matters should receive no less of your thoughts than planning ahead for other problem solving activities.

10.1 Definition

Ethics is an area of study involving what is acceptable and what is not acceptable in association with moral duty and personal and professional obligation. Ethical conflicts can cause as much anxiety as other societal pressures such as obtaining food, clothing, and shelter for survival. Engineers are not immune to ethical pressures and in some ways may be more exposed to them than other professionals. Engineers are involved with using resources, many non-replaceable, to produce goods and services for the public. Engineers must be concerned about the uses of the raw materials and the uses of the end products. Are they being used in an ethical manner and for ethical purposes? Should an engineer get involved with the design and construction of a building that is going to be used for the offices of a tobacco company? What if the engineer is convinced that tobacco is harmful to the health of humans? Should an engineer help design planned obsolescence into a product knowing there are limited natural resources to use for the product?

There are no easy answers to many of the ethical questions that engineers have to face in their career. Personal conflicts arise because of the pressure triangle created by wanting to be a responsible engineer, by wanting to fulfill an obligation to the public, and by wanting to be loyal to an employer or client. It is often difficult to balance all three and not sacrifice one's self interest. Conflict of interest within oneself, or between alternatives, exists when personal desires and professional constraints interfere with an employment obligation or when two or more desires and conditions cannot be met at the same time.

10.2 Sources of Problems

Ethical and legal questions can arise from a number of sources. Generally it is a question of the action to take, or not to take, under a given set of circumstances.

Is it ethical, or legal, to use someone else's trade secret if it is used in an unrelated industry? What if your company will go bankrupt if the secret is not used?

Is it ethical to use the company's computer for your own jobs when it is not being used for company jobs? What is the "cost" to the company when someone does this?

Is it ethical to "borrow" a patented design for use if the patent owner is dead? What if your idea is "somewhat" different from the actual patent?

Is it ethical to advertise yourself as available for engineering jobs?

Is it ethical to bid competitively on jobs, thinking that you can do this job for less than standard rates to get your foot in the door? Will competitive bidding guarantee a safe and reliable product for the consumer? How can contract negotiations be impartially centered on the merits of the contract alone?

Is a kickback ethical? Legal? What is the difference between a kickback and an earned commission? Should an engineer accept a commission from a contractor or manufacturer?

Is it ethical to let others, including the public, know when another engineer is doing something that you consider to be bad engineering? Suppose you are wrong in your accusation? When can freedom of speech interfere with ethical behavior within the confines of your profession? Must you always speak out?

How ethical is it to design a product or service system when you are not licensed as an engineer?

Is it ethical, or legal, to agree with another company on the selling price of a product?

Must you go along with your employer when you are asked to perform an activity you consider unethical, such as altering test data? What rights do you have to balance the responsibility between yourself and your employer?

Engineers cause trouble for themselves and the company they work for when they have personal beliefs about their actions that in practice are not true. Engineers often rationalize their behavior by believing an action taken is really not illegal or unethical.

Activities in this category include providing favors for clients and customers, and industrial spying. Some will carry this belief to the point of saying "If my actions are wrong, there would be a law against them." In many cases a law is passed only after someone has done a "bad" thing and society passes a law to make a public statement about the activity. The passage of a law, of course, does not change the attitude and actions of all people. If it did we wouldn't need police to enforce the law. Laws cannot be passed to include every possible situation in which an engineer might be involved.

A second belief engineers can have that can cause trouble is the belief they won't get caught, even though they know an activity they engage in is illegal or unethical. From copying homework from a fellow student, exceeding the speed limit while driving, exaggerating contributions for tax deductions, to falsifying product test data—if we knew we were going to get caught, we probably wouldn't do it. Ethical behavior requires actions to be taken assuming someone will find out, whether they do or not.

Another way engineers can get into trouble is to believe that the company they work for expects unethical behavior, as long as the results benefit the company. Examples include using low cost material that does not meet specifications, creating product test data without running the tests, paying for insider information, or misrepresenting a product to a customer to get a sale. A second aspect of this behavior is to believe that the company will support and protect employees who engage in these activities if they are caught. Company managers set the environment for employees to follow by their actions or inactions, but ethical engineers cannot allow themselves to be led or forced into a compromising situation. To help engineers in questionable cases, codes of ethics have been created.

10.3 Codes

Codes of ethics have been developed to help engineers answer questions that arise, to avoid potential problems, and to offer guidelines for expected behavior. Codes are established in many professional areas. Oaths of public office or of organizational leaders are a form of ethical code. Other professionals such as doctors and lawyers also have codes of ethics to guide them in their activities. Engineering societies have written and adopted codes to guide them in engineering activities, but there are over 150 different engineering societies. This presents a small problem because there is no single code comparable to the doctors' AMA or lawyers' state bar. However, all the engineering codes attempt to perform the same service, to help guide the engineer. Codes of ethics for engineers have been around for a long time: the first engineering code in the United States was adopted in 1912 by the American Institute of Electrical Engineers.

A code of ethics has several purposes. One is to give guidance and inspiration to engineers. A code will specify actions to take that engineers should consider when making decisions about their work. The code also is a support to an engineer. It specifies rationale to use for a particular decision. A code also can act as a deterrent to certain activities that otherwise might be considered. The code adds discipline to the profession. The code can also be used to educate people about the role of the engineer. Education promotes mutual understanding and contributes to the level of professionalism of engineers. Codes also protect the status quo and promote consistent business practices.

Codes not only contain professional restrictions but also specify protections for an individual's rights. Among the rights an engineer has is the right to pursue legitimate interests freely without the fear of discrimination. An engineer has the right to receive payment for work performed. An engineer has the right to refuse to carry out illegal and unethical work.

Discussion is a key ingredient when deciding which actions to take to avoid unethical actions and behavior. The study topics and the exercises are meant for discussion and the sharing of personal perspectives. Any number of sample codes could be included for reference and discussion. The following code is the Code of Ethics of Engineers adopted by The American Society of Mechanical Engineers.

FIGURE 10.1

Code of Ethics

SOCIETY POLICY

ETHICS

ASME requires ethical practice by each of its members and has adopted the following Code of Ethics of Engineers as referenced in the ASME Constitution, Article C2.1.1.

CODE OF ETHICS OF ENGINEERS

The Fundamental Principles

Engineers uphold and advance the integrity, honor and dignity of the engineering profession by:

I. using their knowledge and skill for the enhancement of human welfare;
II. being honest and impartial, and serving with fidelity the public, their employers and clients, and
III. striving to increase the competence and prestige of the engineering profession.

The Fundamental Canons

1. Engineers shall hold paramount the safety, health and welfare of the public in the performance of their professional duties.
2. Engineers shall perform services only in the areas of their competence.
3. Engineers shall continue their professional development throughout their careers and shall provide opportunities for the professional development of those engineers under their supervision.
4. Engineers shall act in professional matters for each employer or client as faithful agents or trustees, and shall avoid conflicts of interest.
5. Engineers shall build their professional reputation on the merit of their services and shall not compete unfairly with others.
6. Engineers shall associate only with reputable persons or organizations.
7. Engineers shall issue public statements only in an objective and truthful manner.

THE ASME CRITERIA FOR INTERPRETATION OF THE CANONS

The ASME criteria for interpretation of the Canons are advisory in character and represent the objectives toward which members of the engineering profession should strive. They constitute a body of principles upon which an engineer can rely for guidance in specific situations. In addition, they provide interpretive guidance to the ASME Board on Professional Practice and Ethics in applying the Code of Ethics of Engineers.

1. Engineers shall hold paramount the safety, health and welfare of the public in the performance of their professional duties.
    a) Engineers shall recognize that the lives, safety, health and welfare of the general public are dependent upon engineering judgments, decisions and practices incorporated into structures, machines, products, processes and devices.
    b) Engineers shall not approve or seal plans and/or specifications that are not of a design safe to the public health and welfare and in conformity with accepted engineering standards.
    c) Whenever the Engineers' professsional judgment is over-ruled under circumstances where the safety, health, and welfare of the public are endangered, the Engineers shall inform their clients and/or employers of the possible consequences and notify other proper authority of the situation, as may be appropriate.
        c.1 Engineers shall do whatever possible to provide published standards, test codes, and quality control procedures that will enable the public to understand the degree of safety or life expectancy associated with the use of the designs, products, or systems for which they are responsible.
        c.2 Engineers shall conduct reviews of the safety and reliability of the designs, products, or systems for which they are responsible before giving their approval to the plans for the design.
        c.3 Whenever Engineers observe conditions which they believe will endanger public safety or health, they shall inform the proper authority of the situation.
    d) If engineers have knowledge or reason to believe that another person or firm may be in violation of any of the provisions of these Canons, they shall present such information to the proper authority in writing and shall cooperate with the proper authority in furnishing such further information or assistance as may be required.
2. Engineers shall perform services only in areas of their competence.
    a) Engineers shall undertake to perform engineering assignments only when qualified by education or experience in the specific technical field of engineering involved.
    b) Engineers may accept an assignment requiring education or experience outside of their own fields of competence, but their services shall be restricted to other phases of the project in which they are qualified. All other phases of such project shall be performed by qualified associates, consultants, or employees.
3. Engineers shall continue their professional development throughout their careers, and should provide opportunities for the professional development of those engineers under their supervision.
4. Engineers shall act in professional matters for each employer or client as faithful agents or trustees, and shall avoid conflicts of interest.

a) Engineers shall avoid all known conflicts of interest with their employers or clients and shall promptly inform their employers or clients of any business association, interests, or circumstances which could influence their judgment or the quality of their services.

b) Engineers shall not undertake any assignments which would knowingly create a potential conflict of interest between themselves and their clients or their employers.

c) Engineers shall not accept compensation, financial or otherwise, from more than one party for services on the same project, or for services pertaining to the same project, unless the circumstances are fully disclosed to, and agreed to, by all interested parties.

d) Engineers shall not solicit or accept financial or other valuable considerations, for specifying the products or material or equipment suppliers, without disclosure to their clients or employers.

e) Engineers shall not solicit or accept gratuities, directly or indirectly, from contractors, their agents, or other parties dealing with their clients or employers in connection with work for which they are responsible.

f) When in public service as members, advisors, or employees of a governmental body or department, Engineers shall not participate in considerations or actions with respect to services provided by them or their organization(s) in private or product engineering practice.

g) Engineers shall not solicit an engineering contract from a governmental body on which a principal, officer, or employee of their organization serves as a member.

h) When, as a result of their studies, Engineers believe a project(s) will not be successful, they shall so advise their employer or client.

i) Engineers shall treat information coming to them in the course of their assignments as confidential, and shall not use such information as a means of making personal profit if such action is adverse to the interests of their clients, their employers or the public.

  i.1 They will not disclose confidential information concerning the business affairs or technical processes of any present or former employer or client or bidder under evaluation, without his consent, unless required by law.

  i.2 They shall not reveal confidential information or finding of any commission or board of which they are members unless required by law.

  i.3 Designs supplied to Engineers by clients shall not be duplicated by the Engineers for others without the express permission of the client(s).

j) The Engineer shall act with fairness and justice to all parties when administering a construction (or other) contract.

k) Before undertaking work for others in which the Engineer may make improvements, plans, designs, inventions, or other records which may justify copyrights or patents, the Engineer shall enter into a positive agreement regarding the rights of respective parties.

l) Engineers shall admit and accept their own errors when proven wrong and refrain from distorting or altering the facts to justify their decisions.

m) Engineers shall not accept professional employment outside of their regular work or interest without the knowledge of their employers.

n) Engineers shall not attempt to attract an employee from another employer by false or misleading representations.

5. Engineers shall build their professional reputation on the merit of their services and shall not compete unfairly with others.
   a) Engineers shall negotiate contracts for professional services on the basis of demonstrated competence and qualifications for the type of professional service required and at fair and reasonable prices.
   b) Engineers shall not request, propose, or accept professional commissions on a contingent basis under circumstances under which their professional judgments may be compromised.
   c) Engineers shall not falsify or permit misrepresentation of their, or their associates, academic or professional qualification. They shall not misrepresent or exaggerate their degrees of responsibility in or for the subject matter of prior assignments. Brochures or other presentations incident to the solicitation of employment shall not misrepresent pertinent facts concerning employers, employees, associates, joint venturers, or their past accomplishments.
   d) Engineers shall prepare articles for the lay or technical press which are only factual, dignified and free from ostentations or laudatory implications. Such articles shall not imply other than their direct participation in the work described unless credit is given to others for their share of the work.
   e) Engineers shall not maliciously or falsely, directly or indirectly, injure the professional reputation, prospects, practice or employment of another engineer, nor shall they indiscriminately criticize another's work.
   f) Engineers shall not use equipment, supplies, laboratory or office facilities of their employers to carry on outside private practice without consent.
6. Engineers shall associate only with reputable persons or organizations.
   a) Engineers shall not knowingly associate with or permit the use of their names or firm names in business ventures by any person or firm which they know, or have reason to believe, are engaging in business or professional practices of a fraudulent or dishonest nature.
   b) Engineers shall not use association with non-engineers, corporations, or partnerships as "cloaks" for unethical acts.
7. Engineers shall issue public statements only in an objective and truthful manner.
   a) Engineers shall endeavor to extend public knowledge, and to prevent misunderstandings of the achievements of engineering.
   b) Engineers shall be completely objective and truthful in all professional reports, statements or testimony. They shall include all relevant and pertinent information in such reports, statements, or testimony.
   c) Engineers, when serving as expert or technical witnesses before any court, commission, or other tribunal, shall express an engineering opinion only when it is founded upon adequate knowledge of the facts in issue, upon a background of technical competence in the subject matter, and upon honest conviction of the accuracy and propriety of their testimony.
   d) Engineers shall issue no statements, criticisms, or arguments on engineering matters which are inspired or paid for by an interested party, or parties, unless they preface their comments by identifying themselves, by disclosing the identities of the party or parties on whose behalf they are speaking, and by revealing the existence of any pecuniary interest they may have in matters under discussion.

> *e*) Engineers shall be dignified and modest in explaining their work and merit, and shall avoid any act tending to promote their own interest at the expense of the integrity, honor and dignity of the profession or another individual.
>
> 8. Any Engineer accepting membership in The American Society of Mechanical Engineers by this action agrees to abide by this Society Policy on Ethics and procedures for implementation.
>
> Responsibility: Board on Professional Practice and Ethics/Council on Member Affairs

The American Society of Mechanical Engineers 345 East 47th Street New York, NY 10017, Reprinted with permission.

10.4 Specific Situations

Many ethics violations are centered around money. A person might receive money, or other goods of value, for work not performed or for special favors. Or a person might receive kickbacks for awarding a contract, or tender gifts and favors before the issuing of a contract. Even if a company will receive a contract because of its competitive bid, the giving and receiving of gifts or special entertainment should not be condoned. Even seemingly trivial things such as paying for coffee and rolls, or a round of golf, cannot occur without some form of partiality taking place. If vendor A buys coffee on Monday, what feeling exists for vendor B on Tuesday when the same gesture is not made? What's fair for one must be fair for all; the easiest and most consistent policy is to avoid all such situations. Many corporations have established guidelines against receiving gifts of any kind from companies they do business with. It is against company policy to receive a gift whatever its value, regardless of whether the person receiving the gift has influence on company purchases or not. If all such giving and receiving is curtailed, then no one has to decide what a gift of "nominal value" is.

The following excerpt from an ethical business practices notice sent to employees of a Fortune 500 corporation (name withheld by request) illustrates the degree of importance that some companies attach to policies of professional ethics.

> Our Company policy does not permit the acceptance of any gifts by our employees or members of their immediate families from organizations doing business with the Company, or their representatives. Further, our employees must never become involved in any situation which might place them under obligation to any organization with which the Company does business.
>
> Acceptance of gifts, including advertising novelties, by employees or family members from suppliers or business associates is prohibited. The personal use of supplier owned or provided offices, hotels or other residences and facilities is prohibited. An employee or family member is prohibited from using any special discounts or privileges from a supplier that are not available to all employees. Planned social relationships with supplier representatives are discouraged.

It may be regretful that such policy statements are necessary, but business dealings must be conducted on an ethical basis if equity and fairness is to prevail.

An ethic that is violated most often, without regard to type of business or job status, is spending company money improperly. Unwise spending includes using company property for personal use (often called stealing), including the use of company property such as paper, pencils, and other supplies for personal use, as well as making personal phone calls on company time, photocopying personal papers, and using company mail for personal correspondence.

Another category of improper spending involves the purchase of luxury equipment for use on the job when standard equipment would do, such as a $10,000 mahogany desk, or extraordinary plush carpeting for an office.

Impropriety in expense account reporting is another abuse of company funds. If you wouldn't stay at an expensive hotel if you were spending your own money, why would it be considered proper to stay there just because the company is paying for it? Don't use the excuse that everyone else expects it. Just because someone else does something doesn't make it right. Be your own judge of ethical behavior; the real payoff is personal professional integrity. Some case studies will show how difficult a conflict may become for a particular situation, and why ethics can never be neglected.

CASE STUDY #16:
Personal Conflict

A mechanical engineering graduate took a job with one of the larger companies in the United States. Her first assignment was to try to correct product reliability problems on a plastic crash helmet design. The problems seemed to be of a manufacturing nature because of the wide variance in test results of production models. In the process of doing her investigation she learned about the company's facilities, previous helmet design problems that had been solved, how they had been solved, and the structure of the company in general. She worked with other engineers who were still making design changes to try to improve the helmet's performance, including changes to a microphone and earphone system that were part of the helmet. It seemed quite natural to ask about the market that the helmet sold in, its cost, and other marketing concerns. She learned the helmet was used by military pilots, that many were sold to countries other than the United States, and one of the countries was presently at war. TV coverage and news reports indicated that the country had been restricting the civil rights of many of its inhabitants, and this was disturbing to her.

Work continued on improving the quality of the helmet, but at the same time the situation began to add stress to her life; she had misgivings about how her work was contributing to situations that she opposed in other countries. Should she ask for a transfer of assignments? Quit her job? Was she adequately informed of the effect of her work on the civilians in the other country? Who could she talk to about her misgivings? Would she be ridiculed for her feelings just because of her sex? Could she in good conscience continue on the project? Would any action taken by just one person make any difference? She was sure that others would continue the work on improving the helmet. Did older engineers have similar misgivings? How did they rationalize their activities? What should she do?

As mentioned earlier in the chapter, each of us has our own level of tolerance to situations such as this. Some would not even question the significance of the work, responding, "Someone has to do it." Others might say, "If our company doesn't do the work some other company will," or "I can't be responsible for other people."

Can a situation like this have a single answer? No, the situation is in that grey area, not right or not wrong for the majority of people. The young engineer did talk to her superiors, and after working on the helmet project for over a year, and performing well, she was transferred to another project, one with no political overtones.

CASE STUDY #17:
Corporate Conflict

In 1987 the Justice Department filed charges against Chrysler Corporation and two of its executives for selling cars that had been driven with the odometers disconnected, and for selling damage repaired cars as new. Car companies, and other companies, often let employees try out new products to expose them to situations that may have been overlooked during normal testing procedures; Chrysler admitted this is what they were doing. The public exposure of the situation caused faithful Chrysler customers to wonder if they had been shortchanged, and some felt that Chrysler had been unethical in its procedures. Over a ten-year period 72 of the test loaner cars had been involved in accidents, and 40 of those had been repaired and sold as new cars. Thousands of cars had been driven with the odometer disconnected, but only three had been driven over 100 miles.

The company policy was not a secret plot to bilk the customers. No one said, "Let's drive cars with the odometers disconnected so we can sell used cars as new cars." But who is to judge whether the action is ethical or not? If an action violates a law, a person can be tried, convicted, and fined or sent to jail. If a building code is violated, an engineer may lose a job or lose the license to practice engineering, and the out-of-code structure may have to be torn down and rebuilt. When no law or code applies, the customer judges.

In this case the Justice Department considered a law had been broken and acted accordingly; in addition, customers considered the behavior unethical. Chrysler executives wanted to restore faith, make amends to the customers, and not go to jail. Chrysler offered to either replace the cars involved or extend the warranty at no cost to the customer, and pleaded nolo contendere to the criminal charge. The plea was accepted and the charges were dropped. No trials, no lawsuits, and no one lost their job. The consumer and the Justice Department made public an activity that was judged unethical, and the company responded, not with excuses or alibis, but with action.

CASE STUDY #18:
Government Ethics.

Government is big business and it involves not only elected officials, but government agencies, contractors, and the voting public. Special interest groups apply pressure in different forms to influence the laws that are passed, the regulations established, and the contracts that are awarded. What constitutes unethical behavior? Where is the line that separates information and data gathering from prejudicial solicitation? When is what is best for the country as a whole not the best for a specific locale?

Engineers deal with the government and its agencies, and must be alert to possible ethical problems, whether on the giving or receiving end of information flow and contract awarding. An editorial in the *Des Moines Register* on March 21, 1989 summarizes problems and concerns.

JUST SAY NO

As inducements go, they aren't much: free tickets to a football game, free cable TV, free movie passes. The perquisites of office say the defenders; the little things to make up for the modest salaries Iowa pays legislators.

And rank hath privileges. If the little freebies are OK for the rank and file, there's nothing wrong with Senate Majority Leader Bill Hutchins and Senate Minority Leader Calvin Hultman accepting $500 payments to listen to a pitch by chemical pusher E.I. du Pont de Nemours and Co., which has a rather large interest in environmental legislation.

There's an old story that fits: The cocktail-hour conversation turns to the price of integrity, and the dowager announces that some things are beyond monetary value. Her honor, for instance. Not even for $1 million? she is asked. Well, perhaps. How about $100,000? What do you think I am? she asks indignantly. We've established that, she is told; now we're just arguing price.

While accepting two-bit freebies may seem far removed from the issue of bribery, recipients have established that they'll accept favors. Now the only question is the price at which such favors merit a return.

If $500 won't buy a crucial vote or a leader's influence regarding a chemical-pollution issue, will $5000? If a choice ticket to the Iowa-Iowa State game won't influence a vote on a majority appropriation, will a whole row of seats?

You're not for sale for any price, right? Then why allow the question to be raised at all by accepting favors?

The larger the favor the more serious the questions. The smaller the favor the easier it is to forego it. Those penny-ante perks may seem laughably insignificant to the recipients. But they're a conditioner, a preparation.

If legislators are so dense as to think the favors emanate solely from the goodness of a lobbyist's soul, their handicap is not shared by their constituents.

It's so easy to avoid even the appearance of a conflict of interest.

Just say no.

Copyright 1989, the *Des Moines Register*, Reprinted with permission.

10.5 Study Cases

1. Kepone, an insecticide ($C_{10}Cl_{10}O$) production in Hopewell, Virginia, Allied Chemical Corporation, 1949–1975
2. Hooker Chemical and Love Canal, 1975
3. B. Everett Jordan Lake and dam, 1967–1976
4. DC–10 airplane, 1972–1974
5. Lockheed bribe payments to Japan, 1972–1973
6. Kermit Vandiver and B.F.Goodrich, 1968
7. Bay Area Rapid Transit System, 1973
8. Ernest Fitzgerald and Lockheed, November, 1968
9. Ford Pinto 1967–1978
10. Three Mile Island nuclear power plant, March 28, 1979
11. Challenger disaster, January 1986
12. Exxon oil spill in Alaska, March 24, 1989
13. Jim Wright, Speaker of The House of Representatives, 1989
14. Robert T. Morris and computer security, 1990

10.6 References

1. Andrews, Kenneth R., ed. *Ethics In Practice*. Cambridge, Mass.: Harvard Business School Press, 1989.
2. Asch, Peter. *Consumer Safety Regulation*. New York: Oxford University Press, 1988.
3. Baum, Robert J. *Ethics and Engineering Curricula*. New York: Institute of Society, Ethics and the Life Sciences, 1980.
4. Berube, Bertrand G. "A Whistle-blower's Perspective of Ethics in Engineering." *Engineering Education*, Feb. 1988, 294–295.
5. Bick, Patricia Ann. *Business Ethics and Responsibility: An Information Sourcebook*. Phoenix, Ariz.: Oryx Press, 1988.
6. Blinn, Keith W. *Legal and Ethical Concepts in Engineering*. Englewood Cliffs, N.J.: Prentice Hall, 1989. Case study approach.
7. (The) Business Round Table. *Corporate Ethics: A Prime Business Asset*. New York: 1988.
8. Constance, John Dennis. *How To Become a Professional Engineer*. New York: McGraw-Hill, 1978.
9. Davenport, William H. and Daniel Rosenthal. *Engineering: Its Role and Function in Human Society*. New York: Pergamon Press, 1967.

10. Dimond, Diane. "Know-How or No Way?" *Business and Industry,* January, 1988, 108–115.
11. Florman, Samuel. *The Civilized Engineer.* New York: St. Martin's Press, 1987.
12. Florman, Samuel. *The Existential Pleasures of Engineering.* New York: St. Martin's Press, 1976.
13. Gould, Leroy C., et al. *Perceptions of Technological Risks and Benefits.* New York: Russell Sage Foundation, 1988.
14. Gunn, Alastair S. and P. Aarne Vesilind. *Environmental Ethics for Engineers.* Chelsea, Mich.: Lewis Publishers, 1986.
15. Iacocca, Lee, with Sonny Kleinfield. *Talking Straight.* New York: Bantam Books, 1988.
16. King, W.J. "The Unwritten Laws of Engineering." *Mechanical Engineering,* Vol. 66, No. 5,6,7, 1944.
17. Martin, Mike. *Ethics in Engineering.* New York: McGraw-Hill, 1983. Many case summaries.
18. Red, W. Edward. *Engineering, The Career and the Profession.* Monterey, Cal.: Brooks/Cole, 1982.
19. Schaub, James H. *Engineering Profession and Ethics.* New York: Wiley-Interscience, 1983.
20. Straub, Joseph T. *Applied Management.* Boston: Winthrop Publishers, 1979.
21. Vesilind, P. Aarne. "Rules, Ethics and Morals In Engineering Education." *Engineering Education,* Feb. 1988, 289–293.

10.7 Exercises

1. If a person is ethically against taking the life of another person, does it follow that the person should also be against:
    a) Capital punishment for premeditated murder?
    b) Killing someone in self defense?
    c) War?
    d) Capital punishment for manslaughter?
    e) Mercy killing of the terminally ill?

2. A proposed ethic often heard is, "No one should cause physical suffering to another living thing." If you agreed with that ethic, could you participate in:
    a) Boxing?
    b) Football?
    c) Hunting?
    d) Golf?
    e) Baseball?
    f) Branding of cattle?

3. If all project planners and designers were ethical, would the passage of the National Environmental Policy Act (NEPA) have been required? The NEPA requires an environmental impact statement for some projects.

4. Does the fact that NEPA only covers some, but not all, government projects seem ethical? Weapons testing, for example, is not covered.

5. Is it ethical for an engineer to ignore NEPA guidelines just because the project being worked on does not fall under its jurisdiction?

6. Is it ethical for an engineer to ignore the recommended safety factors in the steam boiler code even though they are not law?

7. An engineer signed a pre-employment agreement that gives the company the rights to all patentable ideas created on the job. During the course of performing product research, the engineer thinks of a novel solution to a problem encountered during previous employment. The previous job was in the food industry; the present job is in the steel making industry. The engineer takes the idea to the former employer and tries to sell it. Discuss the ethics of the situation.

8. A student who was uncertain of a major learned of a scholarship available only to English majors. The student had a 3.90 grade point average and recognized financial need, and felt confident about winning the scholarship. The student declared English as a major and applied for the scholarship. Discuss the ethics of the situation.

9. A computer operator runs a program for private purposes at work when the computer is not running company jobs. The computer is on all the time. Discuss the ethics of the situation.

10. A student gives a non-student friend the password to the school computer so the friend can access some programs. Discuss the ethics of the situation. Does it matter if the school is a private or public school?

11. A student breaks the campus computer security code and reviews student grades, but makes no changes. Discuss the ethics of the situation.

12. A programmer is asked to write a program that will give inaccurate information to the company's stockholders. When she discusses this with her supervisor she is told to write the program or be fired. Discuss the ethics of the situation.

13. An engineer makes personal phone calls at work and does not charge them to a home phone number. Each month the calls amount to $10–15. Discuss the ethics of the situation. Would it make a difference if the calls were charged to the home phone?

14. The recording secretary of a faculty committee that reviewed confidential faculty records and made confidential recommendations on tenure and promotion was a member of a department that had a member up for promotion. The committee recommended promotion but was overruled by the president and the board of regents. In a letter of appeal by the non-promoted faculty member, the phrase "and overruled the committee's recommendation" was used. If all deliberations and recommendations were confidential, how was this information known? Discuss possible unethical behavior.

15. Read and discuss each section of the Code of Ethics for Engineers. Discuss possible situations you might find yourself in where the code would help you decide on a course of action.

16. Find real cases that the Code would have jurisdiction over and discuss if the results were in agreement with the code.

17. A company offered an engineer a raise if the engineer agreed to contribute half the raise to the political campaign fund of a congressional candidate. Consider the ethics of the situation:
    a) If the engineer liked the candidate, or
    b) If the engineer did not like the candidate.

18. How concerned should an engineer be about using raw materials or products from countries whose governments violate "human rights?"

19. Consider an engineer who is against the use of tobacco and whose company has just been awarded a large contract to design and build a factory that will produce cigarettes. What are some personal implications when a supervisor tells the engineer "If our company doesn't do the work some other company will," or "I can't be responsible for other people."

20. Discuss the ethical pressures on an engineer who knows that poor product quality on a project may lead to customer dissatisfaction, but that if the project is brought in on schedule there will be a large bonus handed out.

21. Some business managers strive for short term gains in lieu of long term stability. Quick success often leads to promotion and bonus pay. Discuss the ethics of a manager who knowingly performs this way on the job.

22. Is maximum profit always an ethical goal?

23. Discuss situations where a decision might lead to lower annual profits and yet be good for the company. How could you handle such a situation ethically?

24. How would the answer to Exercise 9.50 change if a manager wanted maximum income rather than maximum profit? Can maximum income be used as a measure of success rather than maximum profit? How?

Appendices

APPENDIX A1

Basic quantities and units.

Quantity	Symbol	Dimensions	Metric	English
Acceleration, linear	a	LT^{-2}	meters/sec^2	ft/sec^2
Acceleration, angular	α	T^{-2}	radians/sec^2	radians/sec^2
Acceleration, gravity	g	LT^{-2}	9.80 m/sec^2	32.2 ft/sec^2
Area	A	L^2	meter^2	ft^2
Coefficient of compressibility	β	LT^2M^{-1}	—	—
Damping factor	C	MT^{-1}	N·sec/m	lbf·sec/inch
Displacement, linear	r, d	L	meter	ft
Displacement, angular	$\emptyset$	—	radian	radian
Energy	K, U	ML^2T^{-2}	joule = kg·m^2/sec^2	ft·lbf
Force	F	MLT^{-2}	newton	lbf = lbm·g
Frequency, linear	ν	T^{-1}	hertz	hertz
Frequency, angular	ω	T^{-1}	radians/sec	radians/sec
Length	L	L	meter	ft
Impulse	i	MLT^{-1}	N·sec	lbf·sec
Inertia, rotational	I	ML^2	kg·m^2	lbm/ft^2
Mass	m	M	kilogram	lbm = lbf/g
Mass density	ρ	ML^{-3}	kg/m^3	lbm/ft^3
Mass flow rate	dm/dt	MT^{-1}	kg/sec	lbm/sec
Momentum, linear	**p**	MLT^{-1}	kg·m/sec	lbm·ft/sec
Momentum, angular	L	ML^2T^{-1}	kg·m^2/sec	lbm·ft^2/sec
Period	T	T	second	second
Power	P	ML^2T^{-3}	watt = joule/sec	ft·lbf/sec
Pressure	p	$ML^{-1}T^{-2}$	pascal = N/m^2	lbf/ft^2
Spring constant	k	MT^{-2}	N/m	lbf/inch
Stress	σ	$ML^{-1}T^{-2}$	N/m^2	lbf/ft^2
Surface tension	σ	MT^{-2}	N/m	lbf/inch
Time	t	T	second	second
Torque	τ	ML^2T^{-2}	N·m	ft·lbf
Velocity, linear	v	LT^{-1}	meters/sec	feet/sec
Velocity, angular	ω	T^{-1}	radians/sec	radians/sec
Viscosity, dynamic		$ML^{-1}T^{-1}$	N·sec/m^2	lbm/ft·sec
Viscosity, kinematic		L^2T^{-1}	m^2/sec	ft^2/sec
Volume	V	L^3	meter^3	ft^3
Volume flow rate	Q	L^3T^{-1}	m^3/sec	ft^3/sec
Wavelength	l	L	meter	ft
Work	W	ML^2T^{-2}	joule	ft·lbf

APPENDIX A2

Basic quantities and units.

Quantity	Symbol	Dimensions	Units
Entropy	S	ML^2T^{-2}	joules/°K
Internal energy	U	ML^2T^{-2}	joule
Heat	Q	ML^2T^{-2}	joule
Temperature	T	—	°Kelvin
Capacitance	C	$M^{-1}L^{-2}T^2Q^2$	farad
Charge	q	Q	coulomb = amp·sec
Conductivity	σ	$M^{-1}L^{-3}TQ^2$	1/ohm·meter = mho
Current	i	$T^{-1}Q$	ampere
Current density	j	$L^{-2}T^{-1}Q$	amp/meter^2
Electric dipole moment	**p**	LQ	coulomb·meter
Electric displacement	**D**	$L^{-2}Q$	coulomb/meter^2
Electric polarization	**P**	$L^{-2}Q$	coulomb/meter^2
Electric field strength	**E**	$MLT^{-2}Q^{-1}$	volts/meter
Electric flux	Φ	$ML^3T^{-2}Q^{-1}$	volt·meter
Electric potential	V	$ML^2T^{-2}Q^{-1}$	volt
Electromotive force	V	$ML^2T^{-2}Q^{-1}$	volt
Inductance	L	ML^2Q^{-2}	henry
Magnetic dipole moment	μ	$L^2T^{-1}Q$	amp·meter^2
Magnetic field strength	**H**	$MT^{-1}Q$	amp·meter
Magnetic flux	Φ	$ML^2T^{-1}Q^{-1}$	weber = volt·sec
Magnetic induction	B	$MT^{-1}Q^{-1}$	tesla = webers/meter^2
Magnetization	M	$L^{-1}T^{-1}Q$	amp/meter
Permeability	μ	MLQ^{-2}	henrys/meter
Permittivity	ϵ	$M^{-1}L^{-3}T^2Q^2$	farads/meter
Resistance	R	$ML^2T^{-1}Q^{-2}$	ohm
Resistivity	ρ	$ML^3T^{-1}Q^{-2}$	ohm·meter
Voltage	V	$ML^2T^{-2}Q^{-1}$	volt

APPENDIX B

SI prefixes and multiplying factors.

Prefix Name	Symbol	Multiplying Factor
tera	T	10^{12}
giga	G	10^{9}
mega	M	10^{6}
kilo	k	10^{3}
hecto*	h	10^{2}
deca*	da	10
deci*	d	10^{-1}
centi	c	10^{-2}
milli	m	10^{-3}
micro	μ	10^{-6}
nano	n	10^{-9}
pico	p	10^{-12}
femto	f	10^{-15}
atto	a	10^{-18}

* Not commonly used

APPENDIX C

Selected conversion factors.

To Convert From	To	Multiply By
Btu	gram calorie	252
Btu	joule	1055
Btu	foot·pounds	778
cubic feet	gallon	7.48
foot·pound	newton·meter = joule	1.356
gallon, U.S. liquid	cubic feet	0.134
gallon, U.S. liquid	cubic inches	231
gallon, U.S. liquid	liter	3.79
gallon, U.S. liquid	pounds of water	8.34
gram/cm^3	pounds/ft^3	62.43
horsepower	watts	745.7
horsepower	ft·lbf/second	550
inch	millimeter	25.4
kilogram	pound	2.20
kilowatt·hour	Btu	3414
meter	inch	39.37
millimeters of Hg	atmosphere	0.001316
newton	pound force (lbf)	0.2248
pound	gram	453.6
lbf/in^2	kPascal	6.895
watt	Btu/hr	3.412

APPENDIX D

Average energy values of common fuels.

Fuel	Energy
Coal	
Anthracite	14,000 Btu/pound
Bituminous	12,000 Btu/pound
Lignite	7000 Btu/pound
Crude oil	143,000 Btu/gallon
Diesel fuel*	135-160,000 Btu/gallon
Ethyl alcohol (Ethanol)	75-84,000 Btu/gallon
Fuel oil	140,000 Btu/gallon
Gasoline*	110-126,000 Btu/gallon
Hydrogen (liquid)	34,000 Btu/gallon
Methanol	56,500 Btu/gallon
Natural gas (@ furnace burner)	1000 Btu/ft^3
Compressed @ 3000 psi	29,000 Btu/gallon
Liquid	73,500 Btu/gallon
Propane (bottled gas)	94,000 Btu/gallon

* A function of crude oil composition and distillation technique.

APPENDIX E

Table of selected elements.

Element	Melting Temp °C	Solid Density Mg/m^3 @ 20°C*	Crystal Structure @ 20°C	Common Valence
Aluminum	660.4	2.699	fcc	+3
Antimony	630.7	6.62	—	+5
Argon	−189.2	—	—	Inert
Arsenic	817	5.72	—	+3
Beryllium	1278±5	1.85	hcp	+2
Boron	2300	2.34	—	+3
Calcium	839±2	1.55	fcc	+2
Carbon	~3550	2.25	hex	—
Chlorine	−101	—	—	−1
Chromium	1857±20	7.19	bcc	+3
Cobalt	1495	8.9	hcp	+2
Copper	1083	8.96	fcc	+1
Fluorine	−220	—	—	−1
Germanium	937	5.32	diamond cubic	+4
Gold	1064	19.32	fcc	+1
Helium	−272.2	—	—	Inert
Hydrogen	−259.14	—	—	+1
Iodine	114	4.94	noncubic	−1
Iron	1535	7.87	bcc	+2, +3
Lead	327.5	11.34	fcc	+2
Lithium	180.5	0.534	bcc	+1
Magnesium	649	1.74	hcp	+2
Manganese	1244	7.4	—	+2
Mercury	−38.87	—	—	+2
Neon	−248.7	—	—	Inert
Nickel	1453	8.90	fcc	+2
Nitrogen	−210	—	—	−3
Oxygen	−218.4	—	—	−2
Phosphorus	44	1.8	—	+5
Potassium	64	0.86	bcc	+1
Silicon	1410	2.33	diamond cubic	+4
Silver	961.9	10.5	fcc	+1
Sodium	97.8	0.97	bcc	+1
Sulfur	113	2.07	—	−2
Tin	232	7.3	—	+4
Titanium	1660±10	4.51	hcp	+4
Tungsten	3410±20	19.3	bcc	+4
Uranium	1132	18.7	—	+4
Zinc	419.6	7.135	hcp	+2

* Divide by 27.68 to get lbf/in^3.

APPENDIX F1

Properties of selected materials.

Material	Specific gravity	Thermal conductivity, $\left[\dfrac{\text{Btu} \cdot \text{in.}}{\text{ft}^2 \cdot \text{sec} \cdot °\text{F}}\right]$ at 68°F*	Thermal expansion, in/in/°F at 68°F†	Electrical resistivity, ohm·cm at 68°F‡	Average modulus of elasticity, psi at 68°F§
Metals					
Aluminum (99.9+)	2.7	0.43	12.5×10^{-6}	2.7×10^{-6}	10×10^6
Aluminum alloys	2.7(+)	0.3(±)	12×10^{-6}	3.5×10^{-6}(+)	10×10^6
Brass (70Cu-30Zn)	8.5	0.235	11×10^{-6}	6.2×10^{-6}	16×10^6
Bronze (95Cu-5Sn)	8.8	0.16	10×10^{-6}	9.5×10^{-6}	16×10^6
Cast iron (gray)	7.15	—	5.8×10^{-6}	—	$12\text{-}25 \times 10^6$
Cast iron (white)	7.7	—	5×10^{-6}	—	30×10^6
Copper (99.9+)	8.9	0.77	9×10^{-6}	1.7×10^{-6}	16×10^6
Iron (99.9+)	7.87	0.14	6.53×10^{-6}	9.7×10^{-6}	30×10^6
Lead (99+)	11.34	0.06	16×10^{-6}	20.65×10^{-6}	2×10^6
Magnesium (99+)	1.74	0.37	14×10^{-6}	4.5×10^{-6}	6.5×10^6
Monel (70Ni-30Cu)	8.8	0.05	8×10^{-6}	48.2×10^{-6}	26×10^6
Silver	10.5	0.79	10×10^{-6}	1.6×10^{-6}	11×10^6
Steel (1020)	7.86	0.096	6.5×10^{-6}	16.9×10^{-6}	30×10^6
Steel (1040)	7.85	0.093	6.3×10^{-6}	17.1×10^{-6}	30×10^6
Steel (1080)	7.84	0.089	6.0×10^{-6}	18.0×10^{-6}	30×10^6
Steel (18Cr-8Ni stainless)	7.93	0.028	5×10^{-6}	70×10^{-6}	30×10^6
Ceramics					
Al_2O_3	3.8	0.057	5×10^{-6}	—	50×10^6
Brick					
Building	2.3(±)	0.0012	5×10^{-6}	—	—
Fireclay	2.1	0.0016	2.5×10^{-6}	1.4×10^8	—
Graphite	1.5	—	3×10^{-6}	—	—
Paving	2.5	—	2×10^{-6}	—	—
Silica	1.75	0.0016	—	1.2×10^8	—
Concrete	2.4(±)	0.002	7×10^{-6}	—	2×10^6
Glass					
Plate	2.5	0.0014	5×10^{-6}	10^{14}	10×10^6
Borosilicate	2.4	0.002	1.5×10^{-6}	$>10^{17}$	10×10^6
Silica	2.2	0.0025	0.3×10^{-6}	$\sim 10^{20}$	10×10^6
Vycor	2.2	0.0025	0.35×10^{-6}	—	—
Wool	0.05	0.0005	—	—	—
Graphite (bulk)	1.9	—	3×10^{-6}	10^{-3}	1×10^6
MgO	3.6	—	5×10^{-6}	10^5(2000°F)	30×10^6
Quartz (SiO_2)	2.65	0.025	—	10^{14}	45×10^6
SiC	3.17	0.025	2.5×10^{-6}	2.5(2000°F)	—
TiC	4.5	0.06	4×10^{-6}	50×10^{-6}	50×10^6

APPENDIX F1

Properties of selected materials. (*Continued*)

Material	Specific gravity	Thermal conductivity, $\left[\dfrac{\text{Btu} \cdot \text{in.}}{\text{ft}^2 \cdot \text{sec} \cdot °F}\right]$ at 68°F*	Thermal expansion, in/in/°F at 68°F†	Electrical resistivity, ohm·cm at 68°F‡	Average modulus of elasticity, psi at 68°F§
Polymers					
Melamine-formaldehyde	1.5	0.00057	15×10^{-6}	10^{13}	1.3×10^6
Phenol-formaldehyde	1.3	0.00032	40×10^{-6}	10^{12}	0.5×10^6
Urea-formaldehyde	1.5	0.00057	15×10^{-6}	10^{12}	1.5×10^6
Rubbers (synthetic)	1.5	0.00025	—	—	600-11,000
Rubber (vulcanized)	1.2	0.00025	45×10^{-6}	10^{14}	0.5×10^6
Polyethylene (LD)	0.92	0.00065	100×10^{-6}	10^{15}-10^{18}	14,000-50,000
Polyethylene (HD)	0.96	0.0010	70×10^{-6}	10^{16}-10^{18}	50,000-180,000
Polystyrene	1.05	0.00016	40×10^{-6}	10^{18}	0.4×10^6
Polyvinyl chloride	1.3	0.0003	75×10^{-6}	10^{14}	100,000-600,000
Polyvinylidene chloride	1.7	0.00025	105×10^{-6}	10^{13}	0.05×10^6
Polytertrafluoroethylene	2.2	0.0004	55×10^{-6}	10^{16}	50,000-100,000
Polymethyl methacrylate	1.2	0.0004	50×10^{-6}	10^{16}	0.5×10^6
Nylon	1.15	0.0005	55×10^{-6}	10^{14}	0.4×10^6

*Multiply by 5.19 to get joule · cm/cm² · sec · °C.
†Multiply by 1.8 to get cm/cm/°C.
‡Divide by 2.54 to get ohm · in.
§Multiply by 6900 to get n/m².

L. H. VanVlack, *A Textbook of Materials Technology*. Copyright © 1973, Addison-Wesley Publishing Co., Reading, Massachusetts. Appendix B pp. 313–315. Reprinted with permission.

APPENDIX F2

Average properties of selected engineering materials (English System of Units).

Materials	Specific Weight (lb/in.³)	Elastic Strength[a] Tension (ksi)	Elastic Strength[a] Comp. (ksi)	Elastic Strength[a] Shear (ksi)	Ultimate Strength Tension (ksi)	Ultimate Strength Comp. (ksi)	Ultimate Strength Shear (ksi)	Endurance Limit[c] (ksi)	Modulus of Elasticity (1000 ksi)	Modulus of Rigidity (1000 ksi)	Percent Elongation in 2 in.	Coefficient of Thermal Expansion (10⁻⁶/°F)
Ferrous metals												
Wrought iron	0.278	30	b		48	b	25	23	28		30[d]	6.7
Structural steel	0.284	36	b		66	b		28	29	11.0	28[d]	6.6
Steel, 0.2% C hardened	0.284	62	b		90	b			30	11.6	22	6.6
Steel, 0.4% C hot-rolled	0.284	53	b		84	b		38	30	11.6	29	
Steel, 0.8% C hot-rolled	0.284	76	b		122	b			30	11.6	8	
Cast iron—gray	0.260				25	100		12	15		0.5	6.7
Cast iron—malleable	0.266	32	b		50	b			25		20	6.6
Cast iron—nodular	0.266	70			100				25		4	6.6
Stainless steel (18-8) annealed	0.286	36	b		85	b		40	28	12.5	55	9.6
Stainless steel (18-8) cold-rolled	0.286	165	b		190	b		90	28	12.5	8	9.6
Steel, SAE 4340, heat-treated	0.283	132	145		150	b	95	76	29	11.0	19	
Nonferrous metal alloys												
Aluminum, cast, 195-T6	0.100	24	25		36		30	7	10.3	3.8	5	12.5
Aluminum, wrought, 2014-T4	0.101	41	41	24	62	b	38	18	10.6	4.0	20	12.5
Aluminum, wrought, 2024-T4	0.100	48	48	28	68	b	41	18	10.6	4.0	19	12.5
Aluminum, wrought, 6061-T6	0.098	40	40	26	45	b	30	13.5	10.0	3.8	17	12.5
Magnesium, extrusion, AZ80X	0.066	35	26		49	b	21	19	6.5	2.4	12	14.4
Magnesium, sand cast, AZ63-HT	0.066	14	14		40	b	19	14	6.5	2.4	12	14.4
Monel, wrought, hot-rolled	0.319	50	b		90	b		40	26	9.5	35	7.8
Red brass, cold-rolled	0.316	60			75				15	5.6	4	9.8
Red brass, annealed	0.316	15	b		40	b			15	5.6	50	9.8
Bronze, cold-rolled	0.320	75			100				15	6.5	3	9.4
Bronze, annealed	0.320	20	b		50	b			15	6.5	50	9.4
Titanium alloy, annealed	0.167	135	b		155	b			14	5.3	13	
Invar, annealed	0.292	42	b		70	b			21	8.1	41	0.6

APPENDIX F2

Average properties of selected engineering materials (English System of Units). (Continued)

Materials	Specific Weight (lb/in.^{3})	Elastic Strength[a]			Ultimate Strength			Endurance Limit[c] (ksi)	Modulus of Elasticity (1000 ksi)	Modulus of Rigidity (1000 ksi)	Percent Elongation in 2 in.	Coefficient of Thermal Expansion ($10^{-6}/°F$)
		Tension (ksi)	Comp. (ksi)	Shear (ksi)	Tension (ksi)	Comp. (ksi)	Shear (ksi)					
Nonmetallic materials												
Douglas fir, green[e]	0.022	4.8	3.4			3.9	0.9		1.6			
Douglas fir, air dry[e]	0.020	8.1	6.4			7.4	1.1		1.9			
Red oak, green[e]	0.037	4.4	2.6			3.5	1.2		1.4			1.9
Red oak, air dry[e]	0.025	8.4	4.6			6.9	1.8		1.8			
Concrete, medium strength	0.087		1.2			3.0			3.0			6.0
Concrete, fairly high strength	0.087		2.0			5.0			4.5			6.0

Exact values may vary widely with changes in composition, heat threatment, and mechanical working. More precise information can be obtained from manufacturers.

[a] Elastic strength may be represented by proportional limit, yield point, or yield strength at a specified offset (usually 0.2 percent for ductile metals).
[b] For ductile metals (those with an appreciable ultimate elongation), it is customary to assume the properties in compression have the same values as those in tension.
[c] Rotating beam.
[d] Elongation in 8 in.
[e] All timber properties are parallel to the grain.

A. Higdon, et al., *Mechanics of Materials*. 4th ed. Copyright © 1985, John Wiley & Sons, Inc., New York. Reprinted with permission of John Wiley & Sons, Inc.

APPENDIX F3

Mechanical properties of selected cast irons.

Grey Irons

ASTM Class	S_{ut} ksi	S_{uc} ksi	S_{us} ksi	S_e ksi	Brinell Hardness	E Mpsi	G Mpsi
20	22	83	26	10	156	9.6–14.0	3.9–5.6
25	26	97	32	11.5	174	11.5–14.8	4.6–6.0
30	31	109	40	14	210	13.0–16.4	5.2–6.6
35	36.5	124	48.5	16	212	14.5–17.2	5.8–6.9
40	42.5	140	57	18.5	235	16.0–20.0	6.4–7.8
50	52.5	164	73	21.5	262	18.8–22.8	7.2–8.0
60	62.5	188	88.5	24.5	302	20.4–23.5	7.8–8.5

Nodular Irons

Grade	S_{yt} ksi	S_{ut} ksi	E Mpsi	Brinell Hardness	Elongation, min. % in 2 inches
Tensile Properties—0.2% offset yield					
60-40-18	47.7	66.9	24.5	167	15.0
65-45-12	48.2	67.3	24.4	167	15.0
80-55-06	52.5	81.1	24.4	192	11.2
120-90-02	125.3	141.3	23.8	331	1.5
Compressive Properties—0.2% offset yield					
60-40-18	52.0	—	23.8	—	—
65-45-12	52.5	—	23.6	—	—
80-55-06	56.0	—	23.9	—	—
120-90-02	133.5	—	23.8	—	—
Torsional Properties—0.0375% offset yield					
60-40-18	28.3	68.5	9.1	—	—
65-45-12	30.0	68.9	9.3	—	—
80-55-06	28.0	73.1	9.0	—	—
120-90-02	71.3	126.9	9.2	—	—

APPENDIX F3

Mechanical properties of selected cast irons. *(Continued)*

Class	S_{yt} ksi	S_{ut} ksi	Brinell Hardness	Elongation, min. % in 2 inches
Malleable Irons				
Ferritic: ASTM A47, A338; ANSI G48.1; FED QQ-I-666C				
32510	32	50	156 max	10
35018	35	53	156 max	18
Ferritic: ASTM A197				
—	30	40	156 max	5
Pearlitic and Martensitic: ASTM A220; ANSI G48.2; MIL-I-11444B				
40010	40	60	149–197	10
45008	45	65	156–197	8
45006	45	65	156–207	6
50005	50	70	179–229	5
60004	60	80	197–241	4
70003	70	85	217–269	3
80002	80	95	241–285	2
90001	90	105	269–321	1
Automotive: ASTM A602; SAE J158				
M3210, Annealed	32	50	156 max	10
M4504, Air quenched & tempered	45	65	163–217	4
M5003, Air quenched & tempered	50	75	187–241	3
M5503, Liquid quenched & tempered	55	75	187–241	3
M7002, Liquid quenched & tempered	70	90	229–269	2
M8501, Liquid quenched & tempered	85	105	269–302	1

Source: *Metals Handbook, Desk Edition*. Copyright © 1985, American Society for Metals, Metals Park, OH.

APPENDIX F4

Properties of selected aluminum alloys.

Alloy designation	Principal alloying elements	Hardening process*	Tensile strength, 10^7 N/m²	Yield strength, 10^7 N/m²	Elongation in 50 mm, %	Endurance strength (5×10^8 cycles), 10^7 N/m²	Typical applications
Wrought alloys							
1100	Commercial purity	Cold-working	9–16	3–15	45–15	3–6	Cooking utensils
5052	2.5% Mg	Cold-working	19–29	9–26	30–8	11–14	Bus and truck bodies
Alclad 2024	4.5% Cu, 1.5% Mg (with protective sheet of pure aluminum)	Precipitation	19–47	8–32	22–19	9–14	Aircraft
6061	1.5% Mg₂Si	Precipitation	12–31	6–28	30–17	6–10	General structural
7075	5.6% Zn, 2.5% Mg, 1.6% Cu	Precipitation	23–57	10–50	16–11	12–16	Aircraft
Cast alloys							
195	4.5% Cu	Precipitation	22–28	11–22	8.5–2	5–6	Sand castings
319	3.5% Cu, 6.3% Si	Precipitation	19–25	12–18	2–1.5	7–8	Sand castings
356	7% Si, 0.3% Mg	Precipitation	23–26	15–19	5–4	8–9	Permanent mold castings

* In addition to alloy hardening.
To convert from N/m² to lb/in², multiply by 1.450×10^{-4}.

Guy, A.G. and J.J. Hren, *Elements of Physical Metallurgy*, 3rd ed. Copyright © 1974, Addison-Wesley Publishing Co., Reading, Mass. Table 10.4, p. 495. Reprinted with permission.

APPENDIX F5

Properties of selected magnesium alloys.

Use	Am. Soc. of Testing Materials designation	Composition, %				Condition	Tensile strength, 10^7 N/m^2	Yield strength, 10^7 N/m^2	Elongation in 50 mm, %
		Al	Zn	Mn	Zr				
Sand and permanent mold casting	AZ92A	9.0	2.0	0.1		As-cast	17	10	2
						Solution treated	27	10	10
						Aged	27	14	3
Die casting	AZ91C	9.0	0.7	0.2		As-cast	23	14	3
Sheet	AZ31B	3.0	1.0	0.45		Annealed	26	15	15
						Hard	29	21	8
						Extruded	26	20	15
Structural shapes	AZ80A	8.5	0.5	0.2		Extruded	34	25	12
						Extruded and aged	38	28	8
	ZK60A		5.5		0.5	Extruded	34	26	12
						Extruded and aged	37	30	10

To convert from N/m^2 to lb/in^2, multiply by 1.450×10^{-4}.

Guy, A.G. and J.J. Hren, *Elements of Physical Metallurgy*, 3rd ed. Copyright © 1974, Addison-Wesley Publishing Co., Reading, Mass. Table 7.7, p. 353. Reprinted with permission.

APPENDIX F6

Properties of selected brasses.

Alloy	Composition, %					Tensile strength, 10^7 N/m²	Yield strength, 10^7 N/m²	Elongation in 50 mm, %	Typical applications
	Cu	Zn	Sn	Pb	Mn				
Wrought alloys (annealed)									
Gilding metal	95	5				21	7	40	Coins, jewelry base for gold plate
Commercial bronze	90	10				23	7	45	Grillwork, costume jewelry
Red brass	85	15				25	7	48	Weatherstrip, plumbing lines
Low brass	80	20				26	8	52	Musical instruments, pump lines
Cartridge brass	70	30				28	7	65	Radiator cores, lamp fixtures
Yellow brass	65	35				30	11	55	Reflectors, fasteners, springs
Muntz metal	60	40				37	14	40	Large nuts and bolts, architectural panels
Low-leaded brass	64.5	35		0.5		30	11	55	Ammunition primers, plumbing
Medium-leaded brass	64	35		1.0		30	11	52	Hardware, gears, screws
Free-cutting brass	62	35		3.0		30	12	51	Automatic screw machine stock
Admiralty metal	71	28	1			37	15	65	Condenser tubes, heat exchangers
Naval brass	60	39	1			40	19	45	Marine hardware, valve stems
Manganese bronze	58.5	39.2	1	1 Fe	0.3	45	21	33	Pump rods, shafting rods
Cast alloys									
Cast red brass	85	5	5	5		23	12	25	Pipe fittings, small gears, bearings
Cast yellow brass	60	38	1	1		28	10	25	Hardware fittings, ornamental castings
Cast manganese bronze	58	39.7	1 Al	1 Fe	0.3	48	19	30	Propeller hubs and blades

Guy, A.G. and J.J. Hren, *Elements of Physical Metallurgy*, 3rd ed. Copyright © 1974, Addison-Wesley Publishing Co., Reading, Mass. Table 7.4, p. 348. Reprinted with permission.

APPENDIX F7

Properties of selected bronzes.

Alloy	Condition	Composition, % Cu	Zn	Sn	Pb			Tensile strength, 10^7 N/m²	Yield strength, 10^7 N/m²	Elongation in 50 mm, %	Typical applications
Tin-Bronzes											
5% phosphor-bronze	Wrought, cold-worked	94.8	5				0.2 P	63	61	5	Diaphragms, springs, switch parts
10% phosphor-bronze	Wrought, cold-worked	89.8	10				0.2 P	74	68	10	Bridge bearing plates, special springs
Leaded tin-bronze	Sand cast	88	6	4.5	1.5			26	11	35	Valves, gears, bearings
Gun metal	Sand cast	88	10	2				28	11	30	Fittings, bolts, pump parts
Aluminum-Bronzes (eutectoid hardening)											
5% aluminum-bronze	Wrought, annealed	95					5 Al	41	17	66	Corrosion-resistant tubing
10% aluminum-bronze	Sand cast	86				3.5 Fe	10.5 Al	52	24	20	
Same	Sand cast and hardened							69	31	12	Gears, bearings, bushings
Silicon-Bronze											
Silicon-bronze, type A	Wrought, annealed	95				1 Mn	3 Si	39	14	63	Chemical equipment, hot water tanks
Nickel-Bronzes											
30% cupro-nickel	Wrought, annealed	70					30 Ni	41	17	45	Condenser, distiller tubes
18% nickel-silver	Wrought, annealed	65		17			18 Ni	39	17	42	Table flatware, zippers
Nickel-silver	Sand cast	64	4	8	4		20 Ni	28	17	15	Marine castings, valves
Beryllium-Bronzes (precipitation hardening)											
Beryllium-copper	Wrought, annealed	98				0.3 Co	1.7 Be	48	21	50	Springs, nonsparking tools
Same	Wrought, hardened							121	90	5	

Guy, A.G. and J.J. Hren, *Elements of Physical Metallurgy*, 3rd ed. Copyright © 1974, Addison-Wesley Publishing Co., Reading, Mass. Table 7.6, p. 351. Reprinted with permission.

APPENDIX F8

Properties of selected nickel alloys.

Alloy	Principal alloying elements	Mechanical properties			Endurance strength (10^8 cycles), lb/in^2	Typical applications
		Tensile strength, lb/in^2	*Yield strength,* lb/in^2	*Elongation in 2 inches,* %		
"A" nickel, hot-rolled	none	65,000	30,000	50	30,000	Chemical industry, nickel plating
Duranickel, cold-drawn and precipitation hardened	4.4% aluminum 0.4% titanium	190,000	150,000	20	60,000	Springs, plastics, extrusion equipment
Monel, hot-rolled	30% copper	90,000	55,000	35	42,000	Oil refinery parts
"K" Monel, cold-drawn and precipitation hardened	29% copper 3% aluminum	180,000	160,000	5	46,000	Pump rods, valve stems
Inconel, hot-rolled	14% chromium 6% iron	120,000	65,000	35	45,000	Gas turbine parts
Inconel "X" hot-rolled and aged at 1300°F (704°C)	15% chromium 7% iron 2.5% titanium	180,000	130,000	25	60,000	Springs and bolts subjected to corrosion

Guy, A.G., *Physical Metallurgy For Engineers*. Copyright © 1962, Addison-Wesley Publishing Co., Reading, Mass. Table 9.2, p. 284. Reprinted with permission.

APPENDIX F9

Properties of selected titanium alloys.

Alloy	Heat treatment	Yield strength, lb/in²	Tensile strength, lb/in²	Elongation in 2 inches, %	Reduction of area, %
Commercial titanium	Annealed at 1100 to 1350°F (593 to 732°C)	80,000	95,000	20	42
Ti-6Al-4V	Water quenched from 1750°F (954°C); aged at 1000°F (538°C) for 2 hr	150,000	170,000	13	14
Ti-4Al-4Mn	Water quenched from 1450°F (788°C); aged at 900°F (482°C) for 8 hr	170,000	185,000	13	37
Ti-5Al-2.75Cr-1.25Fe	Water quenched from 1450°F; aged at 900°F for 6 hr	175,000	185,000	10	20
Ti-5Al-1.5Fe-1.4Cr-1.2Mo	Water quenched from 1625°F (885°C); aged at 1000°F for 6 hr	170,000	185,000	15	25

Guy, A.G., *Physical Metallurgy For Engineers*, Copyright © 1962, Addison-Wesley Publishing Co., Reading, Mass. Table 5.8, p. 163. Reprinted with permission.

APPENDIX F10

Properties of selected ceramics @ 20°C.

	Melting Point, °C	Tensile or Bend Strength (σ_t), MPa	Young's Modulus (E), GPa	Fracture Toughness (R), J/m²	Coefficient of Thermal Expansion (α), °C⁻¹ × 10⁻⁶	Thermal Conductivity (k), W/mK	Spalling Index ($k\sigma_t/E\alpha$), kW/m	Toughness/Strength Ratio (K_c/σ_t) = $\sqrt{ER}/\sigma_t$, $\sqrt{m}$
Typical porous insulating firebrick	?	~15	20	? 30	7	1	0.1	0.05
Alumina (Al₂O₃)	2050	300 (700 whisker)	380	80	8	38	3.8	5.5
Magnesia (MgO)	2850	100	315	70	14.8	44	1	0.5
Silicon carbide (SiC)	2300 (decomposes)	200	350	50	4.5	209	26	0.02
Silicon nitride (Si₃N₄)	1900 (sublimes)	600 (1400 whisker)	320	80	2.5	8	6	0.01
Titanium carbide (TiC), Ni/Mo binder	3250	200	350	400	7	17	1.4	0.6
Tungsten carbide (WC), Co binder	2620	350	700	1000	1	34	1.4	0.8
Common bulk glass	Fuses at 1600	70 (up to 4000 filaments)	70	20	10	1	0.1	0.2
Pyroceram glass ceramic	Fuses at 1600	190	126	? 40	10	3	0.5	0.1
Concrete, reinforced	?	20	35	30	7	0.5	0.1	0.1
Carbon graphite	3600 (sublimes)	30	6	100	3	~100	170	0.3

NOTE: There is much more variation in properties of these types of solid than metals, variations occurring with production methods. Properties also vary with temperature and rate, for example, k for Al₂O₃ is 4 W/mK at 1000°C and is 13 W/mK at 1000°C for SiC.

David K. Felbeck and Anthony G. Atkins, *Strength & Fracture of Engineering Solids*. Copyright © 1984, p. 527. Reprinted by permission of Prentice-Hall, Inc., Englewood Cliffs, N.J.

APPENDIX F11

Properties of concrete.

Property	Value*
Compressive strength	3000–7000 psi
Tensile strength	400 psi
Modulus of elasticity	$3.5–5.0 \cdot 10^6$ psi
Modulus of rupture	700 in·lb/in^3
Density	
Wet	145 lb/ft^3
Dry	96 lb/ft^3

* Values depend on water content, cement : sand : gravel ratio, and cure time.

APPENDIX F12

Selected values of modulus of elasticity @ 20°C.

Material	Mpsi	GPa
Aluminum	10.2	70.5
Beryllium	37.7	260
Brick	1.5–2.5	10.5–17.2
Brass	14.6	101
Ceramics—See Appendix F-10		
Cobalt	30.4	210
Concrete—See Appendix F-1 & F-11		
Chromium	40.5	280
Copper	18.8	130
Glass	9–10	62.0–68.9
Graphite	1.3	9
Irons—See Appendix F-3		
Lead	2.34	16
Magnesium	6.48	45
Molybdenum	46.4	320
Nickel	30.4	210
Nylon	0.17	1.2
Osmium	81.2	560
Polyethylene Low density	0.015–0.040	0.2–0.6
High density	0.060–0.180	0.9–2.6
Rubber	(150 psi)	0.001
Silicon Carbide	68.0	470
Silver	12.0	83
Steel—See Appendix F-2		
Titanium	17.4	120
Tungsten	59.4	410

Selected values of modulus of elasticity @ 20°C. (*Continued*)

Material	Mpsi	GPa
Silica Glass	10.5	72
Sintered Carbide, 94% WC, 6% Co	100	690
Elinvar	24.6	170
36% Ni, 12% Cr & W, 52% Fe: S_{yt} = 65,000 psi		
Invar	20.3	140
36% Ni, 64% Fe: S_{yt} = 51,000 psi		
Ni-Span C	27.6	190
42% Ni, 5.5% Cr, 2.5% Ti, 50% Fe: S_{yt} = 180,000 psi		

APPENDIX F13

Steel alloy numbering system.

AISI or SAE Number	Primary Alloying Elements, weight %
10cc	Plain carbon steels
11cc	Plain carbon, re-sulfurized
13cc	Mn, 1.5–2.0%
23cc	Ni, 3.25–3.75%
25cc	Ni, 4.75–5.25%
31cc	Ni, 1.10–1.40%; Cr, 0.55–0.90%
33cc	Ni, 3.25–3.75%; Cr, 1.40–1.75%
40cc	Mo, 0.20–0.30%
41cc	Cr, 0.40–1.20%; Mo, 0.08–0.25%
43cc	Ni, 1.65–2.00%; Cr, 0.40–0.90%; Mo, 0.20–0.30%
46cc	Ni, 1.40–2.00%; Mo, 0.15–0.30%
48cc	Ni, 3.25–3.75%; Mo, 0.20–0.30%
51cc	Cr, 0.70–1.20%
52cc	Cr, 1.30–1.60%
61cc	Cr, 0.70–1.10%; V, 0.10%
81cc	Ni, 0.20–0.40%; Cr, 0.30–0.55%; Mo, 0.08–0.15%
86cc	Ni, 0.30–0.70%; Cr, 0.40–0.85%; Mo, 0.08–0.25%
87cc	Ni, 0.40–0.70%; Cr, 0.40–0.60%; Mo, 0.20–0.30%
92cc	Si, 1.80–2.20%
93cc	Ni, 3.00–3.50%; Cr, 1.00–1.40%; Mo, 0.08–0.15%

cc = carbon content in hundredths.
A B preceding the carbon content signifies a Boron steel, an L signifies a leaded steel. Examples: A 1020 steel has 0.20 parts of carbon by weight. A 10B18 steel has 0.18 parts carbon and has Boron added. A 12L14 steel has 0.14 parts carbon and has lead added.

APPENDIX F14

Magnetic alloys, in order of coercive force.

Material	Residual Magnetization weber/m²	Coercive Force weber/m²
Soft Magnets		
Supermalloy, 79% Ni, 16% Fe, 5% Mo	0.50	$0.0020 \cdot 10^{-4}$
Hypernik, 50% Ni, 50% Fe	0.80	$0.05 \cdot 10^{-4}$
Puron, 99.99% Fe	0.90	$0.05 \cdot 10^{-4}$
4.25% Silicon steel	0.70	$0.40 \cdot 10^{-4}$
2.5% Silicon steel	0.80	$0.80 \cdot 10^{-4}$
1% Silicon steel	0.90	$0.90 \cdot 10^{-4}$
Ingot iron, 99.9% Fe	1.10	$1.00 \cdot 10^{-4}$
Perminvar, 45% Ni, 25% Co, 30% Fe	0.60	$1.2 \cdot 10^{-4}$
Hard Magnets		
Carbon steel, 0.9% C, 1% Mn	1.00	$50 \cdot 10^{-4}$
Tungsten steel, 5% W, 0.7% C, 0.3% Mn	1.03	$70 \cdot 10^{-4}$
Cobalt steel, 36% Co, 4% Cr, 5% W, 0.7% C	1.00	$240 \cdot 10^{-4}$
Alnico 5B, 8% Al, 14% Ni, 24% Co, 3% Cu, 51% Fe	1.27	$600 \cdot 10^{-4}$
Cunife, 60% Cu, 20% Ni, 20% Fe	0.54	$500 \cdot 10^{-4}$
Fine powder, 30% Co, 70% Fe	0.90	$1000 \cdot 10^{-4}$
Ferroxdur, $BaFe_{12}O_{19}$	0.34	$1800 \cdot 10^{-4}$
Bismanol, MnBi	0.43	$3400 \cdot 10^{-4}$

APPENDIX F15

Aluminum identification coding system.

4 Digit Code	Primary Alloy Elements
1---	Unalloyed, > 99% Al
2---	Copper
3---	Manganese
4---	Silicon
5---	Magnesium
6---	Magnesium and Silicon
7---	Zinc
8---	Other elements

Heat Treatment Suffixes

- -F As fabricated   -O   As annealed
- -H Strain hardened by cold working
  - -H1h Hardened only; h = hardness index, 8 = full hard
  - -H2h Hardened and partially annealed
  - -H3h Hardened and stabilized
- -T Heat treated
  - -T2   Annealed, cast alloys
  - -T3   Solution heat treated and cold worked
  - -T4   Solution heat treated and aged naturally
  - -T5   Artificially aged only
  - -T6   Solution heat treated and artificially aged
  - -T7   Solution heat treated and stabilized
  - -T8   Solution heat treated, cold worked, and aged
  - -T9   Solution heat treated, aged, and cold worked
  - -T10 Artificially aged and cold worked

APPENDIX F16

Selected commercial glasses.

Type	SiO$_2$	Na$_2$O	CaO	B$_2$O$_3$	PbO	Al$_2$O$_3$	MgO	K$_2$O
Fused Silica	96	–	–	–	–	–	–	–
Low thermal expansion, high viscosity.								
Pyrex	81	4	–	12	–	2	–	–
Low thermal expansion, low ion exchange.								
Lamp bulbs	73	16	5	–	–	2	4	–
Easy to work.								
Thermometer	73	10	–	10	–	6	–	–
Dimensional stability.								
Container	72	14	10	–	–	2	1	–
High durability, easy to work, resists chemicals.								
Plate	72	14	10	–	–	1	4	–
High durability.								
Window	72	14	8	–	–	1	4	–
Heat absorbing	71	10	9	–	–	4	4	–
Lead tableware	66	6	1	–	16	1	–	10
High index of refraction.								
Lamp stems	55	12	–	–	32	1	–	–
High resistivity.								
Fiberglass	55	1	16	10	–	15	–	–

APPENDIX F17

Variation in properties of wood.

Property	Hardwoods	Softwoods
Density, lb/ft^3	18–50	18–31
Modulus of elasticity, Mpsi	0.9–2.2	0.5–1.9
Compressive strength, psi		
Parallel to grain	2100–9200	1900–8400
Perpendicular to grain	100–900	200–500
Tensile strength, psi		
Parallel to grain	275–1000	200–450
Shear strength, psi		
Perpendicular to grain	700–1500	700–900
Parallel to grain	1400–2300	850–1500

APPENDIX F18

Ultimate tensile and compressive strength of selected brittle materials.

Material	S_{ut}, psi	S_{uc}, psi
Acrylic plastic	10,600	17,000
Gray cast iron		
Class 20	22,000	83,000
Class 60	62,500	188,000
Alumina ceramics	30,000	300,000
28 day concrete	400	5,300
Window glass	10,000	500,000
Aluminum oxide	20,000–30,000	280,000–380,000
Zirconium oxide	4500–12,000	60,000–100,000
Tungsten carbide	130,000	518,000–800,000
Silicon carbide	15,000	150,000

APPENDIX F19

Selected values of Poisson's ratio @ 20°C.

Material	Poisson's Ratio
Fused quartz	0.17
Concrete	0.19
Tungsten carbide	0.22
Zinc	0.25
Glass	0.25
Tungsten	0.28
Cast iron	0.26–0.31
Magnesium	0.29
Steel	0.30
Nickel	0.31
Titanium	0.36
Aluminum	0.34
Copper	0.35
Molybdenum	0.35
Brass	0.35
Tin	0.36
Silver	0.37
Lead	0.44
Rubber	0.49

APPENDIX F20

Mechanical properties of selected engineering materials at room temperature.

Material	Density kg/m³	Elastic Modulus, GPa	Strength, MPa	Critical Stress Intensity Factor, MPa√m	Nom. Fract. Strain —	Maximum Strength/Density, (km/sec)²
Ductile steel	7850–7870	200	350–800	170	0.2–0.5	0.1
Cast iron	6950–7700	55–140	140–650	5–20	0.0–0.2	0.09
Cast Al alloys	2570–2950	65–72	130–300	—	0.01–0.14	0.1
Wrought Al alloys	2620–2820	68–72	55–650	15–100	0.05–0.30	0.2
Polymers	900–2200	0.1–21	5–190	0.5–3	0.0–8.0	0.1
Low-alloy steel	7800–7900	200	1000–2000	15–170	0.05–0.25	0.25
Glasses, ceramics	2200–4000	40–140	10–140	0.2–5	0.0	0.05
Stainless steels	7600–8000	200	300–1400	55–120	0.10–0.30	0.2
Copper alloys	8200–8900	100–117	300–1400	10–100	0.02–0.65	0.17
Titanium alloys	4400–4800	116	400–1600	50–140	0.05–0.30	0.35
Mg alloys	1740–1840	45	80–370	—	0.01–0.25	0.2
Moldable glass fiber resin	1700–2100	11–17	55–440	20–50	0.003–0.015	0.2
Graphite-epoxy	1500	200	1000	45–120	0–0.02	0.65
Tool steels: Hardness to 68 R_c; stable up to 600°C; strength to 2000 MPa				30–100		
Sintered carbides: Hardness at 20°C, 86 to 93 R_A (69 to well above 70 R_c equivalent); to 85 R_A max. at 760°C				3–21		
Superalloys: Strength is approximately 140 MPa at 650°C, 50 MPa at 750°C, 25 MPa at 900°C				—		

From David K. Felbeck and Anthony G. Atkins, *Strength & Fracture of Engineering Solids*, p. 520. © 1984, Prentice-Hall, Inc., Englewood Cliffs, N.J. Reprinted by permission.

APPENDIX F21

Selected values of quasi-static toughness and yield strength.

Material Type	E, GPa	σ_y, MPa	R, kJ/m^2	K_c, MPa$\sqrt{m}$	K_c/σ_y, $\sqrt{m}$
Glasses	70	138	0.01	1	0.01
Ceramics	300	280	0.1	5	0.02
PMMA	2.8	70	0.5	1	0.02
Cast iron	100	300	1	10	0.03
Boron-fiber epoxy composites	170	2000	30	70	0.04
Tough polymers (Polycarbonate)	2.5	63	4	3	0.05
High-strength aluminum-base alloys	70	600	28	44	0.07
High-strength steels (4340)	210	1400	140	170	0.12
High-strength titanium-base alloys	110	1100	170	136	0.12
Maraging steel	210	960	150	180	0.20
Low-carbon steels	210	280	140	170	0.60

From David K. Felbeck and Anthony G. Atkins, *Strength & Fracture of Engineering Solids*, p. 525. © 1984, Prentice-Hall, Inc. Englewood Cliffs, N.J. Reprinted by permission.

APPENDIX F22

Static and dynamic properties of selected materials.

Material	Condition	Static yield strength psi	Ult. strength, psi Static	Ult. strength, psi Dynamic
Ingot iron	Annealed	16,000	37,100	57,400
SAE 1015	Annealed	29,250	50,600	63,500
1022	Cold rolled	64,500	84,000	105,000
1022	Annealed	41,000	65,000	82,700
1040	Annealed	43,000	78,050	91,800
1045	Normalized	55,500	97,750	105,700
1045	Anneal. (Low)	48,200	81,500	117,000
1045	Anneal. (Hi)	51,750	94,300	132,200
1045	Spheroidized	66,400	76,600	98,400
1045	Quench & T.	132,250	142,900	169,000
1095	Normalized	74,000	144,600	151,000
1095	Quench & T.	136,000	176,500	170,500
2345	Austemp. R_c 35	121,000	155,700	182,000
2345	Q. & T. R_c 52	210,000	265,000	276,000
4140	Austemp. R_c 31	82,000	139,750	150,700
4140	Q. & T. R_c 49	195,000	244,000	258,000
5150	Austemp. R_c 31	102,500	148,500	165,300
5150	Q. & T. R_c 52	205,500	257,000	279,100
8715	Heat treated	103,700	141,300	170,000
8739	Austemp. R_c 37	137,000	169,000	207,400
8739	Q. & T. R_c 52	205,000	272,000	290,000
Copper	Annealed	4,000	29,900	36,700
Copper	Cold rolled	30,000	45,000	60,000
1100 Al.	$\frac{1}{2}$ Hard	12,000	17,200	22,100
1100 Al.	Annealed	1,700	11,600	15,400
2017 Al.	As received	38,000	59,900	63,800
2024 Al.	As received	47,000	65,150	68,600
2024 Al.	Annealed	14,000	33,950	44,980
Magnesium F	As received	25,500	35,920	51,760
Magnesium J	As received	29,870	43,750	51,360
302 Stainless	As received	44,000	93,300	110,800
Zamak II	Die cast	14,000	34,500	50,200

[13]For a more extensive table, see Clark, D. S., and Wood, D. A., "The Tensile Impact Properties of Some Metals and Alloys," *Trans. Amer. Soc. for Metals*, 42 (1950), p. 45. See also *Trans. Amer. Soc. for Metals*, 46 (1954), p. 34.

From M. F. Spotts, *Mechanical Design Analysis*, p. 365. © 1964 Prentice-Hall, Inc., Englewood Cliffs, N.J. Reprinted by permission.

APPENDIX F23

Density and mechanical properties of bulk materials and filaments.

Bulk Material	ρ kg/m³	S_u, MPa	S_u/ρ, (km/sec)²	E, GPa	E/ρ, (km/sec)²
Low-carbon steel	7850	400	0.051	210	27
High-tensile steel	7850	2000	0.25	210	27
Aluminum alloy	2700	550	0.20	74	27
Magnesium alloy	1800	350	0.19	45	25
Nickel	8860	490	0.055	210	24
Copper	8940	240	0.027	120	13
Titanium alloy	4500	1000	0.22	120	27
Wood	600	60	0.1	14	23
Glass	2500	60	0.024	70	28
Filaments					
Glass	2500	1000	0.4	70	28
Iron	7850	14000	1.8	210	27
Nickel	8860	3800	0.43	210	24
Copper	8940	2900	0.32	120	13
Alumina (Al_2O_3)	4000	18700	4.7	390	98
Silicon carbide	3200	24100	7.5	560	175
Boron	2500	3450	1.4	380	152
Graphite	2250	21000	9.3	750	334
Carbon	2250	1600	0.71	360	160
Kevlar 49	1440	2700	1.9	130	90
Nylon	1100	400	0.36	4	3.6
Polypropylene	900	900	1.0	8	9

From David K. Felbeck and Anthony G. Atkins, *Strength & Fracture of Engineering Solids*, p. 528. © 1984, Prentice-Hall, Inc. Englewood Cliffs, N.J. Reprinted by permission.

APPENDIX F24

Selected electrical resistivities @ 20°C, from low to high.

Material	ohm·cm
Silver	$1.6 \cdot 10^{-6}$
Annealed copper	$1.7 \cdot 10^{-6}$
Gold	$2.4 \cdot 10^{-6}$
Aluminum	$2.7 \cdot 10^{-6}$
Brass (70%Cu-30%Zn)	$6.2 \cdot 10^{-6}$
Nickel	$7.8 \cdot 10^{-6}$
Iron	$9.7 \cdot 10^{-6}$
White tin (temp > 13.2°C)	$12.8 \cdot 10^{-6}$
Stainless steel (18-8)	$73 \cdot 10^{-6}$
Nichrome (80%Ni-20%Cr)	$100 \cdot 10^{-6}$
Gray tin (temp < 13.2°C)	$100 \cdot 10^{-6}$
Silicon	$2.0 \cdot 10^{5}$
Electric porcelain and other insulators	10^{13}–10^{15}
Nylon	10^{20}
Polystyrene	10^{24}

APPENDIX F25

Density and mechanical properties of selected composite materials.

Material	ρ, kg/m³	S_u, MPa	S_M/ρ, (km/s)²	E, GPa	E/ρ, (km/s)²
Boron-5505 epoxy	1990	1590	0.80	207	104
HTS graphite-5208 epoxy	1550	1480	0.95	172	111
Kevlar 49-resin	1380	1380	1.00	76	55
E glass-1002 epoxy	1800	1100	0.61	39	22
Boron-6061 Al	2600	1490	0.57	214	82
T50 graphite-2011 Al	2580	566	0.22	160	62

From David K. Felbeck/Anthony G. Atkins, *Strength & Fracture of Engineering Solids*, p. 528. © 1984, Prentice-Hall, Inc., Englewood Cliffs, N.J. Reprinted by permission.

APPENDIX G

Areas under the normal curve.

Proportion of total area under the curve that is under the portion of the curve from $-\infty$ to $(X_i - \overline{X'})/\sigma'$. ($X_i$ represents any desired value of the variable X)

$\dfrac{X_i - \overline{X'}}{\sigma'}$	0.09	0.08	0.07	0.06	0.05	0.04	0.03	0.02	0.01	0.00
−3.5	0.00017	0.00017	0.00018	0.00019	0.00019	0.00020	0.00021	0.00022	0.00022	0.00023
−3.4	0.00024	0.00025	0.00026	0.00027	0.00028	0.00029	0.00030	0.00031	0.00033	0.00034
−3.3	0.00035	0.00036	0.00038	0.00039	0.00040	0.00042	0.00043	0.00045	0.00047	0.00048
−3.2	0.00050	0.00052	0.00054	0.00056	0.00058	0.00060	0.00062	0.00064	0.00066	0.00069
−3.1	0.00071	0.00074	0.00076	0.00079	0.00082	0.00085	0.00087	0.00090	0.00094	0.00097
−3.0	0.00100	0.00104	0.00107	0.00111	0.00114	0.00118	0.00122	0.00126	0.00131	0.00135
−2.9	0.0014	0.0014	0.0015	0.0015	0.0016	0.0016	0.0017	0.0017	0.0018	0.0019
−2.8	0.0019	0.0020	0.0021	0.0021	0.0022	0.0023	0.0023	0.0024	0.0025	0.0026
−2.7	0.0026	0.0027	0.0028	0.0029	0.0030	0.0031	0.0032	0.0033	0.0034	0.0035
−2.6	0.0036	0.0037	0.0038	0.0039	0.0040	0.0041	0.0043	0.0044	0.0045	0.0047
−2.5	0.0048	0.0049	0.0051	0.0052	0.0054	0.0055	0.0057	0.0059	0.0060	0.0062
−2.4	0.0064	0.0066	0.0068	0.0069	0.0071	0.0073	0.0075	0.0078	0.0080	0.0082
−2.3	0.0084	0.0087	0.0089	0.0091	0.0094	0.0096	0.0099	0.0102	0.0104	0.0107
−2.2	0.0110	0.0113	0.0116	0.0119	0.0122	0.0125	0.0129	0.0132	0.0136	0.0139
−2.1	0.0143	0.0146	0.0150	0.0154	0.0158	0.0162	0.0166	0.0170	0.0174	0.0179
−2.0	0.0183	0.0188	0.0192	0.0197	0.0202	0.0207	0.0212	0.0217	0.0222	0.0228
−1.9	0.0233	0.0239	0.0244	0.0250	0.0256	0.0262	0.0268	0.0274	0.0281	0.0287
−1.8	0.0294	0.0301	0.0307	0.0314	0.0322	0.0329	0.0336	0.0344	0.0351	0.0359
−1.7	0.0367	0.0375	0.0384	0.0392	0.0401	0.0409	0.0418	0.0427	0.0436	0.0446
−1.6	0.0455	0.0465	0.0475	0.0485	0.0495	0.0505	0.0516	0.0526	0.0537	0.0548
−1.5	0.0559	0.0571	0.0582	0.0594	0.0606	0.0618	0.0630	0.0643	0.0655	0.0668
−1.4	0.0681	0.0694	0.0708	0.0721	0.0735	0.0749	0.0764	0.0778	0.0793	0.0808
−1.3	0.0823	0.0838	0.0853	0.0869	0.0885	0.0901	0.0918	0.0934	0.0951	0.0968
−1.2	0.0985	0.1003	0.1020	0.1038	0.1057	0.1075	0.1093	0.1112	0.1131	0.1151
−1.1	0.1170	0.1190	0.1210	0.1230	0.1251	0.1271	0.1292	0.1314	0.1335	0.1357
−1.0	0.1379	0.1401	0.1423	0.1446	0.1469	0.1492	0.1515	0.1539	0.1562	0.1587
−0.9	0.1611	0.1635	0.1660	0.1685	0.1711	0.1736	0.1762	0.1788	0.1814	0.1841
−0.8	0.1867	0.1894	0.1922	0.1949	0.1977	0.2005	0.2033	0.2061	0.2090	0.2119
−0.7	0.2148	0.2177	0.2207	0.2236	0.2266	0.2297	0.2327	0.2358	0.2389	0.2420
−0.6	0.2451	0.2483	0.2514	0.2546	0.2578	0.2611	0.2643	0.2676	0.2709	0.2743
−0.5	0.2776	0.2810	0.2843	0.2877	0.2912	0.2946	0.2981	0.3015	0.3050	0.3085
−0.4	0.3121	0.3156	0.3192	0.3228	0.3264	0.3300	0.3336	0.3372	0.3409	0.3446
−0.3	0.3483	0.3520	0.3557	0.3594	0.3632	0.3669	0.3707	0.3745	0.3783	0.3821
−0.2	0.3859	0.3897	0.3936	0.3974	0.4013	0.4052	0.4090	0.4129	0.4168	0.4207
−0.1	0.4247	0.4286	0.4325	0.4364	0.4404	0.4443	0.4483	0.4522	0.4562	0.4602
−0.0	0.4641	0.4681	0.4721	0.4761	0.4801	0.4840	0.4880	0.4920	0.4960	0.5000

APPENDIX G

Areas under the normal curve. *(Continued)*

$\dfrac{X_i - \overline{X}'}{\sigma'}$	0.00	0.01	0.02	0.03	0.04	0.05	0.06	0.07	0.08	0.09
+0.0	0.5000	0.5040	0.5080	0.5120	0.5160	0.5199	0.5239	0.5279	0.5319	0.5359
+0.1	0.5398	0.5438	0.5478	0.5517	0.5557	0.5596	0.5636	0.5675	0.5714	0.5753
+0.2	0.5793	0.5832	0.5871	0.5910	0.5948	0.5987	0.6026	0.6064	0.6103	0.6141
+0.3	0.6179	0.6217	0.6255	0.6293	0.6331	0.6368	0.6406	0.6443	0.6480	0.6517
+0.4	0.6554	0.6591	0.6628	0.6664	0.6700	0.6736	0.6772	0.6808	0.6844	0.6879
+0.5	0.6915	0.6950	0.6985	0.7019	0.7054	0.7088	0.7123	0.7157	0.7190	0.7224
+0.6	0.7257	0.7291	0.7324	0.7357	0.7389	0.7422	0.7454	0.7486	0.7517	0.7549
+0.7	0.7580	0.7611	0.7642	0.7673	0.7704	0.7734	0.7764	0.7794	0.7823	0.7852
+0.8	0.7881	0.7910	0.7939	0.7967	0.7995	0.8023	0.8051	0.8079	0.8106	0.8133
+0.9	0.8159	0.8186	0.8212	0.8238	0.8264	0.8289	0.8315	0.8340	0.8365	0.8389
+1.0	0.8413	0.8438	0.8461	0.8485	0.8508	0.8531	0.8554	0.8577	0.8599	0.8621
+1.1	0.8643	0.8665	0.8686	0.8708	0.8729	0.8749	0.8770	0.8790	0.8810	0.8830
+1.2	0.8849	0.8869	0.8888	0.8907	0.8925	0.8944	0.8962	0.8980	0.8997	0.9015
+1.3	0.9032	0.9049	0.9066	0.9082	0.9099	0.9115	0.9131	0.9147	0.9162	0.9177
+1.4	0.9192	0.9207	0.9222	0.9236	0.9251	0.9265	0.9279	0.9292	0.9306	0.9319
+1.5	0.9332	0.9345	0.9357	0.9370	0.9382	0.9394	0.9406	0.9418	0.9429	0.9441
+1.6	0.9452	0.9463	0.9474	0.9484	0.9495	0.9505	0.9515	0.9525	0.9535	0.9545
+1.7	0.9554	0.9564	0.9573	0.9582	0.9591	0.9599	0.9608	0.9616	0.9625	0.9633
+1.8	0.9641	0.9649	0.9656	0.9664	0.9671	0.9678	0.9686	0.9693	0.9699	0.9706
+1.9	0.9713	0.9719	0.9726	0.9732	0.9738	0.9744	0.9750	0.9756	0.9761	0.9767
+2.0	0.9773	0.9778	0.9783	0.9788	0.9793	0.9798	0.9803	0.9808	0.9812	0.9817
+2.1	0.9821	0.9826	0.9830	0.9834	0.9838	0.9842	0.9846	0.9850	0.9854	0.9857
+2.2	0.9861	0.9864	0.9868	0.9871	0.9875	0.9878	0.9881	0.9884	0.9887	0.9890
+2.3	0.9893	0.9896	0.9898	0.9901	0.9904	0.9906	0.9909	0.9911	0.9913	0.9916
+2.4	0.9918	0.9920	0.9922	0.9925	0.9927	0.9929	0.9931	0.9932	0.9934	0.9936
+2.5	0.9938	0.9940	0.9941	0.9943	0.9945	0.9946	0.9948	0.9949	0.9951	0.9952
+2.6	0.9953	0.9955	0.9956	0.9957	0.9959	0.9960	0.9961	0.9962	0.9963	0.9964
+2.7	0.9965	0.9966	0.9967	0.9968	0.9969	0.9970	0.9971	0.9972	0.9973	0.9974
+2.8	0.9974	0.9975	0.9976	0.9977	0.9977	0.9978	0.9979	0.9979	0.9980	0.9981
+2.9	0.9981	0.9982	0.9983	0.9983	0.9984	0.9984	0.9985	0.9985	0.9986	0.9986
+3.0	0.99865	0.99869	0.99874	0.99878	0.99882	0.99886	0.99889	0.99893	0.99896	0.99900
+3.1	0.99903	0.99906	0.99910	0.99913	0.99915	0.99918	0.99921	0.99924	0.99926	0.99929
+3.2	0.99931	0.99934	0.99936	0.99938	0.99940	0.99942	0.99944	0.99946	0.99948	0.99950
+3.3	0.99952	0.99953	0.99955	0.99957	0.99958	0.99960	0.99961	0.99962	0.99964	0.99965
+3.4	0.99966	0.99967	0.99969	0.99970	0.99971	0.99972	0.99973	0.99974	0.99975	0.99976
+3.5	0.99977	0.99978	0.99978	0.99979	0.99980	0.99981	0.99981	0.99982	0.99983	0.99983

E. L. Grant and R. S. Leavenworth, *Statistical Quality Control*, 5th ed. Copyright © 1980 by McGraw-Hill Publishing Co., New York. Reprinted with permission.

APPENDIX H

The *t* distribution.

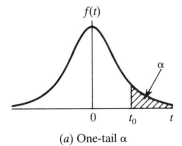

(a) One-tail α

(b) Two-tail α

Given v, the table gives (a) the one-tail t_0 value with α of the area above it, that is, $P(t \geq t_0) = \alpha$, or (b) the two-tail $+t_0$ and $-t_0$ values with α/2 in each tail, that is, $P(t \leq t_0) + P(t \geq +t_0) = \alpha$

	One-tail α					
	0.10	0.05	0.025	0.01	0.005	0.001
	Two-tail α					
v	0.20	0.10	0.05	0.02	0.01	0.002
1	3.078	6.314	12.706	31.821	63.657	318.300
2	1.886	2.920	4.303	6.965	9.925	22.327
3	1.638	2.353	3.182	4.541	5.841	10.214
4	1.533	2.132	2.776	3.747	4.604	7.173
5	1.476	2.015	2.571	3.305	4.032	5.893
6	1.440	1.943	2.447	3.143	3.707	5.208
7	1.415	1.895	2.365	2.998	3.499	4.785
8	1.397	1.860	2.306	2.896	3.355	4.501
9	1.383	1.833	2.262	2.821	3.250	4.297
10	1.372	1.812	2.228	2.764	3.169	4.144
11	1.363	1.796	2.201	2.718	3.106	4.025
12	1.356	1.782	2.179	2.681	3.055	3.930
13	1.350	1.771	2.160	2.650	3.012	3.852
14	1.345	1.761	2.145	2.624	2.977	3.787
15	1.341	1.753	2.131	2.602	2.947	3.733
16	1.337	1.746	2.120	2.583	2.921	3.686
17	1.333	1.740	2.110	2.567	2.898	3.646
18	1.330	1.734	2.101	2.552	2.878	3.611

APPENDIX H

The *t* distribution. *(Continued)*

ν	One-tail α					
	0.10	0.05	0.025	0.01	0.005	0.001
	Two-tail α					
	0.20	0.10	0.05	0.02	0.01	0.002
19	1.328	1.729	2.093	2.539	2.861	3.579
20	1.325	1.725	2.086	2.528	2.845	3.552
21	1.323	1.721	2.080	2.518	2.831	3.527
22	1.321	1.717	2.074	2.508	2.819	3.505
23	1.319	1.714	2.069	2.500	2.807	3.485
24	1.318	1.711	2.064	2.492	2.797	3.467
25	1.316	1.708	2.060	2.485	2.787	3.450
26	1.315	1.706	2.056	2.479	2.779	3.435
27	1.314	1.703	2.052	2.473	2.771	3.421
28	1.313	1.701	2.048	2.467	2.763	3.408
29	1.311	1.699	2.045	2.462	2.756	3.396
30	1.310	1.697	2.042	2.457	2.750	3.385
40	1.303	1.684	2.021	2.423	2.704	3.307
60	1.296	1.671	2.000	2.390	2.660	3.232
80	1.292	1.664	1.990	2.374	2.639	3.195
100	1.290	1.660	1.984	2.365	2.626	3.174
∞	1.282	1.645	1.960	2.326	2.576	3.090

L. Blank, *Statistical Procedures for Engineering, Management, and Science*. Copyright © 1980 by McGraw-Hill Publishing Co., New York. Reprinted with permission.

APPENDIX I

The F distribution ($\alpha = 0.10$, 0.05, and 0.01).

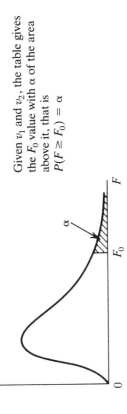

Given v_1 and v_2, the table gives the F_0 value with α of the area above it, that is $P(F \geq F_0) = \alpha$

v_2	α	1	2	3	4	5	6	7	8	9	10	11	12	14	15	19	20	24	30	50	100	500	∞
1	.10	39.9	49.5	53.6	55.8	57.2	58.2	58.9	59.4	59.9	60.2	60.5	60.7	61.1	61.2	61.6	61.7	62.0	62.3	62.7	63.0	63.3	63.3
	.05	161	200	216	255	230	234	237	239	241	242	243	244	245	246	248	248	249	250	252	253	254	254
2	.10	8.53	9.00	9.16	9.24	9.29	9.33	9.35	9.37	9.38	9.39	9.40	9.41	9.42	9.42	9.44	9.44	9.45	9.46	9.47	9.48	9.49	9.49
	.05	18.5	19.0	19.2	19.2	19.3	19.3	19.4	19.4	19.4	19.4	19.4	19.4	19.4	19.4	19.4	19.4	19.5	19.5	19.5	19.5	19.5	19.5
	.01	98.5	99.0	99.2	99.2	99.3	99.3	99.4	99.4	99.4	99.4	99.4	99.4	99.4	99.4	99.4	99.4	99.5	99.5	99.5	99.5	99.5	99.5
3	.10	5.54	5.46	5.39	5.34	5.31	5.28	5.27	5.25	5.24	5.23	5.22	5.22	5.20	5.20	5.18	5.18	5.18	5.17	5.15	5.14	5.14	5.13
	.05	10.1	9.55	9.28	9.12	9.10	8.94	8.89	8.85	8.81	8.79	8.76	8.74	8.71	8.70	8.67	8.66	8.64	8.62	8.58	8.55	8.53	8.53
	.01	34.1	30.8	29.5	28.7	28.2	27.9	27.7	27.5	27.3	27.2	27.1	27.1	26.9	26.9	26.7	26.7	26.6	26.5	26.4	26.2	26.1	26.1
4	.10	4.54	4.32	4.19	4.11	4.05	4.01	3.98	3.95	3.94	3.92	3.91	3.90	3.88	3.87	3.84	3.84	3.83	3.82	3.80	3.78	3.76	3.76
	.05	7.71	6.94	6.59	6.39	6.26	6.16	6.09	6.04	6.00	5.96	5.94	5.91	5.87	5.86	5.81	5.80	5.77	5.75	5.70	5.66	5.64	5.63
	.01	21.2	18.0	16.7	16.0	15.5	15.2	15.0	14.8	14.7	14.5	14.4	14.4	14.2	14.2	14.0	14.0	13.9	13.8	13.7	13.6	13.5	13.5
5	.10	4.06	3.78	3.62	3.52	3.45	3.40	3.37	3.34	3.32	3.30	3.28	3.27	3.25	3.24	3.21	3.21	3.19	3.17	3.15	3.13	3.11	3.10
	.05	6.61	5.79	5.41	5.19	5.05	4.95	4.88	4.82	4.77	4.74	4.71	4.68	4.64	4.62	4.57	4.56	4.53	4.50	4.44	4.41	4.37	4.36
	.01	16.26	13.27	12.06	11.39	10.97	10.67	10.46	10.29	10.16	10.05	9.96	9.89	9.77	9.72	9.58	9.55	9.47	9.38	9.24	9.13	9.04	9.02
6	.10	3.78	3.46	3.29	3.18	3.11	3.05	3.01	2.98	2.96	2.94	2.92	2.90	2.88	2.87	2.84	2.84	2.82	2.80	2.77	2.75	2.73	2.72
	.05	5.99	5.14	4.76	4.53	4.39	4.28	4.21	4.15	4.10	4.06	4.03	4.00	3.96	3.94	3.88	3.87	3.84	3.81	3.75	3.71	3.68	3.67
	.01	13.74	10.92	9.78	9.15	8.75	8.47	8.26	8.10	7.98	7.87	7.79	7.72	7.60	7.56	7.42	7.40	7.31	7.23	7.09	6.99	6.90	6.88
	.10	3.59	3.26	3.07	2.96	2.88	2.83	2.78	2.75	2.72	2.70	2.68	2.67	2.64	2.63	2.60	2.59	2.58	2.56	2.52	2.50	2.48	2.47

APPENDIX I

The F distribution ($\alpha = 0.10$, 0.05, and 0.01). (Continued)

v_2	α	1	2	3	4	5	6	7	8	9	10	11	12	14	15	19	20	24	30	50	100	500	∞
7	.05	5.59	4.74	4.35	4.12	3.97	3.87	3.79	3.73	3.68	3.64	3.60	3.57	3.53	3.51	3.46	3.44	3.41	3.38	3.32	3.27	3.24	3.23
	.01	12.25	9.55	8.45	7.85	7.46	7.19	6.99	6.84	6.72	6.62	6.54	6.47	6.36	6.31	6.18	6.16	6.07	5.99	5.86	5.75	5.67	5.65
8	.10	3.46	3.11	2.92	2.81	2.73	2.67	2.62	2.59	2.56	2.54	2.52	2.50	2.47	2.46	2.43	2.42	2.40	2.38	2.35	2.32	2.30	2.29
	.05	5.32	4.46	4.07	3.84	3.69	3.58	3.50	3.44	3.39	3.35	3.31	3.28	3.24	3.22	3.16	3.15	3.12	3.06	3.02	2.97	2.94	2.93
	.01	11.26	8.65	7.59	7.01	6.63	6.37	6.18	6.03	5.91	5.81	5.73	5.67	5.56	5.52	5.38	5.36	5.28	5.20	5.07	4.96	4.88	4.86
9	.10	3.36	3.01	2.81	2.69	2.61	2.55	2.51	2.47	2.44	2.42	2.40	2.38	2.35	2.34	2.31	2.30	2.28	2.25	2.22	2.19	2.17	2.16
	.05	5.12	4.26	3.86	3.63	3.48	3.37	3.29	3.23	3.18	3.14	3.10	3.07	3.03	3.01	2.95	2.94	2.90	2.86	2.80	2.76	2.72	2.71
	.01	10.56	8.02	6.99	6.42	6.06	5.80	5.61	5.47	5.35	5.26	5.18	5.11	5.00	4.96	4.83	4.81	4.73	4.65	4.52	4.42	4.33	4.31
10	.10	3.28	2.92	2.73	2.61	2.52	2.46	2.41	2.38	2.35	2.32	2.30	2.28	2.25	2.24	2.21	2.20	2.18	2.16	2.12	2.09	2.06	2.06
	.05	4.96	4.10	3.71	3.48	3.33	3.22	3.14	3.07	3.02	2.98	2.94	2.91	2.86	2.85	2.78	2.77	2.74	2.70	2.64	2.59	2.55	2.54
	.01	10.04	7.56	6.55	5.99	5.64	5.39	5.20	5.06	4.94	4.85	4.77	4.71	4.60	4.56	4.43	4.41	4.33	4.25	4.12	4.01	3.93	3.91
11	.10	3.23	2.86	2.66	2.54	2.45	2.39	2.34	2.30	2.27	2.25	2.23	2.21	2.18	2.17	2.13	2.12	2.10	2.06	2.04	2.00	1.98	1.97
	.05	4.84	3.98	3.59	3.36	3.20	3.09	3.01	2.95	2.90	2.85	2.82	2.79	2.74	2.72	2.66	2.65	2.61	2.57	2.51	2.46	2.42	2.40
	.01	9.65	7.21	6.22	5.67	5.32	5.07	4.89	4.74	4.63	4.54	4.46	4.40	4.29	4.25	4.12	4.10	4.02	3.94	3.81	3.71	3.62	3.60
12	.10	3.18	2.81	2.61	2.48	2.39	2.33	2.28	2.24	2.21	2.19	2.17	2.15	2.11	2.10	2.07	2.06	2.04	2.01	1.97	1.94	1.91	1.90
	.05	4.75	3.89	3.49	3.26	3.11	3.00	2.91	2.85	2.80	2.75	2.72	2.69	2.64	2.62	2.56	2.54	2.51	2.47	2.40	2.35	2.31	2.30
	.01	9.33	6.93	5.95	5.41	5.06	4.82	4.64	4.50	4.39	4.30	4.22	4.16	4.05	4.01	3.88	3.86	3.78	3.70	3.57	3.47	3.38	3.36
14	.10	3.10	2.73	2.52	2.39	2.31	2.24	2.19	2.15	2.12	2.10	2.06	2.05	2.02	2.01	1.97	1.96	1.94	1.91	1.87	1.83	1.80	1.80
	.05	4.60	3.74	3.34	3.11	2.96	2.85	2.76	2.70	2.65	2.60	2.57	2.53	2.48	2.46	2.40	2.39	2.35	2.31	2.24	2.19	2.14	2.13
	.01	8.86	6.51	5.56	5.04	4.69	4.46	4.28	4.14	4.03	3.94	3.86	3.80	3.70	3.66	3.53	3.51	3.43	3.35	3.22	3.11	3.03	3.00
15	.10	3.07	2.70	2.49	2.36	2.27	2.21	2.16	2.12	2.09	2.06	2.04	2.02	1.98	1.97	1.93	1.92	1.90	1.87	1.83	1.79	1.76	1.76
	.05	4.54	3.68	3.29	3.06	2.90	2.79	2.71	2.64	2.59	2.54	2.51	2.48	2.42	2.40	2.34	2.33	2.29	2.25	2.18	2.12	2.08	2.07
	.01	8.68	6.36	5.42	4.89	4.56	4.32	4.14	4.00	3.89	3.80	3.73	3.67	3.56	3.52	3.40	3.37	3.29	3.21	3.08	2.98	2.89	2.87
16	.10	3.05	2.67	2.46	2.33	2.24	2.18	2.13	2.09	2.06	2.03	2.01	1.99	1.95	1.94	1.90	1.89	1.87	1.84	1.79	1.76	1.73	1.72
	.05	4.49	3.63	3.24	3.01	2.85	2.74	2.66	2.59	2.54	2.49	2.46	2.42	2.37	2.35	2.29	2.28	2.24	2.19	2.12	2.07	2.02	2.01
	.01	8.53	6.23	5.29	4.77	4.44	4.20	4.03	3.89	3.78	3.69	3.62	3.55	3.45	3.41	3.28	3.26	3.18	3.10	2.97	2.86	2.78	2.75

The degrees of freedom are v_1 for the numerator and v_2 for the denominator.

The F distribution ($\alpha = 0.10, 0.05,$ and 0.01). (Continued)

v_2	α	\multicolumn{17}{c}{v_1 (numerator)}
		1	2	3	4	5	6	7	8	9	10	11	12	14	15	19	20	24	30	50	100	500	∞
18	.10	3.01	2.62	2.42	2.29	2.20	2.13	2.08	2.04	2.00	1.98	1.96	1.93	1.90	1.89	1.85	1.84	1.81	1.78	1.74	1.70	1.67	1.66
	.05	4.41	3.55	3.16	2.93	2.77	2.66	2.58	2.51	2.46	2.41	2.37	2.34	2.29	2.27	2.20	2.19	2.15	2.11	2.04	1.98	1.93	1.92
	.01	8.29	6.01	5.09	4.58	4.25	4.01	3.84	3.71	3.60	3.51	3.43	3.37	3.27	3.23	3.10	3.08	3.00	2.92	2.78	2.68	2.59	2.57
19	.10	2.99	2.61	2.40	2.27	2.18	2.11	2.06	2.02	1.98	1.96	1.94	1.91	1.87	1.86	1.82	1.81	1.79	1.76	1.71	1.67	1.64	1.63
	.05	4.38	3.52	3.13	2.90	2.74	2.63	2.54	2.48	2.42	2.38	2.34	2.31	2.26	2.23	2.17	2.16	2.11	2.07	2.00	1.94	1.89	1.88
	.01	8.18	5.93	5.01	4.50	4.17	3.94	3.77	3.63	3.52	3.43	3.36	3.30	3.19	3.15	3.03	3.00	2.92	2.84	2.71	2.60	2.51	2.49
20	.10	2.97	2.59	2.38	2.25	2.16	2.09	2.04	2.00	1.96	1.94	1.92	1.89	1.85	1.84	1.80	1.79	1.77	1.74	1.69	1.65	1.62	1.61
	.05	4.35	3.49	3.10	2.87	2.71	2.60	2.51	2.45	2.39	2.35	2.31	2.28	2.22	2.20	2.14	2.12	2.08	2.04	1.97	1.91	1.86	1.84
	.01	8.10	5.85	4.94	4.43	4.10	3.87	3.70	3.56	3.46	3.37	3.29	3.23	3.13	3.09	2.96	2.94	2.86	2.78	2.64	2.54	2.44	2.42
24	.10	2.93	2.54	2.33	2.19	2.10	2.04	1.98	1.94	1.91	1.88	1.85	1.83	1.79	1.78	1.74	1.73	1.70	1.67	1.62	1.58	1.54	1.53
	.05	4.26	3.40	3.01	2.78	2.62	2.51	2.42	2.36	2.30	2.25	2.21	2.18	2.13	2.11	2.04	2.03	1.98	1.94	1.86	1.80	1.75	1.73
	.01	7.82	5.61	4.72	4.22	3.90	3.67	3.50	3.36	3.26	3.17	3.09	3.03	2.93	2.89	2.76	2.74	2.66	2.58	2.44	2.33	2.24	2.21
30	.10	2.88	2.49	2.28	2.14	2.05	1.98	1.93	1.88	1.85	1.82	1.79	1.77	1.73	1.72	1.68	1.67	1.64	1.61	1.55	1.51	1.47	1.46
	.05	4.17	3.32	2.92	2.69	2.53	2.42	2.33	2.27	2.21	2.16	2.13	2.09	2.04	2.01	1.95	1.93	1.89	1.84	1.76	1.70	1.64	1.62
	.01	7.56	5.39	4.51	4.02	3.70	3.47	3.30	3.17	3.07	2.98	2.91	2.84	2.74	2.70	2.57	2.55	2.47	2.39	2.25	2.13	2.03	2.01
50	.10	2.81	2.41	2.20	2.06	1.97	1.90	1.84	1.80	1.76	1.73	1.70	1.68	1.64	1.63	1.58	1.57	1.54	1.50	1.44	1.39	1.34	1.33
	.05	4.03	3.18	2.79	2.56	2.40	2.29	2.20	2.13	2.07	2.03	1.99	1.95	1.89	1.87	1.80	1.78	1.74	1.69	1.60	1.52	1.46	1.44
	.01	7.17	5.06	4.20	3.72	3.41	3.19	3.02	2.89	2.79	2.70	2.63	2.56	2.46	2.42	2.29	2.27	2.18	2.10	1.95	1.82	1.71	1.68
100	.10	2.76	2.36	2.14	2.00	1.91	1.83	1.78	1.73	1.70	1.66	1.63	1.61	1.57	1.56	1.50	1.49	1.46	1.42	1.35	1.29	1.23	1.21
	.05	3.94	3.09	2.70	2.46	2.31	2.19	2.10	2.03	1.97	1.93	1.89	1.85	1.79	1.77	1.69	1.68	1.63	1.57	1.48	1.39	1.31	1.28
	.01	6.90	4.82	3.98	3.51	3.21	2.99	2.82	2.69	2.59	2.50	2.43	2.37	2.26	2.22	2.09	2.07	1.98	1.89	1.73	1.60	1.47	1.43
500	.10	2.72	2.31	2.10	1.96	1.86	1.79	1.73	1.68	1.64	1.61	1.58	1.56	1.52	1.50	1.45	1.44	1.41	1.36	1.28	1.21	1.12	1.09
	.05	3.86	3.01	2.62	2.39	2.23	2.12	2.03	1.96	1.90	1.85	1.81	1.77	1.71	1.69	1.61	1.59	1.54	1.48	1.38	1.28	1.16	1.11
	.01	6.69	4.65	3.82	3.36	3.05	2.84	2.68	2.55	2.44	2.36	2.28	2.22	2.12	2.07	1.94	1.92	1.83	1.74	1.56	1.41	1.23	1.16
∞	.10	2.71	2.30	2.08	1.94	1.85	1.77	1.72	1.67	1.63	1.60	1.57	1.55	1.51	1.49	1.43	1.42	1.38	1.34	1.26	1.18	1.08	1.00
	.05	3.84	3.00	2.60	2.37	2.21	2.10	2.01	1.94	1.88	1.83	1.79	1.75	1.69	1.67	1.59	1.57	1.52	1.46	1.35	1.24	1.11	1.00
	.01	6.63	4.61	3.78	3.32	3.02	2.80	2.64	2.51	2.41	2.32	2.25	2.18	2.08	2.04	1.90	1.88	1.79	1.70	1.52	1.36	1.15	1.00

L. Blank, *Statistical Procedures For Engineering, Management, and Science.* Copyright © 1980 by McGraw-Hill Publishing Co., New York. Reprinted with permission.

APPENDIX J

The χ^2 distribution.

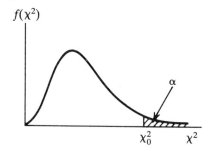

Given v, the table gives the χ_0^2 value with α of the area above it; that is

$$P(\chi^2 \geq \chi_0^2) = \alpha$$

v \ α	0.995	0.990	0.975	0.950	0.900	0.500	0.100	0.050	0.025	0.010	0.005
1	0.00	0.00	0.00	0.00	0.02	0.45	2.71	3.84	5.02	6.63	7.88
2	0.01	0.02	0.05	0.10	0.21	1.39	4.61	5.99	7.38	9.21	10.60
3	0.07	0.11	0.22	0.35	0.58	2.37	6.25	7.81	9.35	11.34	12.84
4	0.21	0.30	0.48	0.71	1.06	3.36	7.78	9.49	11.14	13.28	14.86
5	0.41	0.55	0.83	1.15	1.61	4.35	9.24	11.07	12.83	15.09	16.75
6	0.68	0.87	1.24	1.64	2.20	5.35	10.65	12.59	14.45	16.81	18.55
7	0.99	1.24	1.69	2.17	2.83	6.35	12.02	14.07	16.01	18.48	20.28
8	1.34	1.65	2.18	2.73	3.49	7.34	13.36	15.51	17.53	20.09	21.96
9	1.73	2.09	2.70	3.33	4.17	8.34	14.68	16.92	19.02	21.67	23.59
10	2.16	2.56	3.25	3.94	4.87	9.34	15.99	18.31	20.48	23.21	25.19
11	2.60	3.05	3.82	4.57	5.50	10.34	17.28	19.68	21.92	24.72	26.76
12	3.07	3.57	4.40	5.23	6.30	11.34	18.55	21.03	23.34	26.22	28.30
13	3.57	4.11	5.01	5.89	7.04	12.34	19.81	22.36	24.74	27.69	29.82
14	4.07	4.66	5.63	6.57	7.79	13.34	21.06	23.68	26.12	29.14	31.32
15	4.60	5.23	6.27	7.26	8.55	14.34	22.31	25.00	27.49	30.58	32.80
16	5.14	5.81	6.91	7.96	9.31	15.34	23.54	26.30	28.85	32.00	34.27
17	5.70	6.41	7.56	8.67	10.09	16.34	24.77	27.59	30.19	33.41	35.72
18	6.26	7.01	8.23	9.39	10.87	17.34	25.99	28.87	31.53	34.81	37.16
19	6.84	7.63	8.91	10.12	11.65	18.34	27.20	30.14	32.85	36.19	38.58
20	7.43	8.26	9.59	10.85	12.44	19.38	28.41	31.41	34.17	37.57	40.00
21	8.03	8.90	10.28	11.50	13.24	20.38	29.62	32.67	35.48	38.93	41.40
22	8.64	9.54	10.98	12.34	14.04	21.34	30.81	33.92	36.78	40.29	42.80
23	9.26	10.20	11.69	13.09	14.85	22.34	32.01	35.17	38.08	41.64	44.18
24	9.89	10.86	12.40	13.85	15.66	23.34	33.20	36.42	39.36	42.98	45.56
25	10.52	11.52	13.12	14.61	16.47	24.34	34.38	37.65	40.65	44.31	46.93
26	11.16	12.20	13.84	15.38	17.29	25.34	35.56	38.89	41.92	45.64	48.29
27	11.81	12.88	14.57	16.15	18.11	26.34	36.74	40.11	43.19	46.96	49.65
28	12.46	13.57	15.31	16.93	18.94	27.34	37.92	41.34	44.46	48.28	50.99
29	13.12	14.26	16.05	17.71	19.77	28.34	39.09	42.56	45.72	49.59	52.34
30	13.79	14.95	16.79	18.49	20.60	29.34	40.26	43.77	46.98	50.89	53.67
40	20.71	22.16	24.43	26.51	29.05	39.34	51.80	55.76	59.34	63.69	66.77
50	27.99	29.71	32.36	34.76	37.69	49.33	63.17	67.50	71.42	76.15	79.49
70	43.28	45.44	48.76	51.74	55.33	69.33	85.53	90.53	95.02	100.42	104.22
100	67.33	70.06	74.22	77.93	82.36	99.33	118.50	124.34	129.56	135.81	140.17

L. Blank, *Statistical Procedures For Engineering, Management, and Science.* Copyright © 1980 by McGraw-Hill Publishing Co., New York. Reprinted with permission.

APPENDIX K

Effect of surface conditions on the fatigue strength of steel. Divide N/m² by 6.89×10^3 to get psi

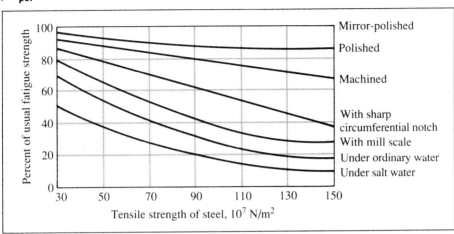

From A. G. Guy and J. J. Hren, *Elements of Physical Metallurgy*, 3rd ed. © 1974 Addison-Wesley Publishing Co., Reading, Mass. Figure 4.27, p. 152. Reprinted with permission.

APPENDIX L

Stress concentration factors.

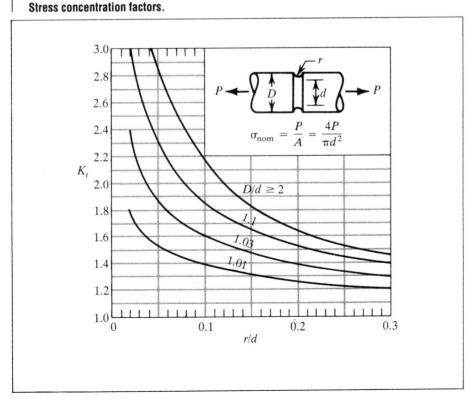

351

APPENDIX L

Stress concentration factors. (*Continued*)

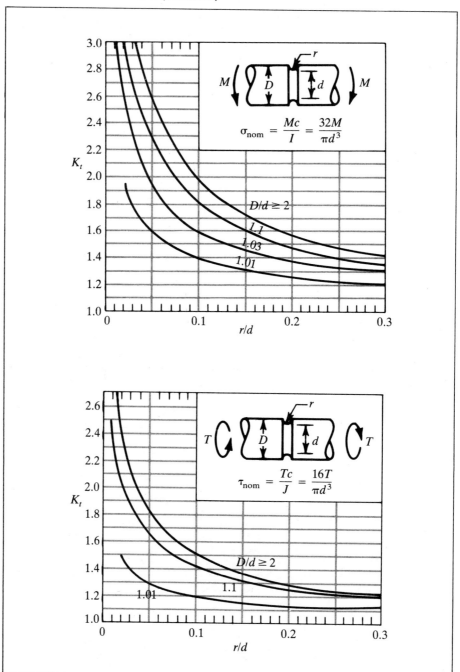

Stress concentration factors. (*Continued*)

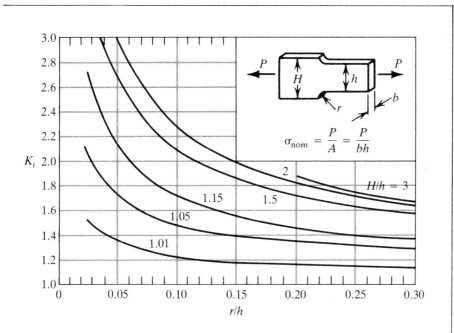

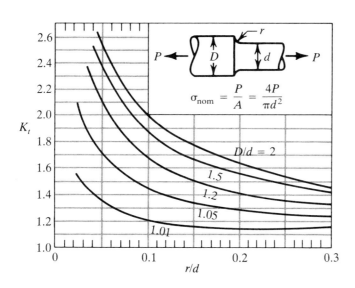

APPENDIX L

Stress concentration factors. (*Continued*)

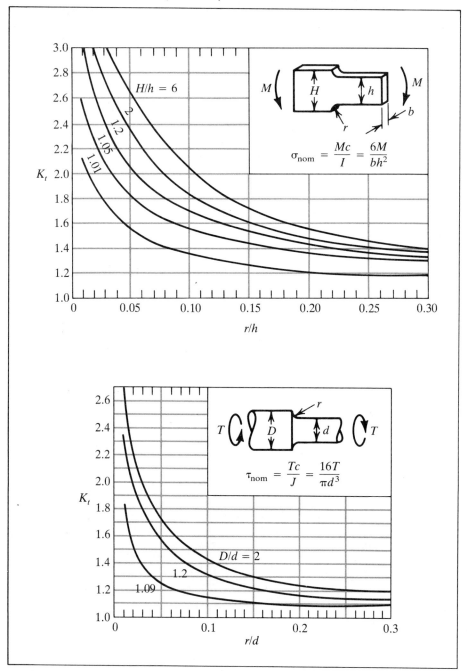

Stress concentration factors. (*Continued*)

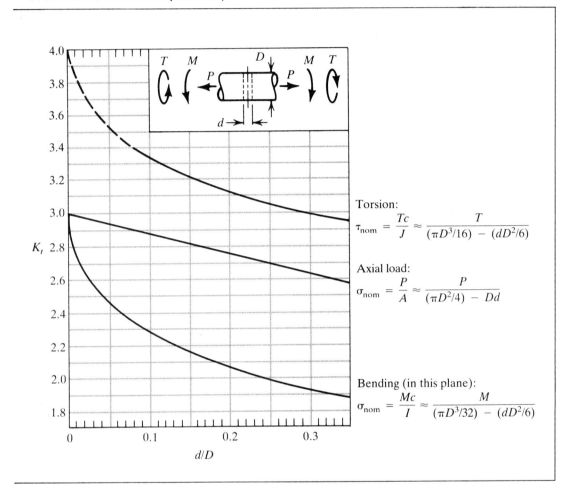

Torsion:
$$\tau_{nom} = \frac{Tc}{J} \approx \frac{T}{(\pi D^3/16) - (dD^2/6)}$$

Axial load:
$$\sigma_{nom} = \frac{P}{A} \approx \frac{P}{(\pi D^2/4) - Dd}$$

Bending (in this plane):
$$\sigma_{nom} = \frac{Mc}{I} \approx \frac{M}{(\pi D^3/32) - (dD^2/6)}$$

APPENDIX M

Stress intensity factors.

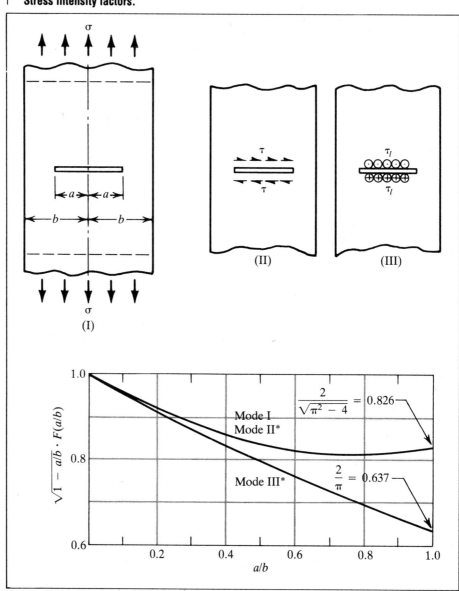

Stress intensity factors. (*Continued*)

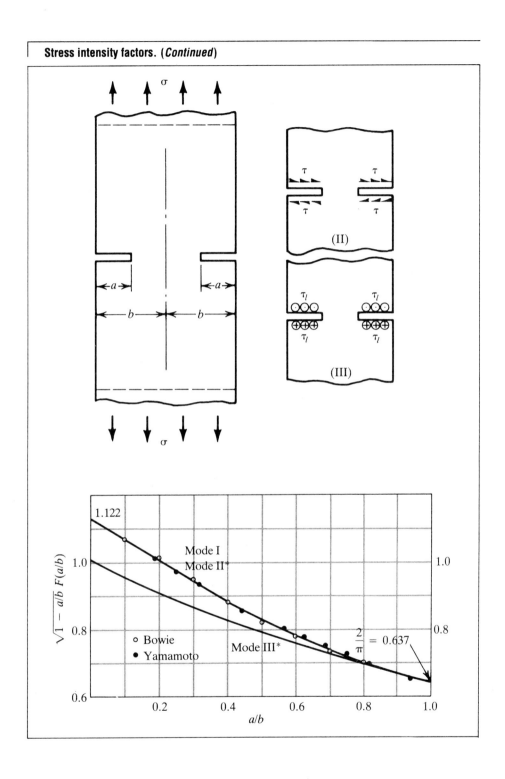

APPENDIX M

Stress intensity factors. (*Continued*)

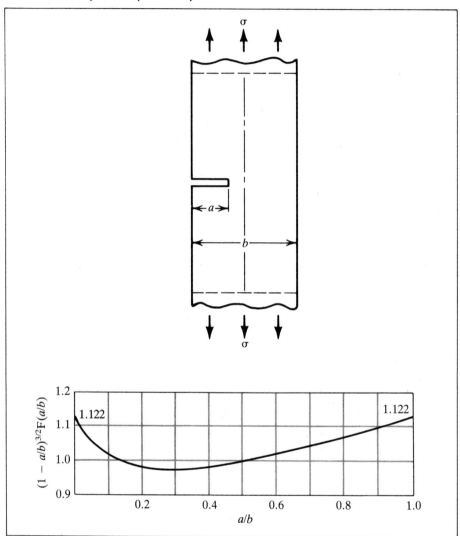

Stress intensity factors. (*Continued*)

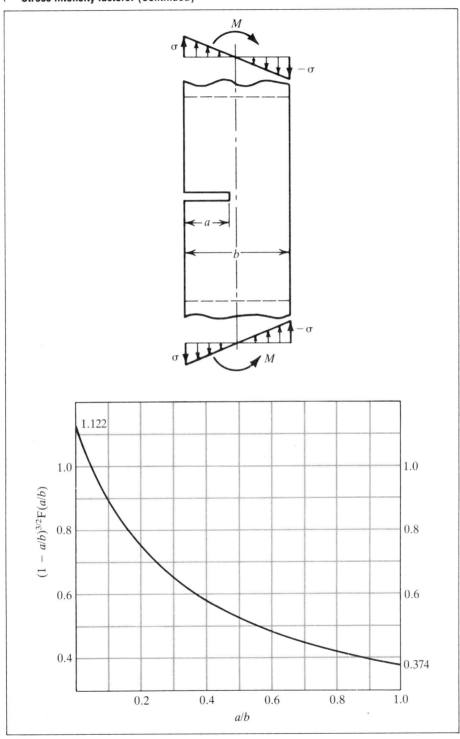

APPENDIX M

Stress intensity factors. (*Continued*)

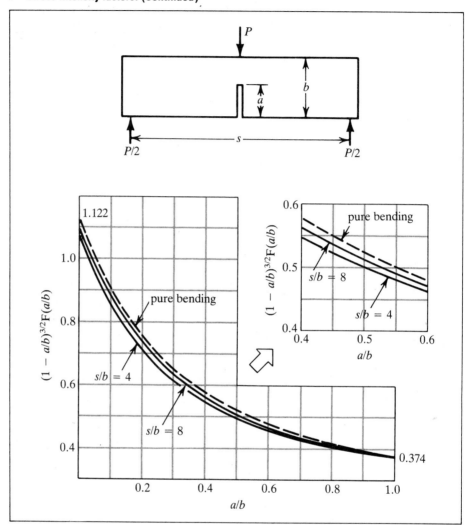

The Stress Analysis of Cracks Handbook, 2nd Ed., by Hiroshi Tada, Paul C. Paris, and George R. Irwin. Paris Productions, Inc., 226 Woodbourne Drive, St. Louis Missouri 63105, 1985.

APPENDIX N

Standard solid copper wire at 20° C.

Wire Gauge	Diameter, mm	Resistance, ohms/km
0000	11.6840	0.1608
000	10.4038	0.2028
00	9.2659	0.2557
0	8.2525	0.3224
1	7.3482	0.4065
2	6.5430	0.5128
3	5.8268	0.6463
4	5.1892	0.8153
5	4.6203	1.0279
6	4.1148	1.2963
7	3.6652	1.6345
8	3.2639	2.0610
9	2.9058	2.5988
10	2.5883	3.2772
12	2.0526	5.2100
14	1.6276	8.2841
16	1.2908	13.1759
18	1.0236	20.9482
20	0.8118	33.3005
22	0.6438	52.9661
24	0.5106	84.2460
26	0.4049	133.9047

Reprinted with permission from *Handbook of Chemistry and Physics*, 67th ed. R. C. Weast, ed. Copyright 1986, CRC Press, Inc. Boca Raton, Fla.

APPENDIX O

Shear, moment, and deflection equations for cantilever beams.

	Slope at Free End	Maximum Deflection	Deflection at Any Point x
1. Concentrated load at end	$\theta = \dfrac{PL^2}{2EI}$	$\delta_{max} = \dfrac{PL^3}{3EI}$	$\delta = \dfrac{Px^2}{6EI}(3L - x)$

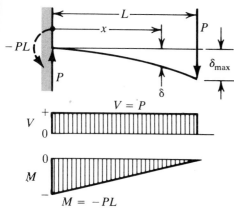

2. Concentrated load at any point	$\theta = \dfrac{Pa^2}{2EI}$	$\delta_{max} = \dfrac{Pa^2}{6EI}(3L - a)$	For $0 \leq x \leq a$: $\delta = \dfrac{Px^2}{6EI}(3a - x)$ For $a \leq x \leq L$: $\delta = \dfrac{Pa^2}{6EI}(3x - a)$

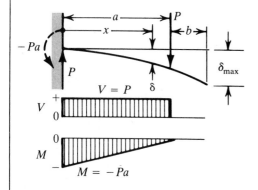

	Shear, moment, and deflection equations for cantilever beams. (*Continued*)		
	Slope at Free End	**Maximum Deflection**	**Deflection at Any Point x**
3. Uniform load	$\theta = \dfrac{wL^3}{6EI}$	$\delta_{max} = \dfrac{wL^4}{8EI}$	$\delta = \dfrac{wx^2}{24EI}(x^2 + 6L^2 - 4Lx)$

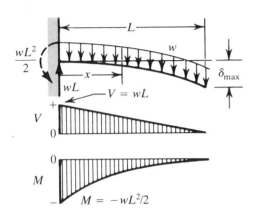

4. Moment load at free end	$\theta = \dfrac{M_b L}{EI}$	$\delta_{max} = \dfrac{M_b L^2}{2EI}$	$\delta = \dfrac{M_b x^2}{2EI}$

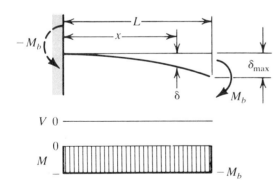

Fundamentals of Machine Component Design, by Robert C. Juvinall. Copyright © 1983, John Wiley & Sons, Inc. Reprinted with permission of John Wiley & Sons, Inc.

APPENDIX O (*Continued*)

Shear, moment, and deflection equations for simply supported beams.

	Slope at Ends, θ	Maximum Deflection, δ_{max}	Deflection at Any Point x, δ
1. Concentrated center load	$\dfrac{PL^2}{16EI}$	At center: $\dfrac{PL^3}{48EI}$	For $0 \leq x \leq L/2$: $\dfrac{Px}{12EI}\left(\dfrac{3L^2}{4} - x^2\right)$

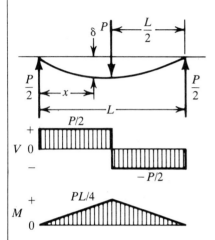

	Slope at Ends, θ	Maximum Deflection, δ_{max}	Deflection at Any Point x, δ
2. Concentrated load at any point	At left end: $\dfrac{Pb(L^2 - b^2)}{6LEI}$	At $x = \sqrt{\dfrac{L^2 - b^2}{3}}$: $\dfrac{Pb(L^2 - b^2)^{3/2}}{9\sqrt{3}LEI}$	For $0 \leq x \leq a$: $\dfrac{Pbx}{6LEI}(L^2 - x^2 - b^2)$

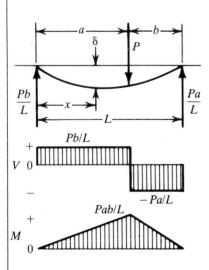

Shear, moment, and deflection equations for simply supported beams. (*Continued*)

	Slope at Ends, θ	**Maximum Deflection, δ_{max}**	**Deflection at Any Point x, δ**
3. Uniform load	$\dfrac{wL^3}{24EI}$	$\dfrac{5wL^4}{384EI}$	$\dfrac{wx}{24EI}(L^3 - 2Lx^2 + x^3)$
4. Overhung load	At left supt: $\dfrac{Pab}{6EI}$ At right supt: $\dfrac{Pab}{3EI}$ At load: $\dfrac{Pb}{6EI}(2L + b)$	$\delta_{max} = \dfrac{Pb^2L}{3EI}$	For $0 \leq x \leq a$: $\dfrac{Pbx}{6aEI}(x^2 - a^2)$ For $0 \leq z \leq b$: $\dfrac{P}{6EI}[z^3 - b(2L + b)z + 2b^2L]$

APPENDIX O (Continued)

Shear, moment, and deflection equations for simply supported beams.

	Slope at Ends, θ	*Maximum Deflection,* δ_{max}	*Deflection at Any Point x,* δ
5. Moment load between supports	At left supt: $\dfrac{-M_0}{6EIL}(2L^2 - 6aL + 3a^2)$ At load: $\dfrac{M_0}{EI}\left(\dfrac{L}{3} + \dfrac{a^2}{L} - a\right)$ At right supt: $\dfrac{M_0}{6EIL}(L^2 - 3a^2)$	At load: $\dfrac{M_0 a}{3EIL}(2a^2 - 3aL + L^2)$	For $0 \leq x \leq a$: $\dfrac{M_0 x}{6EIL}(x^2 + 3a^2 - 6aL + 2L^2)$

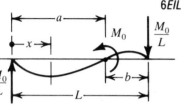

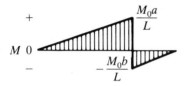

6. Overhung moment load	At left supt: $\dfrac{M_0 a}{6EI}$ At right supt: $\dfrac{M_0 a}{3EI}$ At load: $\dfrac{M_0(a + 3b)}{3EI}$	$\delta_{max} = \dfrac{M_0 b}{6EI}(2L + b)$	For $0 \leq x \leq a$: $-\dfrac{M_0 x}{6aEI}(a^2 - x^2)$ For $0 \leq x' \leq b$: $\dfrac{M_0}{6EI}(2ax' + 3x'^2)$

Shear, moment, and deflection equations for beams with fixed ends *(Continued)*.

	Deflection, δ	*Deflection, δ, at Any Point, x*
1. Concentrated center load	At center: $$\delta_{max} = \frac{PL^3}{192EI}$$	For $0 \le x \le L/2$: $$\frac{Px^2}{48EI}(3L - 4x)$$

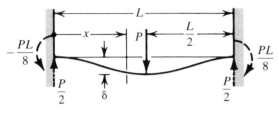

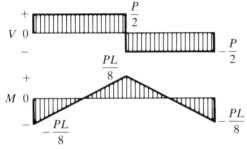

2. Concentrated load at any point   At load:
$$\delta = \frac{Pb^3 a^3}{3EIL^3}$$

For $0 \le a$:
$$\frac{Pb^2 x^2}{6EIL^3} - [3aL - (3a + b)x]$$

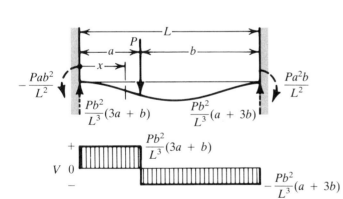

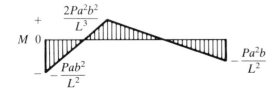

APPENDIX O *(Continued)*

Shear, moment, and deflection equations for beams with fixed ends.

	Deflection, δ	*Deflection*, δ, *at Any Point*, *x*
3. Uniform load	At center: $$\delta_{max} = \frac{wL^4}{384EI}$$	$$\frac{wx^2}{24EI}(L-x)^2$$

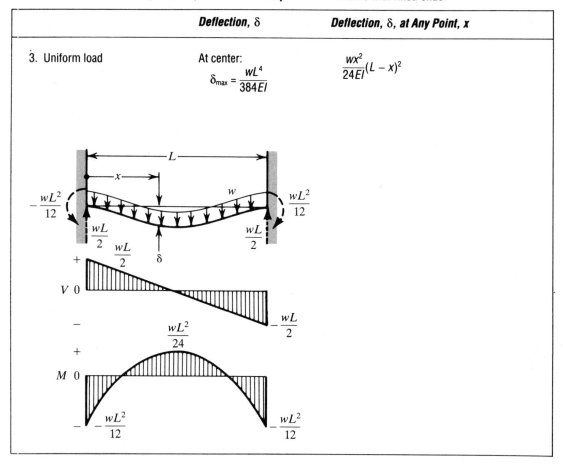

APPENDIX P

Properties of selected geometric shapes.

Rectangle
Area: $A = bh$
Centroid location: $x = 0$, $y = 0$
Area moment of inertia: $I_{xx} = bh^3/12$, $I_{yy} = hb^3/12$
Radius of gyration: $r_x = h/\sqrt{12}$, $r_y = b/\sqrt{12}$

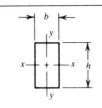

Triangle
Area: $A = bh/2$
Centroid location: $y = h/3$
Area moment of inertia: $I_{xx} = bh^3/36$
Radius of gyration: $r_x = h/\sqrt{18}$

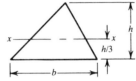

Circle
Area: $A = \pi D^2/4$
Centroid location: $x = 0$, $y = 0$
Area moment of inertia: $I_{xx} = I_{yy} = \pi D^4/64$
Radius of gyration: $r_x = r_y = D/4$

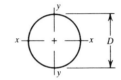

Hollow Circle
Area: $A = \pi(D^2 - D_i^2)/4$
Centroid location: $x = 0$, $y = 0$
Area moment of inertia: $I_{xx} = I_{yy} = \pi(D^4 - D_i^4)/64$
Radius of gyration: $r_x = r_y = \sqrt{(D^2 + D_i^2)/16}$

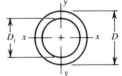

Semicircle
Area: $A = \pi R^2/4$
Centroid location: $x = 0$, $y = 4R/3\pi$
Area moment of inertia: $I_{xx} = \pi R^4/8$, $I_{yy} = \pi R^4[1/8 - 8/9\pi^2]$
Radius of gyration: $r_x = R/\sqrt{2}$, $r_y = (R/2)\sqrt{1/8 - 8/9\pi^2}$

APPENDIX Q1

Wide flange structural sections.

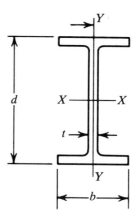

Nominal depth × lbs/ft	t in.	b in.	d in.	Area in²	I_{xx} in⁴	I_{yy} in⁴
W 36 × 300	0.945	16.655	36.74	88.3	20300	1300
Ten sizes exist between 36 × 300 and 36 × 135.						
W 36 × 135	0.600	11.950	35.55	39.7	7800	225
W 33 × 241	0.830	15.860	34.18	70.9	14200	932
Five sizes exist between 33 × 241 and 33 × 118.						
W 33 × 118	0.550	11.480	32.86	34.7	5900	187
W 30 × 211	0.775	15.105	30.94	62.0	10300	757
Six sizes exist between 30 × 211 and 30 × 99.						
W 30 × 99	0.520	10.450	29.65	29.1	3990	128
W 27 × 178	0.725	14.085	27.81	52.3	6990	555
Five sizes exist between 27 × 178 and 27 × 84.						
W 27 × 84	0.460	9.960	26.71	24.8	2850	106
W 24 × 162	0.705	12.955	25.00	47.7	5170	443
Nine sizes exist between 24 × 162 and 24 × 55.						
W 24 × 55	0.395	7.005	23.57	16.2	1350	29.1
W 21 × 147	0.720	12.510	22.06	43.2	3630	376
Eleven sizes exist between 21 × 147 and 21 × 44.						
W 21 × 44	0.350	6.500	20.66	13.0	843	20.7
W 18 × 119	0.655	11.265	18.97	35.1	2190	253
Eleven sizes exist between 18 × 119 and 18 × 35.						
W 18 × 35	0.300	6.000	17.70	10.3	510	15.3
W 16 × 100	0.585	10.425	16.97	29.4	1490	186
Nine sizes exist between 16 × 100 and 16 × 26.						
W 16 × 26	0.250	5.500	15.69	7.68	301	9.6

Wide flange structural sections. (*Continued*)

Nominal depth × lbs/ft	t in.	b in.	d in.	Area in²	I_{xx} in⁴	I_{yy} in⁴
W 14 × 730	3.070	17.890	22.42	215.0	14300	4720
Thirty-four sizes exist between 14 × 730 and 14 × 22.						
W 14 × 22	0.230	5.000	13.74	6.49	199	7.00
W 12 × 336	1.775	13.385	16.82	98.8	4060	1190
Twenty-seven sizes exist between 12 × 336 and 12 × 14.						
W 12 × 14	0.200	3.970	11.91	4.16	88.6	2.36
W 10 × 112	0.755	10.415	11.36	32.9	716	236
Sixteen sizes exist between 10 × 112 and 10 × 12.						
W 10 × 12	0.190	3.960	9.87	3.54	53.8	2.18
W 8 × 67	0.570	8.280	9.00	19.7	272	88.6
Eleven sizes exist between 8 × 67 and 8 × 10.						
W 8 × 10	0.170	3.940	7.89	2.96	30.8	2.09
W 6 × 25	0.320	6.080	6.38	7.34	53.4	17.1
Seven sizes exist between 6 × 25 and 4 × 13.						
W 4 × 13	0.280	4.060	4.16	3.83	11.3	3.86

APPENDIX Q2

Standard flange structural sections.

Nominal depth × lbs/foot	t in.	b in.	d in.	Area in²	I_{xx} in⁴	I_{yy} in⁴
S 24 × 121	0.800	8.050	24.50	35.6	3160	83.3
Four sizes exist between 24 × 121 and 24 × 80.						
S 24 × 80	0.500	7.000	24.00	23.5	2100	42.2
S 20 × 96	0.800	7.200	20.30	28.2	1670	50.2
Two sizes exist between 20 × 96 and 20 × 66.						
S 20 × 66	0.505	6.255	20.00	19.4	1190	27.7
S 18 × 70*	0.711	6.251	18.00	20.6	926	24.1
S 15 × 50*	0.550	5.640	15.00	14.7	486	15.7
S 12 × 50	0.687	5.477	12.00	14.7	305	15.7
Two sizes exist between 12 × 50 and 12 × 31.8.						
S 12 × 31.8	0.350	5.000	12.00	9.35	218	9.36
S 10 × 35*	0.594	4.944	10.00	10.3	147	8.36
S 8 × 23*	0.441	4.171	8.00	6.77	64.9	4.31
S 7 × 20*	0.450	3.860	7.00	5.88	42.4	3.17
S 6 × 17.25*	0.465	3.565	6.00	5.07	26.3	2.31
S 5 × 14.75*	0.494	3.284	5.00	4.34	15.2	1.67
S 4 × 9.5*	0.326	2.796	4.00	2.79	6.79	0.903
S 3 × 7.5*	0.349	2.509	3.00	2.21	2.93	0.586

* A lighter weight section exists for each of these depth sizes.

APPENDIX Q3

Standard channel structural sections.

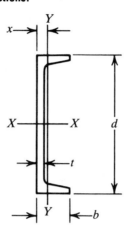

Actual depth in inches × lbs/ft	t in.	b in.	Area in^2	I_{xx} in^4	I_{yy} in^4	x in.
C 15 × 50 (3)	0.716	3.716	14.7	404	11.0	0.798
C 12 × 30 (2)	0.510	3.170	8.82	162	5.14	0.674
C 10 × 30 (3)	0.673	3.033	8.82	103	3.94	0.649
C 9 × 20 (2)	0.448	2.648	5.88	60.9	2.42	0.583
C 8 × 18.75 (2)	0.487	2.527	5.51	44.0	1.98	0.565
C 7 × 14.75 (2)	0.419	2.299	4.33	27.2	1.38	0.532
C 6 × 13 (2)	0.437	2.157	3.83	17.4	1.05	0.514
C 5 × 9 (1)	0.325	1.885	2.64	8.90	0.632	0.478
C 4 × 7.25 (1)	0.321	1.721	2.13	4.59	0.433	0.459
C 3 × 6 (2)	0.356	1.596	1.76	2.07	0.305	0.455

( ) The number of additional smaller weights available for each depth size category.

APPENDIX Q4

Standard angle structural sections.

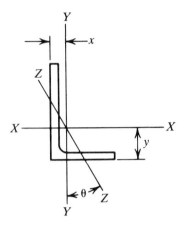

Leg lengths × thickness, inches	Area in²	Weight lbs/ft	I_{xx} in⁴	y in.	I_{yy} in⁴	x in.	I_{zz}* in⁴	Ø deg
L 9 × 4 × 5/8 (2)	7.73	26.3	64.9	3.36	8.32	0.858	5.54	12.2
L 8 × 8 × 1 1/8 (6)	16.7	56.9	98.0	2.41	98.0	2.41	40.64	45.0
L 8 × 6 × 1 (6)	13.0	44.2	80.8	2.65	38.8	1.65	21.30	28.5
L 8 × 4 × 1 (3)	11.0	37.4	69.6	3.05	11.6	1.05	7.87	13.9
L 7 × 4 × 3/4 (3)	7.69	26.2	37.8	2.51	9.05	1.01	5.69	18.0
L 6 × 6 × 1 (8)	11.0	37.4	35.5	1.86	35.5	1.86	15.06	45.0
L 6 × 4 × 7/8 (7)	7.98	27.2	27.7	2.12	9.75	1.12	5.86	22.8
L 6 × 3.5 × 1/2 (2)	4.50	15.3	16.6	2.08	4.25	0.833	2.59	19.0
L 5 × 5 × 7/8 (6)	7.98	27.2	17.8	1.57	5.17	1.57	7.55	45.0
L 5 × 3.5 × 3/4 (6)	5.81	19.8	13.9	1.75	5.55	0.996	3.25	24.9
L 5 × 3 × 5/8 (5)	4.61	15.7	11.4	1.80	3.06	0.796	1.91	19.2
L 4 × 4 × 3/4 (6)	5.44	18.5	7.67	1.27	7.67	1.27	3.29	45.0
L 4 × 3.5 × 5/8 (5)	4.30	14.7	6.37	1.29	4.52	1.04	2.22	36.7
L 4 × 3 × 5/8 (5)	3.98	13.6	6.03	1.37	2.87	0.871	1.61	28.1
L 3.5 × 3.5 × 1/2 (4)	3.25	11.1	3.64	1.06	3.64	1.06	1.52	45.0
L 3.5 × 3 × 1/2 (4)	3.00	10.2	3.45	1.13	2.33	0.875	1.16	35.5
L 3.5 × 2.5 × 1/2 (4)	2.75	9.4	3.24	1.20	1.36	0.705	0.78	25.9
L 3 × 3 × 1/2 (5)	2.75	9.4	2.22	0.932	2.22	0.932	0.94	45.0
L 3 × 2.5 × 1/2 (5)	2.50	8.5	2.08	1.00	1.30	0.750	0.68	33.7
L 3 × 2 × 1/2 (5)	2.25	7.7	1.92	1.08	0.67	0.583	0.41	22.5
L 2.5 × 2.5 × 1/2 (4)	2.25	7.7	1.23	0.806	1.23	0.806	0.53	45.0
L 2.5 × 2 × 3/8 (3)	1.55	5.3	0.91	0.831	0.51	0.581	0.27	31.5
L 2 × 2 × 3/8 (4)	1.36	4.7	0.48	0.636	0.48	0.636	0.21	45.0

( ) Number of additional thinner angles available at the given leg sizes.
* Minimum moment of inertia; maximum moment axis is perpendicular to ZZ through the same center.
$Z = I/c = r^2 A/c$, where Z = section modulus, in³, r = radius of gyration, in., A = section area, in², and c = the distance to the extreme fiber from the centroid, in.
If I_e = moment of inertia relative to an axis parallel to the c axis and e units away, then $I_e = I_c + Ae^2$, where I_c = moment of inertia relative to an axis passing through the centroid of the area A.
Courtesy American Institute of Steel Construction, reprinted with permission.

APPENDIX R

Insulation values of common building materials.

Material	Thickness, inches	R-value
Air space	0.75 or more	0.91
Exterior surface resistance		0.17
Interior surface resistance		0.68
Sand and gravel concrete block	8	1.11
Filled with concrete	8	1.93
Sand and gravel concrete block	12	1.28
Lightweight concrete block	8	2.00
Lightweight concrete block	12	2.27
Face brick	4	0.44
Common brick	4	0.80
Concrete cast in place	1	0.08
Wood sheathing	0.75	0.94
Fiber board insulating sheathing	25/32	2.06
Polyisocyanurate rigid foam board	0.75	5.4
Plywood	1	1.24
Bevel lapped wood siding	0.50 × 8	0.81
Bevel lapped wood siding	0.75 × 10	1.05
Vertical tongue and groove siding	0.75	0.94
Asbestos board	0.25	0.21
3/8 gypsum lath and 3/8 plaster	0.75	0.42
Gypsum board	1	0.90
Interior wood paneling	0.25	0.31
Building paper		0.06

APPENDIX R *(Continued)*

Insulation values of common building materials.

Material	Thickness, inches	R-value
Vapor barrier		0.00
Asphalt shingles	0.125	0.44
Wood shingles	0.375	0.87
Wood 2 × 4 (1.625 × 3.50)	1.625	2.03
	3.50	4.38
Aluminum siding	0.375	0.61
Aluminum siding (insulated)	0.375	1.82
Rock wool batts	1	3.70
Rock wool, loose	1	2.90
Fiberglass batts	1	3.15
Cellulose fiber	1	3.70
Pearlite	1	2.70
Vermiculite	1	2.08
Urethane, aged	1	6.25
Polystyrene, extruded low density	1	4.00
Polystyrene, extruded high density	1	5.00
Polystyrene, beadboard	1	3.57
Acoustical ceiling tile	0.75	1.78
Carpet and fiber pad		2.08
Carpet and rubber pad		1.23
Linoleum and floor tile	0.125	0.05
Stone	1	0.08

Answers To Selected Exercises

1-6. a) Bones in the arms and legs

b) Muscles in the arms making the forearm move. The elbow joint is the fulcrum.

c) The heart

d) Swimming fish, flying birds

e) Duck feathers, some plant leaves

f) Bats in flight, sounds of whales and other fish, dog whistles that humans cannot hear.

g) Lightning bugs, fluorescent fish, luminescent bacteria, fluorescent minerals under ultraviolet light, chlorophyll fluoresces red under certain conditions.

h) Chameleon skin, insects and bacteria that are attracted to light, plants grow toward the light.

1-7. 13°, 6° 21', 3° 1'

1-10. Approximately one ton per day

1-12. 0.432 inches

1-14. The design as specified will not work.

1-15. a) Approximately 12 feet   c) 41 psi   d) 6.11 inches

2-1. c) Ships have progressed from wooden, sail and oar powered, to fiberglass, steel, and other metal alloy hulls powered by steam, diesel, gasoline, and nuclear power. Designs have progressed to include submarines and hydrofoil hull designs that skim the water rather than cutting a wake. Ship design changes have reduced land travel and opened frontiers, but have not made anything obsolete.

f) Firearms started with muzzle loading, single shot pistols and rifles. Powder was carried separately from the bullets which were nearly spherical in shape. The powder was ignited with steel and flint. The powder and bullet became a single unit loaded from the breach ignited with a pressure firing pin. Single shot weapons became multiple shot, and had automatic shell ejection and new bullet loading. Barrels became

Answers To Selected Exercises

rifled to make the bullet rotate for better accuracy and distance. The bow and arrow, and sling shot were made less important as weapons by firearms.

i) Airplanes have progressed from lightweight, single engine, single person, low altitude, short flight machines to multiple engine, multiple passenger, long distance machines with pressurized cabins for high altitude flight. Jet engines have replaced internal combustion engines. Materials have changed to keep weight down and strength up. Navigational devices have been added to fly at night and in the fog. Airplanes reduced long distance travel by automobile and ocean travel by boat.

2-2. *a*) Mopeds, carnival ride cars, farm tractors, golf carts

b) Highway information signs by law, bricks and other building materials by convention and use efficiency

d) Products sold by weight (1 pound of butter, 5 pounds of sugar, 50 pounds of fertilizer)

e) Gas can (red), traffic signal lights (red, yellow, green), highway mile markers (green)

2-4. Use of automatic seat belts in cars (get in the way of passengers entering and leaving auto), kickstand on motorcycles (digs into soft surfaces), vending machines (jam and do not return money), the storage and removal of car jacks and spare tires from the trunks of cars (awkward to remove from storage).

2-6. See Figure 2.14 for sample.

2-11. A knife, football, tennis ball, golf club, bicycle rack, gasoline, oil, kitchen utensils, calculator

2-14. Most companies purchase items that have been standardized such as nuts, bolts, washers, and other hardware components; light bulbs and other electrical devices; rubber O-rings, tires, and other formed rubber parts; and specialty operations they are not equipped to handle such as chrome plating, die casting, and galvanizing.

2-16. Reduced weight from mean = 360 lbs

2-18. Fire extinguisher, match, firecracker, camera flash bulb, CO_2 cartridge, bullet, garbage bag

2-20. Light bulb, grease or oil surface rust protection, food, medicine, paint

2-22. Electric fuse, shear pin, collapsible crash barrier

2-24. Manual can opener, put crank on top; scissors, must have cutting action force on center line; manual pencil sharpener, place crank on top; TV remote control, place most used buttons—channel and volume—in center.

2-26. The interior of a car is not an easily defined geometric shape so the volume is hard to calculate. A room has perpendicular walls and the volume is easier to calculate.

2-27. Range is from $0.30/pound to $2.50/pound.

Answers To Selected Exercises

2-29. Suitcases can hold enough to become too heavy to be easily carried, weights used in exercise machines, door sizes related to people size, clothes must be sized to people dimensions, pole vault pole diameter must be sized to the hand grip of people.

2-31. Oil and air filters must be changed regularly or they will become clogged and not allow the passage of oil or air. No economical alternate method exists to remove the unwanted material from the fluid. *Note:* Electrostatic air filters are available, but most air furnace systems are not using them, perhaps because they are too expensive or noisy.

2-33. Rubber tires, clothes washers, driers, television sets, electric batteries, plastic containers, Styrofoam cups

2-35. One subjective comparison.

	Metal	Glass	Plastic	Coated paper	Uncoated paper
Raw material use	5	1	4	3	2
Fabrication	5	4	3	2	1
Disposal	4	5	3	2	1
Reuse	2	1	5	5	5
Litter life	5	5	5	2	1
Cleanup ease	2	4	2	4	4
By-product effect	3	3	3	2	2
Ease of use	2	2	1	1	1
Shelf life	2	1	3	4	5

1 = good, 5 = poor

2-37. *a)* The critical path is $A-D-F-G-H-I-J-K$ which has a total time of 30 weeks. The project cannot be completed in 27 weeks.

b) Event B has an eight week slack time, event C has a one week slack time, and event E has a ten week slack time.

c) Use four people for the first 15 weeks and two people the second 15 weeks.

2-39. *a)* This analysis has a critical path of: $A-B-C-E-F-G-H-I-L-M-Q-N-R-T-V$, with an estimated completion time of 26.5 working days depending on whether T, the three-day concrete cure time, can be scheduled for a week end.

b) An additional 15–17 days are required depending on whether or not the rain starts on the first or second day of outside work, and whether or not concrete curing time falls on the weekend.

2-40. A computer program that accounts for all the variations of user and environment cannot be written because all variations cannot be identified qualitatively and quantitatively.

2-41. *a)* Implies that other things are hard to obtain, such as technical help, maintenance manuals.

    b) Other problems exist such as light weight, damaged in shipment.

2-43 and 44.

    a) What does "best" mean? Best taste? Best price?

    b) What does "very hard" mean? How many hours? What work was done, number of problems completed?

    c) Car engine stops running, car stops moving. Did the engine run out of fuel? Did the car run into a wall? Did all the tires go flat?

    d) What is "overworked?" Sixty hours a week? Twelve hours a day? A hundred gear boxes assembled before going home?

    e) How heavy is "too heavy?" What weight can a person lift? How many times a day can that weight be lifted and carried a given distance?

    f) "Unbearable?" Backlog can't be filled in "x" weeks? Is the weight of the paper for the orders making something collapse?

    g) "Good?" What size part can be machined? What dimension tolerances can be held?

    h) "Crowded?" Crowded may just mean filled. Does the warehouse contain more than it was designed to hold? How much was it designed to hold? Volume? Weight?

    i) "Dangerous?" Are the flames apt to burn someone? Are the sheets of metal too heavy for the table and it may collapse?

    j) "Beautiful?" Beauty is in the eye of the beholder. More beautiful because of color? Compared to a Van Gogh painting? To another car?

    k) "Better than?" What tested better? Torque-in torque-out ratio? Temperature of operation?

2-46. If the pipes are truly inaccessible, then no amount of cost or time can get them repaired. A more accurate description is probably "Not easily accessed."

3-1. Walk, run, car, taxi, skateboard, bicycle, helicopter, hitch hike

3-3.   *a*) How can you dry your hands?

    b) We need a loaf of bread for supper in 30 minutes.

    c) I need to remember the message I am about to receive.

    d) Where can we be comfortable to eat our lunch?

    e) How can we get across this river?

3-5. Trial and error means test, learn from the test, and iterate to improve the design. This cycle may be necessary several times in engineering design projects.

3-6. Design an automobile that carries six people, but will fold up so it can be carried to an office.

3-7.   *c*) Chernobyl is located approximately 30°E and 50°N on the Earth, so directly opposite would be 150°W and 50°S, somewhere in the Pacific Ocean east of New Zealand.

3-9. The next letter is F. The pattern is the first letter of the counting numbers *o*ne, *t*wo, *t*hree, *f*our, *f*ive,...,*t*hirteen, *f*ourteen, *f*ifteen.

3-10.   *a*) He is short and he cannot reach a floor button higher than 10.

     b) The elevator was being repaired in the morning and could only go to the tenth floor.

     c) He wanted some exercise in the morning walking up stairs, but not at night walking down.

     d) He forgot where he was going in the morning, his office used to be on the tenth floor.

     e) The coffee shop is on the tenth floor and he stopped to get a cup on the way to work and then walked up the last flights.

3-11. Contingency plans in a preliminary schedule might include extra lead time and money, or plans to add more people, time, and material to particular activities if schedule is not maintained.

3-14. Development projects such as medicines and hybrid plants, that need a lot of test time, cannot be speeded up by adding people and money as much as projects such as timber cutting, road building, and brick laying.

3-16. Motor rated too low. Too much friction between belt and rollers. Not enough friction between drive belts and pulleys. Weight rating of system doesn't include all possible product combinations.

3-17. Need curb and gutter to control water flow. Collect the water at low points for sewer openings. Need to know the highest rainfall rate in inches per hour; one-half inch per hour is equivalent to 13,577 gallons per acre per hour; system must be capable of handling the runoff. Consider sewer openings to be safe so children will not be washed into them. How can the system be cleaned out if plugged with debris? What is the shortest route for the system?

3-20. Measure amount of fuel used to produce the same output of heat, or measure the amount of heat output based on a constant input.

3-23. Testing a part outside its normal environment requires a great deal of confidence in knowing what the actual loads are. A test of the shaft all by itself is advantageous if only one aspect of the design is being studied, such as rotating fatigue. Assembled in the transmission will be a better test of all the required functions of the shaft. Tested in an automobile will be best, because outside forces coming from vibration and mounting stress will also be involved.

3-26. A traffic counter could be replaced by a person counting cars. An electronic door opener could be replaced by a door-person. A lane presence device could be replaced with a "traffic-cop." An escalator could be replaced by a stairway. A windmill used to grind grain could be replaced with an animal driven grinder.

3-30. Start on the inside and work out rather than starting on the outside and working in.

3-32. Even a simple job like changing an automobile tire has many variations. Is a bumper jack available, an axle jack, no jack? Is the tire flat, or just being rotated? Flat tires often cause the car to settle so low that there is no room for standard jacks. What kind of wrench is available? Impact? Ratchet? Spinner? Is the tire on a compact car, a full size car, a truck?

3-33. Case 1: Flip either switch up. If a switch is flipped up and the lights go out, that switch is a three-way switch. If a switch is flipped up and more lights go on, that switch is a two-way switch.

Case 2: Flipping one switch will not allow an identification.

Case 3: Flipping one switch will not allow an identification.

Case 4: Flip either switch down. If the lights go off the switch is a two-way switch. If the other lights go on it is the three-way.

4-1. Was the hood release accidentally pulled? How fast was the car moving? How far did the hood open? Does the hood hinge at the windshield, at the side, or at the front? Was there any vibration before the hood came open? Has this ever happened before?

4-3. Is it windy? Does the door have a lock open position? Were you carrying a large load? Did someone keep closing the door as you were carrying packages? Does the door swing both ways? Can you disable the automatic closer if there is one? Will someone hold the door open for you? Is there another entrance you can use?

4-5. Do you own land? Is there land available to rent? What do you want to grow? Vegetables? Flowers? What size garden do you anticipate having? Is water available at the garden site? Do you have any books on gardening? Any other reference material? Have you ever gardened before? Do you own any gardening tools?

4-7. How much will you get? Do you have any outstanding bills to pay? Have you considered investing in stocks? Bonds? Certificates of deposit? Regular savings account? Have you considered donating the money to charity? Do you have a personal need or want?

4-9. Is there a kitchen nearby? What color smoke? Do you see the smoke or smell it? Do you feel heat? Have you looked for the source yet? Is there a fireplace in the house? What time of the year is it? Is the furnace running?

4-13. The order of ranking is from easy-to-check to hard-to-check.

Is the bulb screwed in all the way? Is the bulb burned out? Is this the right switch for that light? Is the fuse blown? Is power out on our street? Has power been disconnected from the house? Is the switch defective? Is the house wiring defective?

4-14. Is the door locked? Do I have the right key? Do I have the right car? Is the lock frozen? (If wintertime.) Is the lock defective?

4-15. Did I pull fast enough? Is there fuel in the fuel tank? Is the key turned on? (If a key ignition.) Is the spark plug wire attached? Is the mower in heavy grass? Is the choke open or closed as necessary for the conditions? (Hot engine or cold engine.) Is the engine flooded with fuel? Is the spark plug bad? Is the magneto or another ignition part operating improperly?

4-18. The hitch on the two units must be compatible, same style and same height. The power of the pulling vehicle must be adequate for the additional load. The units must be compatible for stop and signal light connection. The pulling vehicle must have proper rear view mirrors. Heavy duty rear tires and shock absorbers are also critical if the trailer is very large.

4-20. Public buildings are more likely to have multiple uses, and anticipated equipment and people loads are more difficult to control. Residential homes usually have fixed usage and do not have the high volume traffic that a public building would have.

4-22. Automobiles to carry six passengers and yet be efficient enough to get 50 miles per gallon. A toaster that would not get hot enough to burn a user. A hammer that would not hurt a user's thumb if hit. A lawn mower that would not cut a user's hand even if the user cleaned grass from under the mower while it was running.

4-25. Metals were heated and hammered together. Inertia welding heats two pieces of metal by friction and then forces them together with pressure. Parts are rotated at high speeds in a lathe-like machine.

4-27. *a*) Student should leave earlier or choose a different mode of transportation.

b) Rider should wear pants that are tighter to the leg and cannot get tangled with the chain.

c) Student should learn to type better or review work to correct errors before running the program.

d) The problem could be poor assembly procedure or improper use by customer.

5-2. *a*) Flooding and hard to start, or no starting.

b) A choke that is stuck closed. Dirt in the gasoline.

c) The float was too heavy. The sound of sloshing fluid in the float. A puddle of gas by the float as it sat on the workbench.

d) (1) Drill a hole in the float. Fast, but the hole will need repair.
(2) Heat the float and boil the gas out. Effective, but hot gas fumes might explode.
(3) Wait for the gas to evaporate. Effective, but slow.
(4) Suck the gas out with a vacuum. Effective, but slow; need a vacuum source.
(5) Separate halves and dry out and resolder. Effective, but costly in time. Resolder job is critical.

e) Ranking is subject to facilities available and skill of repair person.

f) Repair as indicated in part (d). Replace float with a new one. Use a material for the float that does not absorb, or is attacked by, gasoline. Fill the existing float with a substance less dense than gasoline.

5-4. Connecting belt, spline shaft, mating gears, friction wheel, chain.

5-7. Alarm signals for fire, computer access to bank accounts, fixed response to TV questions, order products.

5-9. Issue bonus credit if turned in early. Require step by step checking and credit. Decrease credit if turned in late.

5-10. Compass, stars, prevailing winds, position of sun, aurora borealis.

5-12. Traffic person, signal lights, stop and yield signs, gate barriers, overpass, underpass.

5-14. Record license plate numbers and send bills. Prepay and only prepayers can travel road. A coin collector that matches the car speed so paying can be done without slowing down.

5-15. Number the pages. Letter the pages. Notch the sides of the pages. Use different size paper for each page. Place a diagonal line on the edges of the stack of pages.

5-17. Playground apparatus, sandbox, line driveways so cars cannot drive on grass, boat dock bumpers, truck loading dock bumpers, tree protectors, car bumpers, edge of gardens.

5-19. *a*) A buzzer sounding if the head drops, the eyelids stay shut for more than two seconds, or the tires leave the pavement.

b) Crawl on floor to locate. Vacuum area and sift through debris. Make contacts infrared or ultraviolet sensitive and use special light to locate.

c) Car monitoring system that keeps track of both miles traveled and gallons of fuel used. Less exact system could calculate based on miles since last fill up; miles per gallon would continue to increase until the next fill up.

d) Remote control solenoid plungers, pull a wire that allows gravity plungers to operate.

e) Have a person walk the aisles and check each seat belt. Use an electric circuit that uses buckle connections as switches; check only seats with weight on them. Use seat belt units that beep if not buckled whenever takeoff and landing is in progress. Install automatic bucklers, those that would buckle whenever a person sits in the seat.

5-21. Need dense material: lead, gold, concrete, stone, brick.

5-22. Place a net over the area and pull the corners in like a fish net. Make the area a moving surface like people conveyors at large airports. Train dogs to collect the balls. Make special balls with metal centers and use a magnet. Slope area to a trough. Pay someone to walk the area picking up balls. Use a machine with tires that picks up balls as it rolls over them. Make customers pick up all the balls they have hit.

5-24. Horizontal: bubble level, non-rolling of a ball, pressure sensitive altimeter, water or other fluid level.

Vertical: perpendicular to horizontal, hanging weight, falling weight, fluid over a ledge.

5-27. Make A'' an overpass. Place a yield sign at intersection of E' and A'' for A'' traffic. Reroute traffic and close A''. Erect a barrier and eliminate E'. Place a yield sign at E'. Don't allow E' and A'' and make B a one way street.

5-28. Use a fluid in a tube, see Figure 6.10. Use a hanging weight and measure the angle from vertical, see Figure 6.11. Use a mass and a spring to be compressed or stretched in the direction of acceleration, see Figure 6.12.

5-30. Use an ice cube, a lighted candle, a bowl of any solid or liquid that will evaporate. Use any substance that will absorb water such as granular fertilizer, salt, or silica gel.

5-33. Use a Styrofoam container, parachute, hard boil and drop into water, pack egg in carton with "lots" of cushion material, use tube of fluid with density just less than that of an egg and let egg settle slowly in fluid.

5-35. Keep pouring and hope enough hits the bottle until it is full. Use an extension of straws. Use a garden hose.

5-36. Rubber tipped rod attached to head. Wind from a fan. Use your nose. Tip the book and let gravity turn pages.

5-39. Long handled hammer. Throw rocks at the nail. Shoot bullets at the nail. Catapult rocks at the nail.

6-2. See Exercise 6.32, Figure 6.18.

6-4.

Favorable	Unfavorable
a. Stereo system	Car without a muffler
b. Spotlight in theater	Glare from setting sun
c. Sound speakers	Car with bad shocks
d. Night light	Changing a car tire at night
e. Chemotherapy	Nuclear waste
f. Refrigeration of food	Winter weather
g. Balloons	Low water pressure in house
h. Cooking food	Danger of burning skin
i. Cattle prod	Explosions
j. Cooling fan	Hurricane

6-6. *b*) 44 cars per minute

c) 6.9 mph

6-8. *a*) acceleration = $\Delta h/L$.

b) The area of the vertical tubes does not matter, only the difference in heights.

6-13. Two units mounted in perpendicular planes could be used, and the actual acceleration would be the vector sum of the two readings.

6-14. Force in Gs = 0.40

6-16. Approximately 5420 feet

6-17. While accelerating, 0.45 Gs; while braking, –0.95 Gs

6-20. The lowest cost = $16,825P

6-24. *a*) Common available material thicknesses are 1.250, 1.375, or 1.500, not 1.40.

   b) Flame cut is appropriate. Cut variations from mean profile will give size variations. Matching plates for good looking assembly will be difficult.

   c) The only rivets needed are to hold the laminates together. The entire shear load should not occur at any laminate boundary. Twenty-six 1-inch holes will reduce the effective cross section of material. Drilling holes is added labor cost that should be minimized.

   d) Welding makes a one piece assembly as far as crack propagation is concerned.

   e) Hook cannot hang straight unless mass on both sides of centerline are the same.

   f) There are three critical sections to consider for possible maximum stress, at the upper hole, at the shaft cross section, and at the curved beam section at the bottom.
   (1) At the top hole, section A-A, SF is approximately 6.4.
   (2) At the mid section B-B, SF is approximately 10.
   (3) At section C-C, SF is approximately 4.9.

6-27. The cost of required Btu's at 65% efficiency is still less than 100% electric efficiency. The inefficiency of using electricity is the loss when oil, gas, or coal is converted to electric power at the power generating plant, and the power loss during distribution through the wires.

6-29. *a*) Shortened

6-30. The cross sectional area of any component is proportional to F for a given material, so the material cost is proportional to the product of the force F and length L.

   a) Take first moments around point C: the minimum ($F_{AE} L_{AE}$) product = 40,000.

   b) Take first moments around point C: the minimum ($F_{DF} L_{DF}$) product = 50,000.

   c) Take first moments around point C: the minimum ($F_{DF} L_{DF}$) product = 40,000.

   The most widely used solution is DF or BD. The members are in tension and cables can be used. Cable is easier to deliver to the job site than rigid members which are needed for compression loads. The actual answer will depend on the cable sizes available for the tension loads applied.

7-1. *a*) Material, type of bottom (V or flat), how many passengers, weight, dimensions, maximum load, price, safety record, oars, comfort.

   b) Size of books it will hold, required strength, mounting, style, how many books it will hold, other items it will hold, backed or open through, book ends built in, adjustable shelves, number of shelves, material, height.

   c) Material, table surface, weight, leg stability, dimensions, style, ease of setup and take-down.

d) Capacity, color, dimensions, weight, type of loading, price.

e) Available software, type of disk used, type of disk drives, weight, dimensions, price, available printers, memory capacity, delivery, service.

f) Type of seat, type of brakes, color, height, number of speeds, type of chain guard, price.

g) Weight, capacity in number of people, material used, waterproofing, cost, appearance, ease of setup and take-down, durability, compactness for storage, fire resistance, screen doors/windows, type of tie down anchors, odor containment.

h) Location, number and type of rooms, number of levels, type of foundation, condition of plumbing, heating, and electrical systems, appearance, cost, size of yard.

7-5. Sales, warranty claims, safety records, percent of market penetration.

7-7. Have the machine tool vendor machine parts. Have an outside vendor machine parts. Change the processing of the part to make it some other way.

7-10. *a*) D is the highest value choice.

b) D is still the choice.

c) C or D are equal in value.

7-12. *a*) The evaluation is somewhat subjective; high density would be important if a building is being built as a fallout shelter, thermal resistance is critical in a very hot area for keeping cool and in a cold area for keeping warm, and cost is always an important item. Use values of 6, 9, and 10.

b) If the best value in each category is given a 10, and there is no cutoff value that is restrictive, then all other evaluations can be calculated. If the value for density of 169 and 170 is 10, then for 120: $(120/170)10 = 7.1$, for 140: $(140/170)10 = 8.2$; for 62: $(62/170)10 = 3.6$. If the value for heat resistivity of 0.25 is 10, for 0.20: $(0.20/0.25)10 = 8$, for 0.10: $(0.10/0.25)10 = 4$, for 0.02: $(0.02/0.25)10 = 0.8$, and for 0.0007 a value of zero. If the value for cost of 0.50 is 10, then for 3: $(0.50/3)10 = 1.7$, for 5: $(0.50/5)10 = 1$, for 17: $(0.50/17)10 = 0.3$, and for 30: $(0.50/30)10 = 0.2$. With these numbers the ranking chart looks like this.

Item	Value	Normalized	Material				
			1	2	3	4	5
Density	6	0.24	10/2.4	7.1/1.7	8.2/2.0	10/2.4	3.6/0.9
Heat *R*	9	0.36	0/0	8/2.9	4/1.4	0.8/0.3	10/3.6
Cost	10	0.40	0.3/0.1	1/0.4	1.7/0.7	0.2/0.1	10/4.0
Totals			2.5	5.0	4.1	2.8	8.5

c) Material #5 is judged the best.

d) Durability, color, ease of installation, availability

7-14. For Case Study #6:

	1 vs					
	2	1	2 vs			
	3	3	3	3 vs		
	4	4	4	4	4 vs	
	5	5	5	5	5	
Idea		1	2	3	4	5
Number of "wins"		1	0	2	3	4

Proposal 5 is the "best."

7-17.

	Advantages	**Disadvantages**
Wooden pencil	Less expensive. Always know if there is lead available.	Wears out with each sharpening. Handling changes as size and weight changes. Need access to a sharpener.
Mechanical pencil	Lead sharpness is consistent. Handling is constant. Erasers are generally changeable.	More expensive. If lead advance does not work, pencil is scrap.

7-18. An example of a checklist: Items to check when buying a used house.

 a) Attic insulation, affects heating and cooling costs.

 b) Roof, age and general condition, 20 year "normal" life.

 c) Chimney, if brick check mortar joints, does it have a clay liner? If metal check for rust and tight joints.

 d) Roof flashing, check for rust and tight joints, caulking may be required.

 e) Plaster, any cracks, lose or crumbly plaster. Check source of problem: sagging timbers, settling foundation.

 f) Bath walls, any wet plaster around pipes and drains?

 g) Wall insulation, affects cooling and heating costs.

 h) Storm doors and windows, are they well sealed, do any moving sashes not slide or swing as they should? Test them.

 i) Wiring, is the service entrance large enough for all appliances and future need? Has updating to circuit breakers been done?

 j) Rain gutters and downspouts, any rust? Are they working properly? Check by running water off roof with a hose.

 k) Siding
 Brick: Check mortar joints and caulking around doors and windows.
 Wood: Check for cracked or water soaked boards. Does it need paint? Check caulking around doors and windows.

l) Exterior doors, check weather stripping and latch mechanisms.

m) Foundation, look for recent cracks, water stains, and out of plumb walls.

n) Water supply, if well and pump, have water tested for contaminants. Check for rusty joints. Check all shutoff valves. Turn more than one faucet on and check water supply.

o) Furnace, check for rust. If questionable have inspected.

p) Water heater, check for rust, check age.

q) Floors, check for sagging and squeaks.

r) Check surface water runoff for potential basement problems.

8-3. *a*) $R = (L^2 + 4h^2)/8h$

b) $R = 4.87$

8-6. *a*) y in inches, W in pounds, L in inches, b in inches, and h in inches

b) No difference

8-8. $1 = kv^{(-2a_3)}\rho^{(-a_3)}p^{(a_3)}$, $\Pi_1 = p/(v^2\rho)$ and $v = c\sqrt{p/\rho}$

8-10. $1 = kf^{(2a_4-a_3)}\Delta p^{(a_4)}D^{(a_3)}\rho^{(a_4)}v^{(2a_4-a_3)}$, $\Pi_1 = f^2\Delta p\rho v^2$, $\Pi_2 = D/fv$

8-12. The model dropping through 30 feet has the same amount of energy to dissipate as the real object falling 7.5 feet.

8-14. Start with $C = 2\pi R$. Let C' and R' be the new circumference and radius, and X = the change in circumference. Then $C' = C + X = 2\pi(R + 0.50'')$ and $X = 2\pi R + \pi - 2\pi R = \pi$.

8-17. The offset data is:

Year	0	1	2	3	4	5	6	7
Production	0.9	1.0	1.1	1.3	1.2	1.3	1.4	1.4

Linear coordinate regression results in $P = 0.07(Y - 1934) + 0.950$, and $r^2 = 0.893$.

8-19. The offset data is:

Year	0	20	30	40	50	60
Population	62.9	76.0	92.0	105.7	122.8	131.7

Linear coordinate regression results in $P = 1.22(Y - 1880) + 57.85$, and $r^2 = 0.975$.

Semilog regression gives the model $P = 61.65(1.013^{Y-1880})$, and $r^2 = 0.987$. Using the semilog model, $P(1950) = 152.3$, compared to 151.3 actual; $P(1960) = 173.3$ compared to 179.3 actual; $P(1970) = 197.1$ compared to 203.2 actual; and $P(1980) = 224.3$ compared to 226.5 actual.

8-21. Log-log regression gives the "best" fit with the model: $V = 40.14(P^{-0.7045})$, $r^2 = 0.9997$, a sum of the residuals = -0.997 and a sum of the squares of the residuals = 0.006.

8-23. Log-log regression gives the "best" fit with the model: $G = 0.1153(S^{0.1571})$, $r^2 = 0.975$, a sum of the residuals = -1.0000 and a sum of the squares of the residuals = 0.000006.

8-25. Use Equations 12.22 for a quadratic least squares analysis, and $Y = -0.817X^2 + 8.39X - 0.0000003$ with a sum of the squares of the residuals = 32.3.

8-27. Yes

8-29. A value of $r = 0.70$ is acceptable for a 95 percent confidence, $0.70 > 0.632$. A value of 0.70 is not acceptable for a 99 percent confidence level, $0.70 < 0.765$.

8-32. The expanded equations are:

$$a\Sigma x_i^3 + b\Sigma x_i^2 + c\Sigma x_i + nd = \Sigma y_i$$
$$a\Sigma x_i^4 + b\Sigma x_i^3 + c\Sigma x_i^2 + d\Sigma x_i = \Sigma y_i^2$$
$$a\Sigma x_i^5 + b\Sigma x_i^4 + c\Sigma x_i^3 + d\Sigma x_i^2 = \Sigma y_i^3$$
$$a\Sigma x_i^6 + b\Sigma x_i^5 + c\Sigma x_i^4 + d\Sigma x_i^3 = \Sigma y_i^4$$

8-33. a) log-log

  b) propane

8-36. a) Cost is the highest importance because it is to the third power. Keep cost low for a high V.

  b) Look at cost as the most important and distance of control as the second most important.

  c) Everything except the $S^{0.5}$. How can increased size make a control device more attractive to the customer?

8-37. Cost of purchase. Cost of installation. Cost of maintenance. Appearance and styling. Ease of operation if moving parts are required. Delivery time. Effective R-value. Product life.

8-40. Step function: Use of a three speed fan. Lights on or off. A manual five-speed transmission.

Continuous function: Rheostat control of lights. Throttle control on an auto. Pitch settings on an airplane propeller.

8-41. Calculate the average temperature at each pressure and the data becomes: (67,91), (71,90), (75,87.5), (79,85.5), and (83,84). This models well with linear coordinate regression as $T = -0.46P + 122.23$, $r^2 = 0.983$, and a sum of the squares of the residuals = 0.607.

8-43. a) $1 = kP^{(-a_4)}L^{[a_5/(a_5+2a_3-2a_4)]}A^{(a_5/a_3)}E^{(a_5/a_4)}\delta^{a_5}$

  $\Pi_1 = L^2E/P$, $\Pi_2 = \delta/L$, $\Pi_3 = A/L^2$

  b) The value of k will not be constant over the range of the experiment.

8-45. For the uniformly loaded beam $\emptyset = wL^3/24EI$.
For the single center load $\emptyset = PL^2/16EI$.
For slope to be the same, $wL^4/24EI = PL^2/16EI$, and $P = 2wL/3$.

8-47. The maximum shear on the uniformly loaded beam = $wL^2/2$. For the model, the maximum shear = $P/2 = (5wL/8)/2 = 5wL/16$, less than the uniformly loaded beam.

8-49. Information required to conduct an experiment should include: length of the wall; thickness of the wall; height of the wall; properties of the wall material ($E, S_{yt}, S_{ut}, S_{yc}, S_{uc}$); height and width of door openings, and their location from the end of the wall; height, width, distance off the ground for window openings, and their location from the end of the wall.

8-51. Even though the data shows a high correlation, whether the diameter and length are actually related depends on the processes used in the manufacture. If the length is created on a lathe and the diameter created on a grinder it is doubtful if the dimensions are truly related. If the length and diameter are being created at the same time in the same process then perhaps there is some degree of correlation. A good statistical correlation does not prove there is a real correlation between the data. Don't fall into the statistical trap.

8-53. $Y = mX + b$, $Y = mX + [\Sigma y_i - m\Sigma x_i]/n$, $Y = mX + \Sigma y_i/n - m\Sigma x_i/n$; but $\Sigma y_i/n = \mathbf{y}$ and $\Sigma x_i/n = \mathbf{x}$, so $Y = mX + \mathbf{y} - m\mathbf{x}$, and $Y - \mathbf{y} = m(X - \mathbf{x})$.

8-56. Actual voltage = 122.8

9.3. d^2I/dh^2 is approximately equal to -124, so $I(10.39)$ is a maximum.

9.7. Other variables being equal, the higher the exponent value, the lower the tool life.

9.9. $F(2,2) = 12$ is a minimum.

9.10. No maximum or minimum

9.11.   *a*) $(0,f(0)) = (0,6)$ is a minimum.

       b) $(2,f(2)) = (2,0)$ is a minimum; $(3,f(3)) = (3,1)$ is a maximum; $(4,f(4)) = (4,0)$ is a minimum.

       c) $(2.125,f(2.125)) = (2.125,-0.515625)$ is a minimum.

9.12. $(\pi/2, f(\pi/2)) = (\pi/2, \pi/2)$ is an inflection point.

9.13. The maximum value of $I = 2978$ in^4 when $x = 3.078$ in.

9.14. The maximum value of $I = 3843$ in^4 when $x = 3.760$ in.

9.20. The cylindrical tank model is lower in cost than the rectangular model.

9.21. The curves move up and show a slight change in spacing.

9.22.   *a*) $h = 5.46$ inches

       b) Material cost = $0.01/0.20 = $0.05/in^3

       c) Total cost = 28($0.05) = $1.40

9.24. The diameter for lowest cost decreases compared to Case Study #15 because more material is attributed to the diameter D.

9.25. $I_{max} = 546.33$ when $b = 6.86$ and $h = 9.85$

9.26. $I_{max} = 560.81$ when $b = 6.00$ and $h = 10.39$

9.31.   *a*) $f(x) = x^4 + 6$: Eighth interval = $[-0.1114, 0.1642]$
Eighth interval division = $(0.618)(0.1642 + 0.1114) = 0.1703$
$x_9 = 0.1642 - 0.1703 = -0.0061$, $f(-0.0061) = 6.0000$

b) $f(x) = x^4 - 12x^3 + 52x^2 - 96x + 64$: Eighth interval = [3.9409, 4.2174]
Eighth interval division = $(0.618)(4.2174 - 3.9409) = 0.1703$
$x_9 = 4.2174 - 0.1703 = 4.0461, f(4.0461) = 0.0089$

c) $f(x) = x^2 - 4.25x + 4$: Eighth interval = [1.9476, 2.2232]
Eighth interval division = $(0.618)(2.2232 - 1.9476) = 0.1703$
$x_9 = 1.9476 + 0.1703 = 2.1179, f(2.1179) = -0.5156$

9.32. A minimum value of 12.16 occurs at (1.67, 2.5) and (2.5, 1.67).

9.33. Fibonacci search. $x_6 = y_6 = 2.381 - 0.238 = 2.101$
$f(1.905, 1.905) = 12.208, f(1.905, 2.101) = 12.010, f(2.101, 1.905) = 12.010$, and $f(2.101, 2.101) = 12.030$
Estimate for $x_7 = y_7$ = the middle of the intervals = $(1/2)(1.905 + 2.101) = 2.003$
$f(2.003, 2.003) = 12.000$, so the minimum is $f(2.003, 2.003) = 12.000$.

9.34. Golden section search. $x_6 = y_6 = 1.636 + 0.447 = 2.083$
$f(1.910, 1.910) = 12.025, f(1.910, 2.083) = 12.008, f(2.083, 1.910) = 12.008$, and $f(2.083, 2.083) = 12.020$
Estimate for $x_7 = y_7$ = the middle of the interval = $(1/2)(2.083 + 1.910) = 1.996$
$f(1.996, 1.996) = 12.000$, so the minimum is $f(1.996, 1.996) = 12.000$.

9.36. A minimum value of C is 1484 when $y = 10$ and $z = 9$.

9.37. $C(9.692, 8.308) = 1480.5$ represents the minimum.

9.38. $C(9.502, 8.491) = 1481.4$ represents the minimum.

9.40. c) $x \leq -2$ or $x \geq 3$

9.41. $R_1 < 2r$

9.42. The box dimensions are $1.26 \times 1.26 \times 0.63$, and $A = 4.763$.

9.44. $H = \sqrt{KV/2L}$

9.46. $P_{max}(8/9, 32/9) = \$23.11$, and leaves 2.78 hours of machine time available.

9.48. The possible winning combinations (R, P) are: (3,2), (3,3), (2,3), (2,4), (1,3), (1,4), (1,5), (1,6), (0,4), (0,5), (0,6), and (0,7).

9.51. $P(22, 35) = \$1314$

9.53. $P(52, 13) = \$1014$

9.54. Use 18.7 ounces of alloy A, 21.3 ounces of alloy B, and 46.7 ounces of alloy C.

9.57. The four variable Yates algorithm format is:

Experimental code					Experimental response	(1)	(2)	(3)	(4)	Effect estimate
(1)	−	−	−	−						
a	+	−	−	−						
b	−	+	−	−						
ab	+	+	−	−						
c	−	−	+	−						
ac	+	−	+	−						
bc	−	+	+	−						
abc	+	+	+	−						
d	−	−	−	+						
ad	+	−	−	+						
bd	−	+	−	+						
abd	+	+	−	+						
cd	−	−	+	+						
acd	+	−	+	+						
bcd	−	+	+	+						
abcd	+	+	+	+						

Four columns of addition-subtraction are required before the effect estimate column can be completed. The rest of the analysis follows the three variable format.

9.60. The density of air = $1.2929[(273.13/T)(B - 0.3783e)/760]$ kg/m^3, where
T = temperature in degrees centigrade plus 273.13, (°C +273.13)
B = atmospheric pressure in mm
e = vapor pressure of air

9.62. a) $L = H/\sin \theta$

b) $V = 2FH/S_{ut} \sin 2\theta$

c) $D = \sqrt{4F/\pi S_{ut} \cos \theta}$

9.63. 0.375 diameter is the most economical choice.

9.64. At 5000 lbf use 1.000 inch diameter.

At 6000 lbf, not possible under the restrictions of the problem.

9.65. $\sin^4 \theta (C^2 + H^2) - 4C^2 \sin^3 \theta + \sin^2 \theta (6C^2 - H^2) - 4C^2 \sin \theta + C^2 = 0$

This equation is not practical to solve unless C and H are known, or at least the ratio between them known, and even then computer search is most likely the preferred solution method.

9.70. $\theta = 2.637 = 151.1°$

10-4. It would seem that if NEPA was developed to protect the land and the people, that it should have jurisdiction and control over all matters. Why should any activity be exempt? The overused excuse of national security is often a copout.

10-6. Many codes exist to guide project design. They are not law in the sense that they have passed through legislation. If, however, the code has not been followed and a failure occurs, violation of the code is considered a negligent act and those responsible can be successfully sued and their licenses lost. Some cases exists where following the code was not good enough because the state of the art had advanced beyond code requirements. The engineer was successfully sued even though the code was followed. Codes are guidelines. Sound decisions must be made whether they are followed or not.

10-8. If the scholarships are offered to encourage students into a particular field, then the student reacted exactly as desired. If a scholarship is offered as a reward for past performance then the student was not responding as desired. The action would be unethical if the student only declared English as a major for the time it took to get the scholarship, and then changed to another major. At this point the school could take steps to rescind the scholarship.

10-10. Another form of stealing. The student has paid for the right to use the computer through tuition or lab fees, not the right to give the privelege to someone else. To allow someone else to use the computer is not ethical.

10-12. It is unfortunate when a person is threatened with the loss of a job to become an accomplice to a wrongdoing. It is a case of the supervisor being guilty of two counts of unethical behavior. The first is wanting to mislead the stockholders. The second is to ask someone else to join in the deception. The programmer has the choice of joining the charade and being equally guilty of deception, or of writing the program correctly and risking the loss of the job. One of the problems of lying is that usually more lies are required to keep the first lie from being found out.

10-14. Several conditions could exist. 1. The recording secretary could have told the person. 2. Someone else on the committee could have told the person. 3. The person may have assumed the actions were taken and written the statement for effect. In all three cases a breach of ethics occurred.

10-17. If the engineer liked the candidate the temptation to go along with the scheme is more enticing, but it is still wrong. Raises should be related to the worth of the engineer to the company objectives. If the engineer did not like the candidate, refusal would be easier, but alienation of a supervisor might occur. The person who makes such a request is being unethical, and to draw someone else into the picture just makes it worse.

10-20. The pressure is highest if others are involved in the bonus because the pressure comes not only from oneself but from fellow workers. Pressure for a short term gain at the expense of covering a problem is not unusual. Sometimes the engineer is the only one who knows the problem. Sometimes it is known that someone else is aware of a problem. The incentive system (the reward for high output) without a check on quality encourages poor judgement and action for short term gain over long term sound reputation.

10-23. Money might be spent on product improvement or new product research. If an improved product can be sold in later years for a higher price, or more units can be sold, then money spent now will help future profits.

Index

Numbers in ( ) are exercise numbers. Bold face entries are Appendices.

Accelerometer, 132 (5.28), 156 (6.8), 157 (6.11, 12, 13)
Activity chart, 53-57
Airplanes, 29
  Boeing 737-200, 8
  DC-10, 7
  de Havilland Comet, 6
  Voyager, 27
Alexander L. Kieland oil rig, 7
Aluminum, property values of, **F-1, F-2, F-4**
Analysis
  beam, **O**
  data
    experimental, 216-217
    graphical, 218-219
  force, 142-143
  functional, 135
  human factors, 136-137
  least squares, 217
  linear coordinate, 220-224
  idea, 81-82
  market, 96, 138, 176
  safety, 136
  service requirements, 137-138
  using drawings, 140-144
Aswan Dam, 32-33
Automobile(s), 27-29
  springs, 35
Aviation, 13, 17, 27, 29

Beam analysis, **O**
Bill of material, 43
Boundary conditions, 205-206
Brainstorming, 119-120
Brass, property values of, **F-1, F-2, F-6**
Bridges, 5
  arch, 36
  arch truss, 38
  Brooklyn, 10
  draw, 35
  Golden Gate, 5, 35
  Hackensack River, 5
  Mackinac Suspension, 10
  Mianus River, 7
  Queensboro truss, 34
  South Carolina Canadian, 7
  suspension, 22 (1.14)
  Tacoma Narrows, 5, 35
  Walt Whitman suspension, 37
Brittle material data, **F-10**
Bronze, property values of, **F-1, F-2, F-7**
Buckingham Pi Theorem, 210-211
Buildings, 5, 6, 7

Case Study #1: Door Holder-opener, 79-80
Case Study #2: Parts Moving Problem, 89-91, 120-121, 148-153
Case Study #3: Idea Generation, 121-123, 176
Case Study #4: Drawing Board Table Adjustment, 141-145
Case Study #5: Room Cooling, 147-148
Case Study #6: Ball Field Revision, 171-174
Case Study #7: Tree Age Model, 202-205
Case Study #8: Wind Force Model, 212-215
Case Study #9: Empirical Formula Refinement, 222-228
Case Study #10: Tool Life Optimization, 246-249
Case Study #11: Pipe Packing Optimization, 251-253, 289 (9.18)
Case Study #12: Cost Ratio Analysis, 253-255
Case Study #13: Balcony Seating, 272-275
Case Study #14: Spherical Storage Tank, 276-279
Case Study #15: Experimental Optimization, 281-286
Case Study #16: Personal Conflict, 306-307
Case Study #17: Corporate Conflict, 307
Case Study #18: Government Ethics, 308
Cast iron, 5, 31
  property values of, **F-1, F-2, F-3**
Ceramics, property values of, **F-1, F-10**
Chain saw, 30
Chi-squared distribution data, **J**

Code
  building, 48-50, 106 (4.20)
  of ethics, 301-305
Coefficient
  of correlation, 220
  of determination, 220
Communication, 45, 57-64, 174. *See also* Reports
  oral, 61-64
  written, 58-61
Compromise. *See* Design, compromise
Computer(s), 7, 86, 147, 274-275
Concrete, 6
  property values of, **F-1, F-2, F-11**
Consumer, 12, 136, 167, 176
Conversion factors, **C**
Copper
  property values of, **F-1**
  wire, **N**
Cost, 42, 43, 146
Creativity, 76-80, 108
  roadblocks, 78-79
Crack(s), 7
Crystal structure, **E**

Dams
  and locks, 10
  Aswan, 32-33
  Vaiont, 6
Data
  analysis, 11, 216-228
  collection, 8, 11
  error, 11
Databases, 99
Decision, 166-173. *See also* Engineering, decisions
Density, material values of, **E**
Design
  compromise, 4, 7, 45, 74, 135
  constraints, 46-52
  definition, 24, 155 (6.1)
  codes, 47-50, 106 (4.20)
  failure, 10-13
  human aspects, 136-137
  problems, 182-195
  side effects, 31
Dimensional analysis, 209-216
Dimensionless groups, 215-216

Drawing(s), 174
  errors, 11-12
Dynamic damper, 131 (5.26)

Earthquake, 10
Edison, 30, 37
Electricity, 30
Energy, 15-16, 21
  nuclear, 10
  solar, 16
  wind, 16
Energy values of fuel, **D**
Engineer(ing)
  activities, 24-65
  advancements, 26-37
  constraints, 46-52
  day-to-day concerns, 44-46
  decisions, 5, 45, 83
  design. *See* Design
  developments, 26-37
  experiences, 5-10
  failures. *See* Failure
  mistakes, 44
  necessary skills, 25-26
  personal characteristics and abilities, 64-65
  primary job, 24-25
  responsibilities, 38-44
Environment. *See* Problems, pollution
Equal division search technique, 256-258, 265-266
Ethics, 45, 298-308
  code of, 301-305
  definition of, 298
  sources of problems, 299-300
Expansion joint, 7
Experimentation, 11, 37, 228-232, 279-286
Experimental data analysis, 216-228

F distribution data, **I**
Failure. *See also* specific modes of failure
  building, 5-7
  bridge, 5-7
  engineering, 5
  roof, 7
Fatigue, 8
Feasibility study, 125-127
Ferrous alloys, property values of, **F-1, F-2**
Fibonacci search technique, 258-260, 265, 266
Fire, 6, 8
Forecast uncertainty, 56
Four-color map problem, 86 (3.30)

Fracture mechanics, 6, 34
  stress intensity factors, **M**

Gamma rays, 130 (5.21)
Gantt chart, 54-55
Golden section search technique, 260-263, 265, 266
Government
  agencies, 99-100
  regulations, 47-50
Graphical analytical methods, 218-219
Grumman bus, 7

Hartford Civic Center, 7
Heat flow, 291 (9.41)
Hyatt Regency Hotel skywalk collapse, 7
Hydrogen embrittlement, 35

Idea(s)
  alternate word, 113-114
  analysis, 81-82
  combination, 112-113
  comparison, 170-173, 180 (7.14)
  evaluation, 138
  generation, 80-81, 108-127
  implementation, 82-83, 174-176
  question and answer, 114-116
  recording, 114
  refinement, 81
  selection, 168-171
Inertia formulas, **P**
Information, 10, 16
  sources of, 96-101, 113

John Hancock Center, 6

Lagrange multipliers, 250-251
Laws
  legal, 47
  physical, 117-119
Lead, property values of, **F-1**
Least square analysis, 217
Liberty Ships, 6, 34
Light bulbs, 37, 229-232, 239 (8.56)
Linear programming, 266-272
Lock and dam, 10

Magnesium, property values of, **F-1, F-2, F-5**
Maintenance, 7, 51, 176
Manufacturing, 27, 47
  process selection, 144-145
Materials, 144. *See also* specific material name or material category.
  density values, **E**

crystal structure, **E**
properties, thermal, **E**
Medicine, 17
Melting temperature values, **E**
Metallurgy, 30
Methane, 5
Metric prefixes, **B**
Model(s), 146, 198-220
  analog, 199
  boundary conditions, 205-206
  full size, 206-207
  graphical, 218
  iconic, 199
  mathematical, 216-220
    log-log, 216, 217, 218, 219
    linear coordinate, 216, 217, 218, 220-223
    polynomial, 219-220
    semilog, 216, 217, 218, 224-228
  scale, 207-209
  symbolic, 200
  testing, 135
Modulus of elasticity, values of, **F-12**. *See also* specific material or material category

Natural events, 18
Natural gas, 5
Neutron radiation, 9, 34
Nickel, property values of, **F-8**
Normal distribution data, **G**
Nuclear power, 9, 34
Nylon, property values of, **F-1**

Optimization, 46, 240-286
  calculus techniques, 241-256
    Lagrange multipliers, 250-251
  experimental, 279-286
  linear programming, 266-272
  search techniques, 256-266
    equal division, 256-258, 265-266
    Fibonacci, 258-260, 265, 266
    golden section, 260-263, 265, 266
    multi-dimensional, 265-266
    slope, 263-264, 265

Panama Canal, 10
Particulate settling, 239 (8.55)
Patent records, 97
Pendulum, 163 (6.29)
PERT, 56-57
Petroleum, 15, 27
Plexiglass™, 136
Plutonium oxide, 10

Pollution, 8, 13
　waterway, 18
Polyethylene, property values of, **F-1**
Polymers, property values of, **F-1**
Problem solving activities, 74-83
Problem definition, 80, 88-104
　cautions, 102-103
　methods of, 88-96
　related activities, 96-101
Problems
　information, 16
　medicine, 17
　pollution, 13, 21 (1.4), 111-112
　sources of, 74-75
　system, 103-104
　transportation, 17
　unsolved, 13-18
Product
　cost, 42, 43
　distribution, 18
　life, 43, 50
　servicing, 44, 137-138
　testing, 45, 101, 175
Professional societies, 101
Project planning, 52-57, 175
　critical path network, 54, 56
　Gantt chart, 54-55
　implementation, 82-83
　PERT, 56-57

R value of materials, **R**
Refinement, 138-147, 151-154
Reliability, 135

Reports
　written, 58-61
　feasibility, 125-127
　project proposal, 59
　project status, 59-60
　oral, 61-64
　visual aids, 62-63

Safety, 17, 167
　analysis, 136
　features, 12, 44
Shipping, 12, 51
Slope search technique, 263-264, 265
Sound exposure levels, 239
Specifications, 40-41
State of the art, 5, 7, 167
Steel, 35
　property values of, **F-1, F-2**
　stainless, 9
Storage, 12, 138
Strength, fatigue, 8
　values of, **F-1**
Stress
　concentration, 5, 6
　factors, **L**
　intensity factors, **M**
Structural sections
　angles, **Q-4**
　channels, **Q-3**
　standard flange beams, **Q-2**
　wide flange beams, **Q-1**
Styling, 50

Survey
　land, 10
　market, 96
Suspension. *See also* Bridges
　formula, 22 (1.14)
Systems, 103-104

t distribution data, **H**
Teflon™, **F-9**
Titanium, property values of, **F-2, F-9**
Transistors, 36
Transportation, 17

Vaiont Dam, 6
Value analysis, 110-112
Vector analysis, 142-143
Visual aids, 62-63
Voyager (airplane), 27
Voyager 2 (spacecraft), 10

Wind
　testing, 5
　modeling, 212-215, 293 (9.58)
Wood, 5, 185 (7.42)
　material properties, values of, **F-2**
Work simplification, 124-125

Xerox, 167

Yates algorithm, 279-285
Yield strength, values of, **F2-F10**